HANDBOOK OF
FATIGUE FRACTURE IN PRESSURE VESSELS AND PIPING

HANDBOOK OF FATIGUE FRACTURE IN PRESSURE VESSELS AND PIPING

Jaylen Lewis

Oxford Book Company
Jaipur ▪ New Delhi

Published By

Oxford Book Company

Regd. Off. "Bony Residency", Gate No. 2
Opp. Tilak Public School, Vishweswariya Nagar,
Gopalpura Road, Jaipur - 302018, India
Ph: 0141 - 2761280, Telefax: 2761381
email - oxfordbookjpr@gmail.com
website - oxfordbookcompany.com

Delhi Office: 102, 1st Floor, Satyam House,
4327/3, Ansari Road, Darya Ganj,
New Delhi - 110002, India
Ph: 011-45652440

Distributed by
KEYA INTERNATIONAL PRESS LLP
E-mail: keyainternationalpressllp@gmail.com

First published in India, 2024

ISBN : 9789355241566

Typeset by Shivangi Computers, Jaipur

Printed and bound in India by Replika Press Pvt. Ltd.

Contents

Preface

In physics, elasticity is the physical property of a material that returns to its original shape after the stress that made it deform or distort is removed. The relative amount of deformation is called the strain. Above a certain stress known as the elastic limit or the yield strength of an elastic material, the relationship between stress and strain becomes nonlinear. Beyond this limit, the solid may deform irreversibly, exhibiting plasticity. A stress-strain curve is one tool for visualising this transition. Furthermore, not only solids exhibit elasticity. Some non-Newtonian fluids, such as viscoelastic fluids, will also exhibit elasticity in certain conditions. In response to a small, rapidly applied and removed strain, these fluids may deform and then return to their original shape. Under larger strains, or strains applied for longer periods of time, these fluids may start to flow like a liquid, with some viscosity.

In materials science, fatigue is the progressive and localised structural damage that occurs when a material is subjected to cyclic loading. The nominal maximum stress values are less than the ultimate tensile stress limit, and may be below the yield stress limit of the material. Fatigue occurs when a material is subjected to repeated loading and unloading. If the loads are above a certain threshold, microscopic cracks will begin to form at the surface. Eventually a crack will reach a critical size, and the structure will suddenly fracture. The shape of the structure will significantly affect the fatigue life; square holes or sharp corners will lead to elevated local stresses where fatigue cracks can initiate. Round holes and smooth transitions or fillets are therefore important to increase the fatigue strength of the structure.

In science, buckling is a mathematical instability, leading to a failure mode. Theoretically, buckling is caused by a bifurcation in the solution to the equations of static equilibrium. At a certain stage under an increasing load, further load is able to be sustained in one of two states of equilibrium: an undeformed state or a laterally-deformed state. In practice, buckling is characterised by a sudden failure of a structural member subjected to highcompressive stress, where the actual compressive stress at the point of failure is less than the ultimate compressive stresses that the material is capable of withstanding. For example, during earthquakes, reinforced concrete members may experience lateral deformation of the longitudinal reinforcing bars. This mode of failure is also described as failure due to elastic instability. Mathematical analysis of buckling makes use of an axial load eccentricity that introduces a moment, which does not form part of the primary forces to which the member is subjected. When load is constantly being applied on a member, such as column, it will ultimately become large enough to cause the member to become unstable. Further load will cause significant and somewhat unpredictable deformations, possibly leading to complete loss of load-carrying capacity. The member is said to have buckled, to have deformed.

Hope this handbook will guide students through the creation of piping symbols for fittings, flanges, valves, and mechanical equipment.

— ***Jaylen Lewis***

Chapter 1

Introduction

Fatigue (Material)

In materials science, fatigue is the progressive and localised structural damage that occurs when a material is subjected to cyclic loading. The nominal maximum stress values are less than the ultimate tensile stress limit, and may be below the yield stress limit of the material.

Fatigue occurs when a material is subjected to repeated loading and unloading. If the loads are above a certain threshold, microscopic cracks will begin to form at the surface. Eventually a crack will reach a critical size, and the structure will suddenly fracture. The shape of the structure will significantly affect the fatigue life; square holes or sharp corners will lead to elevated local stresses where fatigue cracks can initiate. Round holes and smooth transitions or fillets are therefore important to increase the fatigue strength of the structure.

Fatigue Life

ASTM defines fatigue life, N_f, as the number of stress cycles of a specified character that a specimen sustains before failure of a specified nature occurs. One method to predict fatigue life of materials is the Uniform Material Law (UML). UML was developed for fatigue life prediction of aluminium and titanium alloys by the end of 20th century and extended to high-strength steels and cast iron. For some materials there is a theoretical value for stress amplitude below which it will not fail for any number of cycles, called a fatigue limit, endurance limit, or fatigue strength

Elasticity (Physics)

In physics, elasticity (or stretchiness) is the physical property of a material that returns to its original shape after the stress (e.g. external forces) that made it deform or distort is removed. The relative amount of deformation is called the strain.

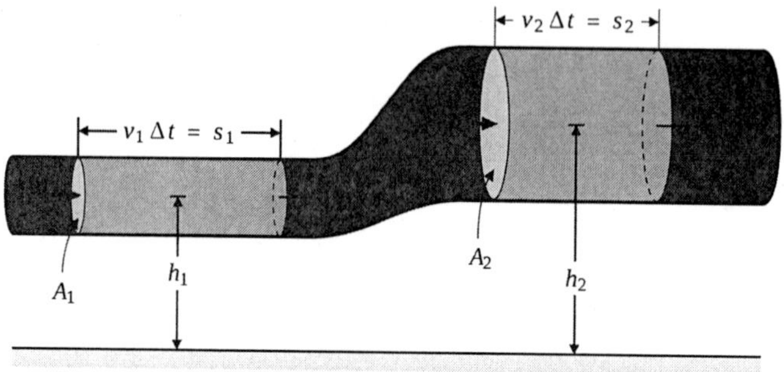

The elastic regime is characterised by a linear relationship between stress and strain, denoted linear elasticity. The classic example is a metal spring. This idea was first stated by Robert Hooke in 1675 as a Latin anagram "ceiiinosssttuu" whose solution he published in 1678 as "Ut tensio, sic vis" which means "As the extension, so the force." This linear relationship is called Hooke's law. The classic model of linear elasticity is the perfect spring. Although the general proportionality constant between stress and strain in three dimensions is a 4th order tensor, when considering simple situations of higher symmetry such as a rod in one dimensional loading, the relationship may often be reduced to applications of Hooke's law.

Because most materials are elastic only under relatively small deformations, several assumptions are used to linearise the theory. Most importantly, higher order terms are generally discarded based on the small deformation assumption. In certain special cases, such as when considering a rubbery material, these assumptions may not be permissible. However, in general, elasticity refers to the linearised theory of the continuum stresses and strains.

Transitions to Inelasticity

Above a certain stress known as the elastic limit or the yield strength of an elastic material, the relationship between stress and strain becomes nonlinear. Beyond this limit, the solid may deform irreversibly, exhibiting plasticity. A stress-strain curve is one tool for visualising this transition.

Furthermore, not only solids exhibit elasticity. Some non-Newtonian fluids, such as viscoelastic fluids, will also exhibit elasticity in certain conditions. In response to a small, rapidly applied and removed strain, these fluids may deform and then return to their original shape. Under larger strains, or strains applied for longer periods of time, these fluids may start to flow like a liquid, with some viscosity.

Characteristics of Fatigue

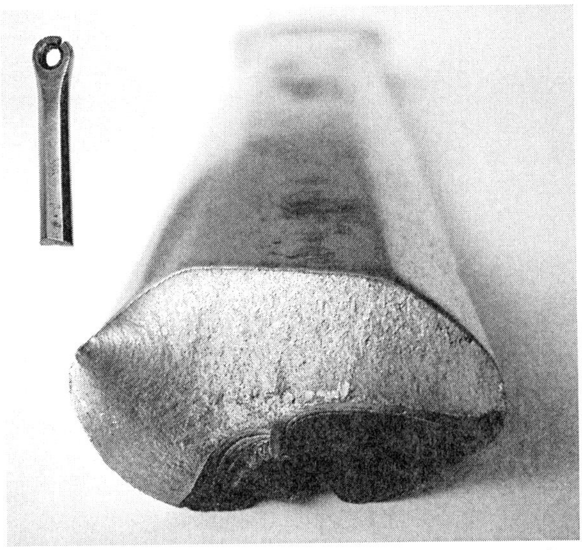

Figure: Fracture of an aluminium crank arm. Dark area of striations: slow crack growth. Bright granular area: sudden fracture.

- In metals and alloys, the process starts with dislocation movements, eventually forming persistent slip bands that nucleate short cracks.
- Fatigue is a stochastic process, often showing considerable scatter even in controlled environments.
- The greater the applied stress range, the shorter the life.
- Fatigue life scatter tends to increase for longer fatigue lives.
- Damage is cumulative. Materials do not recover when rested.
- Fatigue life is influenced by a variety of factors, such as temperature, surface finish, microstructure, presence of oxidising or inert chemicals, residual stresses, contact (fretting), etc.
- Some materials (e.g., some steel and titanium alloys) exhibit a theoretical fatigue limit below which continued loading does not lead to structural failure.
- In recent years, researchers have found that failures occur below the theoretical fatigue limit at very high fatigue lives (10^9 to 10^{10} cycles). An ultrasonic resonance technique is used in these experiments with frequencies around 10–20 kHz.

- High cycle fatigue strength (about 10^3 to 10^8 cycles) can be described by stress-based parameters. A load-controlled servo-hydraulic test rig is commonly used in these tests, with frequencies of around 20–50 Hz. Other sorts of machines—like resonant magnetic machines—can also be used, achieving frequencies up to 250 Hz.
- Low cycle fatigue (typically less than 10^3 cycles) is associated with widespread plasticity in metals; thus, a strain-based parameter should be used for fatigue life prediction in metals and alloys. Testing is conducted with constant strain amplitudes typically at 0.01–5 Hz.

Timeline of Early Fatigue Research History

- 1837: Wilhelm Albert publishes the first article on fatigue. He devised a test machine for conveyor chains used in the Clausthal mines.
- 1839: Jean-Victor Poncelet describes metals as being tired in his lectures at the military school at Metz.
- 1842: William John Macquorn Rankine recognises the importance of stress concentrations in his investigation of railroad axle failures. The Versailles train crash was caused by axle fatigue.
- 1843: Joseph Glynn reports on fatigue of axle on locomotive tender. He identifies the keyway as the crack origin.
- 1848: Railway Inspectorate report one of the first tyre failures, probably from a rivet hole in tread of railway carriage wheel. It was likely a fatigue failure.
- 1849: Eaton Hodgkinson is granted a small sum of money to report to the UK Parliament on his work in ascertaining by direct experiment, the effects of continued changes of load upon iron structures and to what extent they could be loaded without danger to their ultimate security.
- 1854: Braithwaite reports on common service fatigue failures and coins the term fatigue.
- 1860: Systematic fatigue testing undertaken by Sir William Fairbairn and August Wöhler.
- 1870: Wöhler summarises his work on railroad axles. He concludes that cyclic stress range is more important than peak stress and introduces the concept of endurance limit.

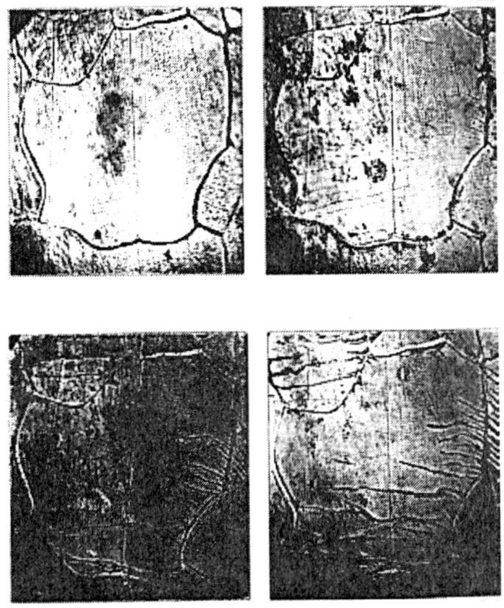

Figure: Micrographs showing how surface fatigue cracks grow as material is further cycled. From Ewing & Humfrey (1903)

- 1903: Sir James Alfred Ewing demonstrates the origin of fatigue failure in microscopic cracks.
- 1910: O. H. Basquin proposes a log-log relationship for SN curves, using Wöhler's test data.
- 1945: A. M. Miner popularises A. Palmgren's (1924) linear damage hypothesis as a practical design tool.
- 1954: L. F. Coffin and S. S. Manson explain fatigue crack-growth in terms of plastic strain in the tip of cracks.
- 1961: P. C. Paris proposes methods for predicting the rate of growth of individual fatigue cracks in the face of initial scepticism and popular defence of Miner's phenomenological approach.
- 1968: Tatsuo Endo and M. Matsuishi devise the rainflow counting algorithm and enable the reliable application of Miner's rule to random loadings.
- 1970: W. Elber elucidates the mechanisms and importance of crack closure in slowing the growth of a fatigue crack due to the wedging effect of plastic deformation left behind the tip of the crack.

High-cycle Fatigue

Historically, most attention has focused on situations that require more than 10^4 cycles to failure where stress is low and deformation primarily elastic.

The S-N Curve

In high-cycle fatigue situations, materials performance is commonly characterised by an S-N curve, also known as a Wöhler curve . This is a graph of the magnitude of a cyclic stress (S) against the logarithmic scale of cycles to failure (N).

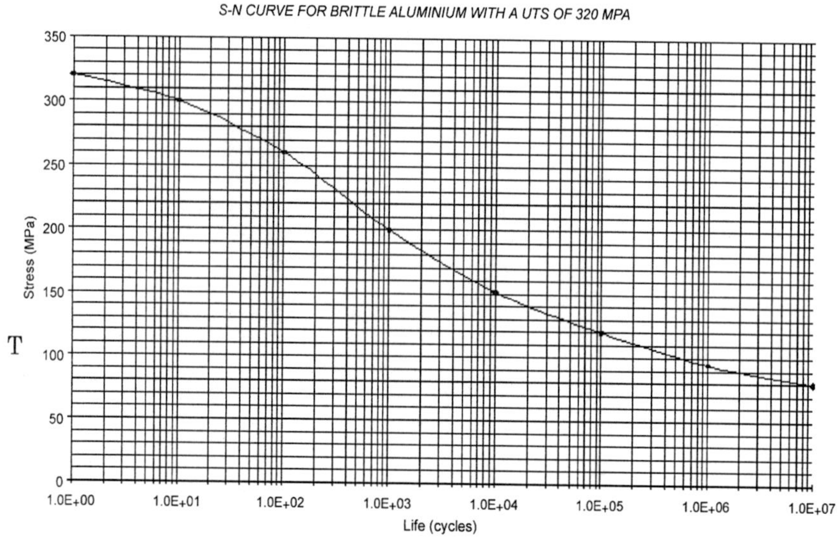

S-N curves are derived from tests on samples of the material to be characterised (often called coupons) where a regular sinusoidal stress is applied by a testing machine which also counts the number of cycles to failure. This process is sometimes known as coupon testing. Each coupon test generates a point on the plot though in some cases there is a runout where the time to failure exceeds that available for the test. Analysis of fatigue data requires techniques from statistics, especially survival analysis and linear regression.

Probabilistic Nature of Fatigue

As coupons sampled from a homogeneous frame will manifest variation in their number of cycles to failure, the S-N curve should more properly be an S-N-P curve capturing the probability of failure after a given number of cycles of a certain stress. Probability distributions that are common in data analysis and in design against fatigue include the lognormal distribution, extreme value distribution, Birnbaum–Saunders distribution, and Weibull distribution.

Complex Loadings

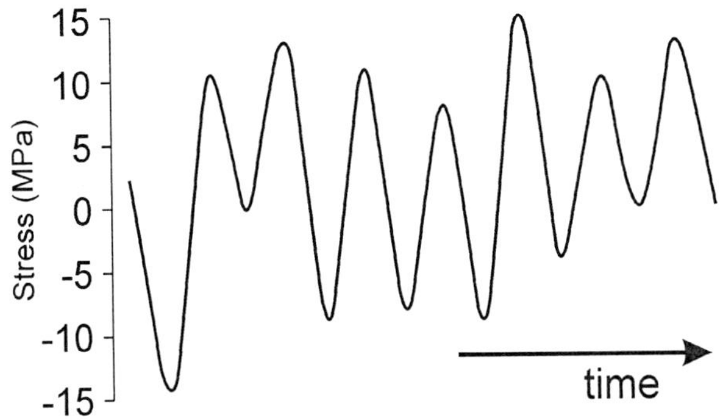

Figure: Spectrum loading

In practice, a mechanical part is exposed to a complex, often random, sequence of loads, large and small. In order to assess the safe life of such a part:

1. Reduce the complex loading to a series of simple cyclic loadings using a technique such as rainflow analysis;
2. Create a histogram of cyclic stress from the rainflow analysis to form a fatigue damage spectrum;
3. For each stress level, calculate the degree of cumulative damage incurred from the S-N curve; and
4. Combine the individual contributions using an algorithm such as Miner's rule.

Miner's Rule

In 1945, M. A. Miner popularised a rule that had first been proposed by A. Palmgren in 1924. The rule, variously called Miner's rule or the Palmgren-Miner linear damage hypothesis, states that where there are k different stress magnitudes in a spectrum, S_i (1 d" i d" k), each contributing $n_i(S_i)$ cycles, then if$N_i(S_i)$ is the number of cycles to failure of a constant stress reversal S_i, failure occurs when:

$$\sum_{i=1}^{k} \frac{n_i}{N_i} = C$$

C is experimentally found to be between 0.7 and 2.2. Usually for design purposes, C is assumed to be 1. This can be thought of as assessing what proportion of life is consumed by stress reversal at each magnitude then forming a linear combination of their aggregate. Though Miner's rule is a useful approximation in many circumstances, it has several major limitations:

1. It fails to recognise the probabilistic nature of fatigue and there is no simple way to relate life predicted by the rule with the characteristics of a probability distribution. Industry analysts often use design curves, adjusted to account for scatter, to calculate $N_i(S_i)$.
2. There is sometimes an effect in the order in which the reversals occur. In some circumstances, cycles of low stress followed by high stress cause more damage than would be predicted by the rule. It does not consider the effect of overload or high stress which may result in a compressive residual stress. High stress followed by low stress may have less damage due to the presence of compressive residual stress.

Paris' Relationship

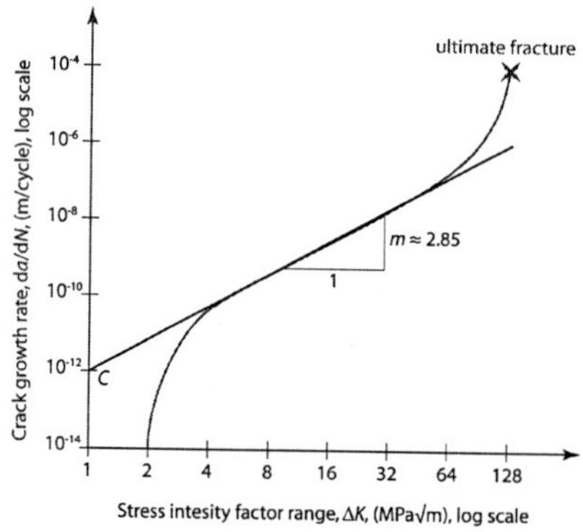

Figure: Typical fatigue crack growth rate graph.

In Fracture mechanics, Anderson, Gomez and Paris derived relationships for the stage II crack growth with cycles N, in terms of the cyclical component ΔK of the Stress Intensity Factor K

$$\frac{da}{dN} = C(\Delta K)^m$$

where a is the crack length and m is typically in the range 3 to 5 (for metals).

This relationship was later modified (by Forman, 1967) to make better allowance for the mean stress, by introducing a factor depending on (1-R) where R = min stress/max stress, in the denominator.

Low-cycle Fatigue

Where the stress is high enough for plastic deformation to occur, the account in terms of stress is less useful and the strain in the material offers a simpler description. Low-cycle fatigue is usually characterised by the Coffin-Manson relation (published independently by L. F. Coffin in 1954 and S. S. Manson 1953):

$$\frac{\Delta \in_p}{2} = \in'_f (2N)^c$$

where:

- $\Delta \in_p / 2$ is the plastic strain amplitude;
- $\in'_f$ is an empirical constant known as the fatigue ductility coefficient, the failure strain for a single reversal;
- 2N is the number of reversals to failure (N cycles);
- c is an empirical constant known as the fatigue ductility exponent, commonly ranging from -0.5 to -0.7 for metals in time independent fatigue. Slopes can be considerably steeper in the presence of creep or environmental interactions.

A similar relationship for materials such as Zirconium, used in the nuclear industry.

Fatigue and Fracture Mechanics

The account above is purely empirical and, though it allows life prediction and design assurance, life improvement or design optimisation can be enhanced using fracture mechanics. It can be developed in four stages.

1. Crack nucleation;
2. Stage I crack-growth;
3. Stage II crack-growth; and
4. Ultimate ductile failure.

Factors That Affect Fatigue-life

- Cyclic stress state: Depending on the complexity of the geometry and the loading, one or more properties of the stress state need to be considered, such as stress amplitude, mean stress, biaxiality, in-phase or out-of-phase shear stress, and load sequence,
- Geometry: Notches and variation in cross section throughout a part lead to stress concentrations where fatigue cracks initiate.
- Surface quality. Surface roughness cause microscopic stress

concentrations that lower the fatigue strength. Compressive residual stresses can be introduced in the surface by e.g. shot peening to increase fatigue life. Such techniques for producing surface stress are often referred to as peening, whatever the mechanism used to produce the stress. Low plasticity burnishing, laser peening, and ultrasonic impact treatment can also produce this surface compressive stress and can increase the fatigue life of the component. This improvement is normally observed only for high-cycle fatigue.

- Material Type: Fatigue life, as well as the behaviour during cyclic loading, varies widely for different materials, e.g. composites and polymers differ markedly from metals.
- Residual stresses: Welding, cutting, casting, and other manufacturing processes involving heat or deformation can produce high levels of tensile residual stress, which decreases the fatigue strength.
- Size and distribution of internal defects: Casting defects such as gas porosity, non-metallic inclusions and shrinkage voids can significantly reduce fatigue strength.
- Direction of loading: For non-isotropic materials, fatigue strength depends on the direction of the principal stress.
- Grain size: For most metals, smaller grains yield longer fatigue lives, however, the presence of surface defects or scratches will have a greater influence than in a coarse grained alloy.
- Environment: Environmental conditions can cause erosion, corrosion, or gas-phase embrittlement, which all affect fatigue life. Corrosion fatigue is a problem encountered in many aggressive environments.
- Temperature: Extreme high or low temperatures can decrease fatigue strength.

Design Against Fatigue

Dependable design against fatigue-failure requires thorough education and supervised experience instructural engineering, mechanical engineering, or materials science. There are three principal approaches to life assurance for mechanical parts that display increasing degrees of sophistication:

1. Design to keep stress below threshold of fatigue limit (infinite lifetime concept);

2. Design (conservatively) for a fixed life after which the user is instructed to replace the part with a new one (a so-called lifed part, finite lifetime concept, or "safe-life" design practice);
3. Instruct the user to inspect the part periodically for cracks and to replace the part once a crack exceeds a critical length. This approach usually uses the technologies of nondestructive testing and requires an accurate prediction of the rate of crack-growth between inspections. This is often referred to as damage tolerant design or "retirement-for-cause".

Stopping Fatigue

Fatigue cracks that have begun to propagate can sometimes be stopped by drilling holes, called drill stops, in the path of the fatigue crack. This is not recommended as a general practice because the hole represents a stress concentration factor which depends on the size of the hole and geometry, though the hole is typically less of a stress concentration than the removed tip of the crack. The possibility remains of a new crack starting in the side of the hole. It is always far better to replace the cracked part entirely.

Material Change

Changes in the materials used in parts can also improve fatigue life. For example, parts can be made from better fatigue rated metals. Complete replacement and redesign of parts can also reduce if not eliminate fatigue problems. Thus helicopter rotor blades and propellers in metal are being replaced by composite equivalents. They are not only lighter, but also much more resistant to fatigue. They are more expensive, but the extra cost is amply repaid by their greater integrity, since loss of a rotor blade usually leads to total loss of the aircraft. A similar argument has been made for replacement of metal fuselages, wings and tails of aircraft.

Infamous Fatigue Failures

Versailles Train Crash

Following the King's fete celebrations at the Palace of Versailles, a train returning to Paris crashed in May 1842 at Meudon after the leading locomotive broke an axle. The carriages behind piled into the wrecked engines and caught fire. At least 55 passengers were killed trapped in the carriages, including the explorer Jules Dumont d'Urville. This accident is known in France as the "Catastrophe ferroviaire de

Meudon". The accident was witnessed by the British locomotive engineer Joseph Locke and widely reported in Britain. It was discussed extensively by engineers, who sought an explanation.

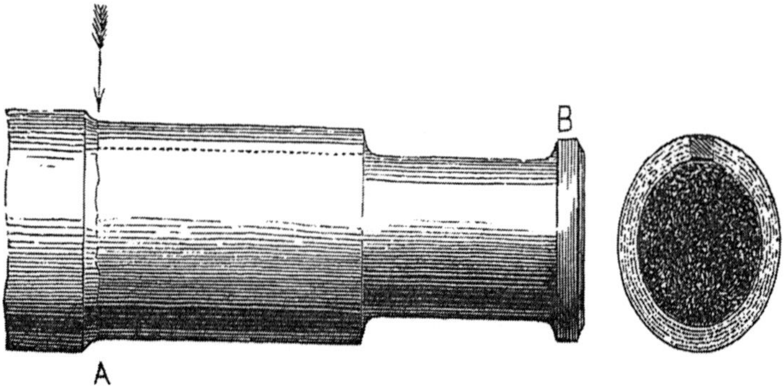

Figure: Drawing of a fatigue failure in an axle by Joseph Glynn, 1843.

The derailment had been the result of a broken locomotive axle. Rankine's investigation of broken axles in Britain highlighted the importance of stress concentration, and the mechanism of crack growth with repeated loading. His and other papers suggesting a crack growth mechanism through repeated stressing, however, were ignored, and fatigue failures occurred at an ever increasing rate on the expanding railway system. Other spurious theories seemed to be more acceptable, such as the idea that the metal had somehow "crystallised". The notion was based on the crystalline appearance of the fast fracture region of the crack surface, but ignored the fact that the metal was already highly crystalline.

De Havilland Comet

Two de Havilland Comet passenger jets broke up in mid-air and crashed within a few months of each other in 1954. As a result systematic tests were conducted on a fuselage immersed and pressurised in a water tank. After the equivalent of 3,000 flights investigators at the Royal Aircraft Establishment (RAE) were able to conclude that the crash had been due to failure of the pressure cabin at the forward Automatic Direction Finder window in the roof. This 'window' was in fact one of two apertures for the aerials of an electronic navigation system in which opaque fibre glass panels took the place of the window 'glass'. The failure was a result of metal fatigue caused by the repeated pressurisation and depressurisation of the aircraft cabin. Another fact was that the supports around the windows were riveted, not bonded, as the original specifications for the aircraft had called for.

The problem was exacerbated by the punch rivet construction technique employed. Unlike drill riveting, the imperfect nature of the hole created by punch riveting caused manufacturing defect cracks which may have caused the start of fatigue cracks around the rivet. The Comet's pressure cabin had been designed to a safety factor comfortably in excess of that required by British Civil Airworthiness Requirements (2.5 times the cabin proof pressure as opposed to the requirement of 1.33 times and an ultimate load of 2.0 times the cabin pressure) and the accident caused a revision in the estimates of the safe loading strength requirements of airliner pressure cabins.

In addition, it was discovered that the stresses around pressure cabin apertures were considerably higher than had been anticipated, especially around sharp-cornered cut-outs, such as windows. As a result, all future jet airliners would feature windows with rounded corners, the curve eliminating a stress concentration. This was a noticeable distinguishing feature of all later models of the Comet. Investigators from the RAE told a public inquiry that the sharp corners near the Comets' window openings acted as initiation sites for cracks. The skin of the aircraft was also too thin, and cracks from manufacturing stresses were present at the corners.

Alexander L. Kielland Oil Platform Capsize

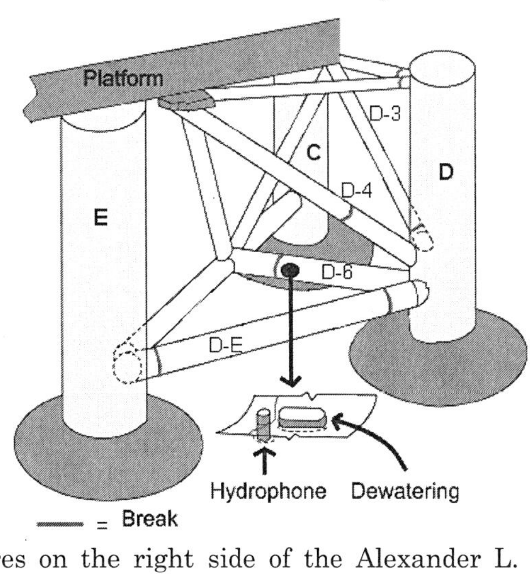

Figure: Fractures on the right side of the Alexander L. Kielland rig

The Alexander L. Kielland was a Norwegian semi-submersible drilling rig that capsized whilst working in the Ekofisk oil field in March 1980 killing 123 people. The capsize was the worst disaster in Norwegian waters since World War II.

The rig, located approximately 320 km east from Dundee, Scotland, was owned by the Stavanger Drilling Company of Norway and was on hire to the U.S.company Phillips Petroleum at the time of the disaster. In driving rain and mist, early in the evening of 27 March 1980 more than 200 men were off duty in the accommodation on the Alexander L. Kielland.

The wind was gusting to 40 knots with waves up to 12 m high. The rig had just been winched away from the Edda production platform. Minutes before 18:30 those on board felt a 'sharp crack' followed by 'some kind of trembling'. Suddenly the rig heeled over 30° and then stabilised. Five of the six anchor cables had broken, the one remaining cable preventing the rig from capsizing. The list continued to increase and at 18.53 the remaining anchor cable snapped and the rig turned upside down.

A year later in March 1981, the investigative report concluded that the rig collapsed owing to a fatigue crack in one of its six bracings (bracing D-6), which connected the collapsed D-leg to the rest of the rig. This was traced to a small 6 mm fillet weld which joined a non-load-bearing flange plate to this D-6 bracing. This flange plate held a sonar device used during drilling operations. The poor profile of the fillet weld contributed to a reduction in its fatigue strength. Further, the investigation found considerable amounts of lamellar tearing in the flange plate and cold cracks in the butt weld. Cold cracks in the welds, increased stress concentrations due to the weakened flange plate, the poor weld profile, and cyclical stresses (which would be common in the North Sea), seemed to collectively play a role in the rig's collapse.

Pressure Vessel

A pressure vessel is a closed container designed to hold gases or liquids at a pressure substantially different from the ambient pressure.

The pressure differential is dangerous and many fatal accidents have occurred in the history of their development and operation. Consequently, their design, manufacture, and operation are regulated by engineering authorities backed by legislation. For these reasons, the definition of a pressure vessel varies from country to country, but involves parameters such as maximum safe operating pressure and temperature.

Uses

Figure: A pressure tank connected to a water well and domestic hot water system

Figure: A few pressure tanks, here used to hold propane

Pressure vessels are used in a variety of applications in both industry and the private sector. They appear in these sectors as industrial compressed air receivers and domestic hot water storage tanks. Other examples of pressure vessels are diving cylinder, recompression chamber, distillation towers, autoclaves, and many other vessels in mining or oil refineries and petrochemical plants, nuclear reactor vessel, habitat of a space ship, habitat of a submarine, pneumatic reservoir, hydraulic reservoir under pressure, rail vehicle airbrake reservoir, road vehicle airbrake reservoir and storage vessels for liquified gases such as ammonia, chlorine, propane, butane, and LPG.

Pressure Vessel Features

Shape of a Pressure Vessel

Pressure vessels may theoretically be almost any shape, but shapes made of sections of spheres, cylinders, and cones are usually employed. A common design is a cylinder with end caps called heads. Head shapes are frequently either hemispherical or dished (torispherical). More complicated shapes have historically been much harder to analyse for safe operation and are usually far more difficult to construct.

Theoretically, a spherical pressure vessel has approximately twice the strength of a cylindrical pressure vessel. However, a spherical shape is tough to manufacture, therefore more expensive, so most pressure vessels are cylindrical with 2:1 semi-elliptical heads or end caps on each end. Smaller pressure vessels are assembled from a pipe and two covers. A disadvantage of these vessels is that greater breadths are more expensive, so that for example the most economic shape of a 1,000 litres (35 cu ft), 250 bars (3,600 psi) pressure vessel might be a breadth of 914.4 millimetres (36 in) and a width of 1,701.8 millimetres (67 in) including the 2:1 semi-elliptical domed end caps.

Construction Materials

Figure: Steel Pressure Vessel

Theoretically almost any material with good tensile properties that is chemically stable in the chosen application could be employed. However, pressure vessel design codes and application standards (ASME BPVC Section II, EN 13445-2 etc.) contain long lists of approved materials with associated limitations in temperature range.

Many pressure vessels are made of steel. To manufacture a cylindrical or spherical pressure vessel, rolled and possibly forged parts would have to be welded together. Some mechanical properties of steel, achieved by rolling or forging, could be adversely affected by welding, unless special precautions are taken. In addition to adequate mechanical strength, current standards dictate the use of steel with a high impact resistance, especially for vessels used in low temperatures. In applications where carbon steel would suffer corrosion, special corrosion resistant material should also be used.

Some pressure vessels are made of composite materials, such as filament wound composite using carbon fibre held in place with a polymer. Due to the very high tensile strength of carbon fibre these vessels can be very light, but are much more difficult to manufacture. The composite material may be wound around a metal liner, forming a composite overwrapped pressure vessel.

Other very common materials include polymers such as PET in carbonated beverage containers and copper in plumbing.

Pressure vessels may be lined with various metals, ceramics, or polymers to prevent leaking and protect the structure of the vessel from the contained medium. This liner may also carry a significant portion of the pressure load.

Scaling

No matter what shape it takes, the minimum mass of a pressure vessel scales with the pressure and volume it contains and is inversely proportional to the strength to weight ratio of the construction material (minimum mass decreases as strength increases).

Scaling of Stress in Walls of Vessel

Pressure vessels are held together against the gas pressure due to tensile forces within the walls of the container. The normal (tensile) stress in the walls of the container is proportional to the pressure and radius of the vessel and inversely proportional to the thickness of the walls. Therefore pressure vessels are designed to have a thickness proportional to the radius of tank and the pressure of the tank and inversely proportional to the maximum allowed normal stress of the particular material used in the walls of the container.

Because (for a given pressure) the thickness of the walls scales with the radius of the tank, the mass of a tank (which scales as the length times radius times thickness of the wall for a cylindrical tank) scales with the volume of the gas held (which scales as length times radius squared). The exact formula varies with the tank shape but depends on the density, ñ, and maximum allowable stress ó of the material in addition to the pressure P and volume V of the vessel.

Spherical Vessel

For a sphere, the mass of a pressure vessel is

$$M = \frac{3}{2} PV \frac{\rho}{\sigma},$$

where

M is mass,

P is the pressure difference from ambient (the gauge pressure),

V is volume,

ρ is the density of the pressure vessel material,

σ is the maximum working stress that material can tolerate.

Other shapes besides a sphere have constants larger than 3/2 (infinite cylinders take 2), although some tanks, such as non-spherical wound composite tanks can approach this.

Cylindrical Vessel with Hemispherical Ends

This is sometimes called a "bullet" for its shape.

For a cylinder with hemispherical ends,

$$M = 2\pi R^3 (R + W) P \frac{\rho}{\sigma}$$

where

- R is the radius
- W is the middle cylinder width only, and the overall width is W + 2R

2.1 Cylindrical Vessel with Semi-elliptical Ends

In a vessel with an aspect ratio of middle cylinder width to radius of 2:1,

$$M = 6\pi R^3 P \frac{\rho}{\sigma}.$$

Gas Storage

In looking at the first equation, the factor PV, in SI units, is in units of (pressurisation) energy. For a stored gas, PV is proportional to the mass of gas at a given temperature, thus

$$M = \frac{3}{2} nRT \frac{\rho}{\sigma}$$

The other factors are constant for a given vessel shape and material. So we can see that there is no theoretical "efficiency of scale", in terms of the ratio of pressure vessel mass to pressurisation energy, or of pressure vessel mass to stored gas mass. For storing gases, "tankage efficiency" is independent of pressure, at least for the same temperature.

So, for example, a typical design for a minimum mass tank to hold helium (as a pressurant gas) on a rocket would use a spherical chamber for a minimum shape constant, carbon fibre for best possibleñ / ó, and very cold helium for best possible M / pV.

Stress in thin-walled pressure vessels

Stress in a shallow-walled pressure vessel in the shape of a sphere is

$$\sigma_\theta = \sigma_{long} = \frac{pr}{2t},$$

where σ_e is hoop stress, or stress in the circumferential direction, σ_{long} is stress in the longitudinal direction, p is internal gauge pressure, r is the inner radius of the sphere, and t is thickness of the cylinder wall. A vessel can be considered "shallow-walled" if the diameter is at least 10 times (sometimes cited as 20 times) greater than the wall depth.

Stress in a shallow-walled pressure vessel in the shape of a cylinder is

$$\sigma_\theta = \frac{pr}{t}$$

$$\sigma_{long} = \frac{pr}{2t},$$

where σ_e is hoop stress, or stress in the circumferential direction, σ_{long} is stress in the longitudinal direction, p is internal gauge pressure, r is the inner radius of the cylinder, and t is thickness of the cylinder wall.

Almost all pressure vessel design standards contain variations of these two formulas with additional empirical terms to account for wall thickness tolerances, quality control of welds and in service corrosion allowances.

For example, the ASME Boiler and Pressure Vessel Code (BPVC) (UG-27) formulas are:

Spherical shells:

$$\sigma_\theta = \frac{pr}{t}$$

$$\sigma_\theta = \sigma_{long} = \frac{p(r+0.2t)}{2tE}$$

Cylindrical shells:

$$\sigma_\theta = \frac{p(r+0.6t)}{tE}$$

$$\sigma_{long} = \frac{p(r-0.4t)}{2tE}$$

where E is the joint efficient, and all others variables as stated above.

Winding Angle of Carbon Fibre Vessels

Wound infinite cylindrical shapes optimally take a winding angle of 54.7 degrees, as this gives the necessary twice the strength in the circumferential direction to the longitudinal.

Design and Operation Standards

Pressure vessels are designed to operate safely at a specific pressure and temperature, technically referred to as the "Design Pressure" and "Design Temperature".

A vessel that is inadequately designed to handle a high pressure constitutes a very significant safety hazard. Because of that, the design and certification of pressure vessels is governed by design codes such as the ASME Boiler and Pressure Vessel Code in North America, the Pressure Equipment Directive of the EU (PED),Japanese Industrial Standard (JIS), CSA B51 in Canada, Australian Standards in Australia and other international standards like Lloyd's, Germanischer Lloyd, Det Norske Veritas, Société Générale de Surveillance (SGS S.A.), Stoomwezen etc.

Note that where the pressure-volume product is part of a safety standard, any incompressible liquid in the vessel can be excluded as it does not contribute to the potential energy stored in the vessel, so only the volume of the compressible part such as gas is used.

List of Standards

- EN 13445: The current European Standard, harmonised with

the Pressure Equipment Directive(97/23/EC). Extensively used in Europe.

- ASME Boiler and Pressure Vessel Code Section VIII: Rules for Construction of Pressure Vessels
- BS 5500: Former British Standard, replaced in the UK by BS EN 13445 but retained under the name PD 5500 for the design and construction of export equipment.
- AD Merkblätter: German standard, harmonised with the Pressure Equipment Directive.
- EN 286 (Parts 1 to 4): European standard for simple pressure vessels (air tanks), harmonised with Council Directive 87/404/EEC.
- BS 4994: Specification for design and construction of vessels and tanks in reinforced plastics.
- ASME PVHO: US standard for Pressure Vessels for Human Occupancy
- CODAP: French Code for Construction of Unfired Pressure Vessel
- AS/NZS 1200: Pressure equipment
- API 510
- ISO 11439
- IS 2825-1969 (RE1977)_code_unfired_Pressure_vessels
- FRP tanks and vessels
- AIAA S-080-1998: AIAA Standard for Space Systems - Metallic Pressure Vessels, Pressurised Structures, and Pressure Components
- AIAA S-081A-2006: AIAA Standard for Space Systems - Composite Overwrapped Pressure Vessels (COPVs)
- B51-09 Canadian Boiler, pressure vessel, and pressure piping code
- Stoomwezen: Former pressure vessels code in the Netherlands, also known as RToD: Regels voor Toestellen onder Druk (Dutch Rules for Pressure Vessels)

Leak Before Burst

Leak before burst describes a pressure vessel designed such that a crack in the vessel will grow through the wall, allowing the contained fluid to escape and reducing the pressure, prior to growing so large as to cause fracture at the operating pressure.

Many pressure vessel standards, including the ASME Boiler and Pressure Vessel Code and the AIAA metallic pressure vessel standard, either require pressure vessel designs to be leak before burst, or require pressure vessels to meet more stringent requirements for fatigue and fracture if they are not shown to be leak before burst.

Alternatives to Pressure Vessels

Depending on the application and local circumstances, alternatives to pressure vessels exist. Examples can be seen in domestic water collection systems, where the following may be used:

- Gravity controlled systems which typically consist of an unpressurised water tank at an elevation higher than the point of use. Pressure at the point of use is the result of the hydrostatic pressure caused by the elevation difference. Gravity systems produce 0.43 pounds per square inch (3.0 kPa) per foot of water head (elevation difference). A municipal water supply or pumped water is typically around 90 pounds per square inch (620 kPa).
- Inline pump controllers or pressure-sensitive pumps.

History of Pressure Vessels

Large pressure vessels were invented during the industrial revolution, particularly in Great Britain, to be used as boilers for making steam to drive steam engines.

Design and testing standards and a system of certification came about as the result of fatal boiler explosions.

In an early effort to design a tank capable of withstanding pressures up to 10,000 psi (69 MPa), a 6-inch (150 mm) diameter tank was developed in 1919 that was spirally-wound with two layers of high tensile strength steel wire to prevent sidewall rupture, and the end caps longitudinally reinforced with lengthwise high-tensile rods.

Piping

Within industry, piping is a system of pipes used to convey fluids (liquids and gases) from one location to another. The engineering discipline of piping design studies the efficient transport of fluid.

Industrial process piping (and accompanying in-line components) can be manufactured from wood, fibreglass, glass, steel, aluminium, plastic, copper, and concrete. The in-line components, known as fittings, valves, and other devices, typically sense and control the pressure, flow rate and temperature of the transmitted fluid, and usually are included in the field of Piping Design (or Piping Engineering). Piping systems are documented in piping and instrumentation diagrams (P&IDs). If necessary, pipes can be cleaned by the tube cleaning process.

Figure: Large-scale piping system in an HVAC mechanical room

"Piping" sometimes refers to Piping Design, the detailed specification of the physical piping layout within a process plant or commercial building. In earlier days, this was sometimes called Drafting, Technical drawing, Engineering Drawing, and Design but is today commonly performed by Designers who have learned to use automated Computer Aided Drawing / Computer Aided Design (CAD) software.

Plumbing is a piping system that most people are familiar with, as it constitutes the form of fluid transportation that is used to provide potable water and fuels to their homes and business. Plumbing pipes also remove waste in the form of sewage, and allow venting of sewage gases to the outdoors. Fire sprinkler systems also use piping, and may transport nonpotable or potable water, or other fire-suppression fluids.

Piping also has many other industrial applications, which are crucial for moving raw and semi-processed fluids for refining into more useful products. Some of the more exotic materials of construction are Inconel, titanium, chrome-moly and various other steel alloys.

Engineering Subfields

Generally, Industrial Piping Engineering has three major subfields:

- Piping Material
- Piping Design
- Stress Analysis.

Stress Analysis

Process piping and power piping are typically checked by Pipe Stress Engineers to verify that the routing, nozzle loads, hangers, and supports are properly placed and selected such that allowable pipe stress is not exceeded under different situation such as sustain, operating, hydro test etc as per the ASME or any other legislative code and local government standards. It is necessary to evaluate the mechanical behaviour of the piping under regular loads (internal pressure and thermal stresses) as well under occasional and intermittent loading cases such as earthquake, high wind or special vibration, and water hammer. This evaluation is usually performed with the assistance of a specialised (finite element) pipe stress analysis computer program such as CAESAR II, ROHR2, CAEPIPE and AUTOPIPE.

Wooden Piping History

Early wooden pipes were constructed out of logs that had a large hole bored lengthwise through the centre. Later wooden pipes were constructed with staves and hoops similar to wooden barrel construction. Stave pipes have the advantage that they are easily transported as a compact pile of parts on a wagon and then assembled as a hollow structure at the job site. Wooden pipes were especially popular in mountain regions where transport of heavy iron or concrete pipes would have been difficult.

Wooden pipes were easier to maintain than metal, because the wood did not expand or contract with temperature changes as much as metal and so consequently expansion joints and bends were not required. The thickness of wood afforded some insulating properties to the pipes which helped prevent freezing as compared to metal pipes. Wood used for water pipes also does not rot very easily. Electrolysis, that bugbear of many iron pipe systems, doesn't affect wood pipes at all, since wood is a much better electrical insulator.

In the Western United States where redwood was used for pipe construction, it was found that redwood had "peculiar properties" that protected it from weathering, acids, insects, and fungus growths. Redwood pipes stayed smooth and clean indefinitely while iron pipe by comparison would rapidly begin to scale and corrode and could eventually plug itself up with the corrosion.

Materials

The material with which a pipe is manufactured often forms as the basis for choosing any pipe. Materials that are used for manufacturing pipes include:

- Carbon Steel (CS)
- Low Temperature Service Carbon Steel (LTCS)
- Stainless Steel (SS)
- Non Ferrous Metals (Inconel, Incoloy, Cupro-nickel, etc.)
- Non Metallic (GRE, PVC, HDPE, tempered glass, etc.)

Standards

There are certain standard codes that need to be followed while designing or manufacturing any piping system. Organisations that promulgate piping standards include:

- ASME - The American Society of Mechanical Engineers
- ASTM - American Society for Testing and Materials
- API - American Petroleum Institute
- AWS - American Welding Society
- AWWA - American Water Works Association
- MSS – Manufacturers' Standardisation Society
- ANSI - American National Standards Institute
- NFPA - National Fire Protection Association
- EJMA - Expansion Joint Manufacturers Association

Fracture Mechanics

Fracture mechanics is the field of mechanics concerned with the study of the propagation of cracks in materials. It uses methods of analytical solid mechanics to calculate the driving force on a crack and those of experimental solid mechanics to characterise the material's resistance to fracture.

In modern materials science, fracture mechanics is an important tool in improving the mechanical performance of materials and components. It applies the physics of stress and strain, in particular the theories of elasticity and plasticity, to the microscopic crystallographic defects found in real materials in order to predict the macroscopic mechanical failure of bodies. Fractography is widely used with fracture mechanics to understand the causes of failures and also verify the theoretical failure predictions with real life failures.

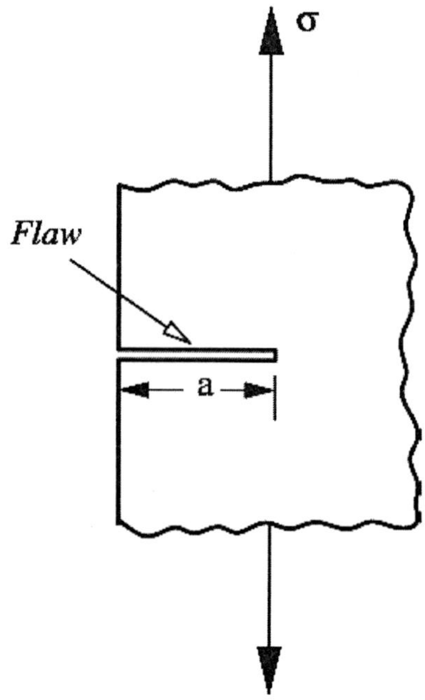

Figure: An edge crack (flaw) of length a in a material.

Linear Elastic Fracture Mechanics

Griffith's Criterion

Fracture mechanics was developed during World War I by English aeronautical engineer, A. A. Griffith, to explain the failure of brittle materials. Griffith's work was motivated by two contradictory facts:

- The stress needed to fracture bulk glass is around 100 MPa (15,000 psi).
- The theoretical stress needed for breaking atomic bonds is approximately 10,000 MPa (1,500,000 psi).

A theory was needed to reconcile these conflicting observations. Also, experiments on glass fibres that Griffith himself conducted suggested that the fracture stress increases as the fibre diameter decreases. Hence the uniaxial tensile strength, which had been used extensively to predict material failure before Griffith, could not be a specimen-independent material property. Griffith suggested that the low fracture strength observed in experiments, as well as the size-dependence of strength, was due to the presence of microscopic flaws in the bulk material.

To verify the flaw hypothesis, Griffith introduced an artificial flaw in his experimental specimens. The artificial flaw was in the form of a surface crack which was much larger than other flaws in a specimen. The experiments showed that the product of the square root of the flaw length (a) and the stress at fracture ($ó_f$) was nearly constant, which is expressed by the equation:

$$\sigma_f \sqrt{a} \approx C$$

An explanation of this relation in terms of linear elasticity theory is problematic. Linear elasticity theory predicts that stress (and hence the strain) at the tip of a sharp flaw in a linear elastic material is infinite. To avoid that problem, Griffith developed a thermodynamic approach to explain the relation that he observed.

The growth of a crack requires the creation of two new surfaces and hence an increase in the surface energy. Griffith found an expression for the constant C in terms of the surface energy of the crack by solving the elasticity problem of a finite crack in an elastic plate. Briefly, the approach was:

- Compute the potential energy stored in a perfect specimen under an uniaxial tensile load.
- Fix the boundary so that the applied load does no work and then introduce a crack into the specimen. The crack relaxes the stress and hence reduces the elastic energy near the crack faces. On the other hand, the crack increases the total surface energy of the specimen.
- Compute the change in the free energy (surface energy " elastic energy) as a function of the crack length. Failure occurs when the free energy attains a peak value at a critical crack length, beyond which the free energy decreases by increasing the crack length, i.e. by causing fracture. Using this procedure, Griffith found that

$$C = \sqrt{\frac{2E\gamma}{\pi}}$$

where E is the Young's modulus of the material and ã is the surface energy density of the material. Assuming E = 62 GPa and ã = 1 J/m^2 gives excellent agreement of Griffith's predicted fracture stress with experimental results for glass.

Irwin's Modification

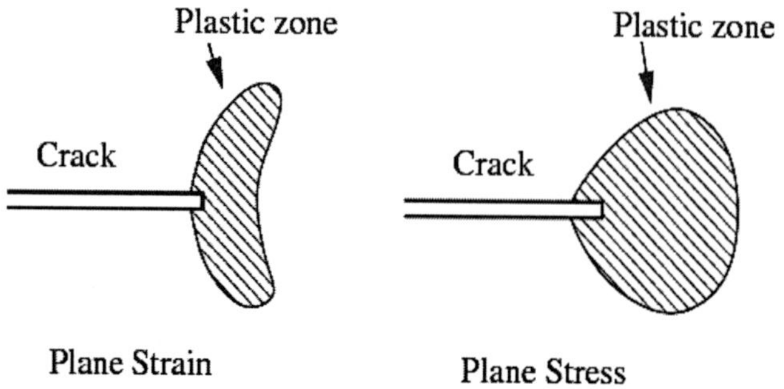

Figure: The plastic zone around a crack tip in a ductile material.

Griffith's work was largely ignored by the engineering community until the early 1950s. The reasons for this appear to be (a) in the actual structural materials the level of energy needed to cause fracture is orders of magnitude higher than the corresponding surface energy, and (b) in structural materials there are always some inelastic deformations around the crack front that would make the assumption of linear elastic medium with infinite stresses at the crack tip highly unrealistic. F. Erdogan (2000)

Griffith's theory provides excellent agreement with experimental data for brittle materials such as glass. For ductile materials such as steel, though the relation still holds, the surface energy (ã) predicted by Griffith's theory is usually unrealistically high. A group working under G. R. Irwin at the U.S. Naval Research Laboratory (NRL) during World War II realised that plasticity must play a significant role in the fracture of ductile materials.

In ductile materials (and even in materials that appear to be brittle), a plastic zone develops at the tip of the crack. As the applied load increases, the plastic zone increases in size until the crack grows and the material behind the crack tip unloads. The plastic loading and unloading cycle near the crack tip leads to the dissipation of energy as heat. Hence, a dissipative term has to be added to the energy balance relation devised by Griffith for brittle materials. In physical terms, additional energy is needed for crack growth in ductile materials when compared to brittle materials.

Irwin's strategy was to partition the energy into two parts:

- the stored elastic strain energy which is released as a crack grows. This is the thermodynamic driving force for fracture.
- the dissipated energy which includes plastic dissipation and the surface energy (and any other dissipative forces that may be at work). The dissipated energy provides the thermodynamic resistance to fracture. Then the total energy dissipated is

  G = 2ã + G_p

where γ is the surface energy and G_p is the plastic dissipation (and dissipation from other sources) per unit area of crack growth.

The modified version of Griffith's energy criterion can then be written as

$$\sigma_f \sqrt{a} = \sqrt{\frac{EG}{\pi}}.$$

For brittle materials such as glass, the surface energy term dominates and . For ductile materials such as steel, the plastic dissipation term dominates and . For polymers close to the glass transition temperature, we have intermediate values of .

Stress Intensity Factor

Another significant achievement of Irwin and his colleagues was to find a method of calculating the amount of energy available for fracture in terms of the asymptotic stress and displacement fields around a crack front in a linear elastic solid. This asymptotic expression for the stress field around a crack tip is

$$\sigma_{ij} \approx \left(\frac{K}{\sqrt{2\pi r}}\right) f_{ij}(\theta)$$

where σ_{ij} are the Cauchy stresses, r is the distance from the crack tip, θ is the angle with respect to the plane of the crack, and f_{ij} are functions that are independent of the crack geometry and loading conditions. Irwin called the quantity K the stress intensity factor. Since the quantity f_{ij} is dimensionless, the stress intensity factor can be expressed in units of $MPa - \sqrt{m}.$.

When a rigid line inclusion is considered, a similar asymptotic expression for the stress fields is obtained.

Strain Energy Release

Irwin was the first to observe that if the size of the plastic zone around a crack is small compared to the size of the crack, the energy required to grow the crack will not be critically dependent on the state of stress at the crack tip. In other words, a purely elastic solution may be used to calculate the amount of energy available for fracture.

The energy release rate for crack growth or strain energy release rate may then be calculated as the change in elastic strain energy per unit area of crack growth, i.e.,

$$G := -\left[\frac{\partial}{\partial a}\right]_p = -\left[\frac{\partial U}{\partial a}\right]_u$$

where U is the elastic energy of the system and a is the crack length. Either the load P or the displacement u can be kept fixed while evaluating the above expressions.

Irwin showed that for a mode I crack (opening mode) the strain energy release rate and the stress intensity factor are related by:

$$G = G_I = \begin{cases} \dfrac{K_1^2}{E} & \textit{plane stress} \\[2ex] \dfrac{\left(1-\nu^2\right)K_1^2}{E} & \textit{plane strain} \end{cases}$$

where E is the Young's modulus, í is Poisson's ratio, and K_I is the stress intensity factor in mode I. Irwin also showed that the strain energy release rate of a planar crack in a linear elastic body can be expressed in terms of the mode I, mode II (sliding mode), and mode III (tearing mode) stress intensity factors for the most general loading conditions. Next, Irwin adopted the additional assumption that the size and shape of the energy dissipation zone remains approximately constant during brittle fracture. This assumption suggests that the energy needed to create a unit fracture surface is a constant that depends only on the material. This new material property was given the name fracture toughness and designated G_{Ic}. Today, it is the critical stress intensity factor K_{Ic} which is accepted as the defining property in linear elastic fracture mechanics.

But a problem arose for the NRL researchers because naval materials, e.g., ship-plate steel, are not perfectly elastic but undergo significant plastic deformation at the tip of a crack. One basic assumption in Irwin's linear elastic fracture mechanics is that the size of the plastic zone is small compared to the crack length. However, this assumption is quite restrictive for certain types of failure in structural steels though such steels can be prone to brittle fracture, which has led to a number of catastrophic failures.

Linear-elastic fracture mechanics is of limited practical use for structural steels for another more practical reason. Fracture toughness testing is very expensive and engineers believe that sufficient information for selection of steels can be obtained from the simpler and cheaperCharpy impact test.

Elastic–plastic Fracture Mechanics

Most engineering materials show some nonlinear elastic and inelastic behaviour under operating conditions that involve large loads. In such materials the assumptions of linear elastic fracture mechanics may not hold, that is,

- the plastic zone at a crack tip may have a size of the same order of magnitude as the crack size
- the size and shape of the plastic zone may change as the applied load is increased and also as the crack length increases.

Therefore a more general theory of crack growth is needed for elastic-plastic materials that can account for:

- the local conditions for initial crack growth which include the nucleation, growth, and coalescence of voids or decohesion at a crack tip.
- a global energy balance criterion for further crack growth and unstable fracture.

R-curve

An early attempt in the direction of elastic-plastic fracture mechanics was Irwin's crack extension resistance curve or R-curve. This curve acknowledges the fact that the resistance to fracture increases with growing crack size in elastic-plastic materials. The R-curve is a plot of the total energy dissipation rate as a function of the crack size and can be used to examine the processes of slow stable crack growth and unstable fracture. However, the R-curve was not widely used in applications until the early 1970s. The main reasons appear to be that the R-curve depends on the geometry of the specimen and the crack driving force may be difficult to calculate.

J-integral

In the mid-1960s James R. Rice (then at Brown University) and G. P. Cherepanov independently developed a new toughness measure to describe the case where there is sufficient crack-tip deformation that the part no longer obeys the linear-elastic approximation. Rice's analysis, which assumes non-linear elastic (or monotonic deformation theory plastic) deformation ahead of the crack tip, is designated the J integral. This analysis is limited to situations where plastic deformation at the crack tip does not extend to the furthest edge of the loaded part. It also demands that the assumed non-linear elastic behaviour of the material is a reasonable approximation in shape and magnitude to the real material's load response. The elastic-plastic failure parameter is designated J_{Ic} and is conventionally converted to K_{Ic} using Equation (3.1) of the Appendix to this article. Also note that the J integral approach reduces to the Griffith theory for linear-elastic behaviour.

Cohesive Zone Models

When a significant region around a crack tip has undergone plastic deformation, other approaches can be used to determine the possibility of further crack extension and the direction of crack growth and branching. A simple technique that is easily incorporated into numerical calculations is the cohesive zone model method which is based on concepts proposed independently by Barenblatt and Dugdale in the early 1960s. The relationship between the Dugale-Barenblatt models and Griffith's theory was first discussed by Willis in 1967. The equivalence of the two approaches in the context of brittle fracture was shown by Rice in 1968. Interest in cohesive zone modelling of fracture has been reignited since 2000 following the pioneering work on dynamic fracture by Xu and Needleman , and Camacho and Ortiz.

Fully Plastic Failure

If the material is so tough that the yielded region ahead of the crack extends to the far edge of the specimen before fracture, the crack is no longer an effective stress concentrator. Instead, the presence of the crack merely serves to reduce the load-bearing area. In this regime the failure stress is conventionally assumed to be the average of the yield and ultimate strengths of the material.

Engineering Applications

The following information is needed for a fracture mechanics prediction of failure:

- Applied load
- Residual stress
- Size and shape of the part
- Size, shape, location, and orientation of the crack

Usually not all of this information is available and conservative assumptions have to be made.

Occasionally post-mortem fracture-mechanics analyses are carried out. In the absence of an extreme overload, the causes are either insufficient toughness (K_{Ic}) or an excessively large crack that was not detected during routine inspection.

Short Summary

Arising from the manufacturing process, interior and surface flaws are found in all metal structures. Not all such flaws are unstable under service conditions. Fracture mechanics is the analysis of flaws to discover those that are safe (that is, do not grow) and those that are liable to propagate as cracks and so cause failure of the flawed structure. Ensuring safe operation of structure despite these inherent flaws is achieved through damage tolerance analysis. Fracture mechanics as a subject for critical study has barely been around for a century and thus is relatively new. There is a high demand for engineers with fracture mechanics expertise—particularly in this day and age where engineering failure is considered 'shocking' amongst the general public.

Appendix: Mathematical Relations

Griffith's Criterion

For the simple case of a thin rectangular plate with a crack perpendicular to the load Griffith's theory becomes:

$$G = \frac{\pi\sigma^2 a}{E} \qquad (1)$$

where G is the strain energy release rate, σ is the applied stress, a is half the crack length, and E is the Young's modulus. The strain energy release rate can otherwise be understood as: *the rate at which energy is absorbed by growth of the crack.* However, we also have that:

$$G_c = \frac{\pi\sigma_f^2 a}{E} \qquad (2)$$

If G e– G_c, this is the criterion for which the crack will begin to propagate.

Irwin's Modifications

Eventually a modification of Griffith's solids theory emerged from this work; a term called stress intensity replaced strain energy release rate and a term called fracture toughness replaced surface weakness energy. Both of these terms are simply related to the energy terms that Griffith used:

$$K_I = \sigma\sqrt{\pi a} \tag{3}$$

and

$$K_c = \sqrt{EG_c} \quad \text{(for plane stress)} \tag{4}$$

$$K_c = \sqrt{\frac{EG_c}{1-\nu^2}} \quad \text{(for plane strain)} \tag{5}$$

where K_I is the stress intensity, K_c the fracture toughness, and ν is Poisson's ratio. It is important to recognise the fact that fracture parameter K_c has different values when measured under plane stress and plane strain

Fracture occurs when $K_I \geq K_c$. For the special case of plane strain deformation, K_c becomes K_{Ic} and is considered a material property. The subscript I arises because of the different ways of loading a material to enable a crack to propagate. It refers to so-called "mode I" loading as opposed to mode II or III: There are three ways of applying a force to enable a crack to propagate:

- Mode I crack – Opening mode (a tensile stress normal to the plane of the crack)
- Mode II crack – Sliding mode (a shear stress acting parallel to the plane of the crack and perpendicular to the crack front)
- Mode III crack – Tearing mode (a shear stress acting parallel to the plane of the crack and parallel to the crack front)

We must note that the expression for K_I in equation 2.1 will be different for geometries other than the centre-cracked infinite plate, as discussed in the article on the stress intensity factor. Consequently, it is necessary to introduce a dimensionless correction factor, Y, in order to characterise the geometry. We thus have:

$$K_I = Y\sigma\sqrt{\pi a} \tag{6}$$

where Y is a function of the crack length and width of sheet given by:

$$Y\left(\frac{a}{W}\right)=\sqrt{\sec\left(\frac{\pi a}{W}\right)} \tag{7}$$

for a sheet of finite width W containing a through-thickness crack of length $2a$, or

$$Y\left(\frac{a}{W}\right)=1.12-\frac{0.41}{\sqrt{\pi}}\frac{a}{W}+\frac{18.7}{\sqrt{\pi}}\left(\frac{a}{W}\right)^2-\cdots \tag{8}$$

for a sheet of finite width W containing a through-thickness edge crack of length a

Elasticity and Plasticity

Since engineers became accustomed to using K_{Ic} to characterise fracture toughness, a relation has been used to reduce J_{Ic} to it: $K_{Ic}=\sqrt{E^{*}J_{Ic}}$

where $E^{*}=E$ for plane stress and $E^{*}=\frac{E}{1-\nu^2}$ for plane strain (9)

The remainder of the mathematics employed in this approach is interesting, but is probably better summarised in external pages due to its complex nature.

Chapter 2

Deep-level Transient Spectroscopy

Deep Level Transient Spectroscopy (DLTS) is an experimental tool for studying electrically active defects (known as charge carrier traps) in semiconductors. DLTS establishes fundamental defect parameters and measures their concentration in the material. Some of the parameters are considered as defect "finger prints" used for their identifications and analysis.

DLTS investigates defects present in a space charge (depletion) region of a simple electronic device. The most commonly used are Schottky diodes or p-n junctions. In the measurement process the steady-state diode reverse polarisation voltage is disturbed by a voltage pulse. This voltage pulse reduces the electric field in the space charge region and allows free carriers from the semiconductor bulk to penetrate this region and recharge the defects causing their non-equilibrium charge state. After the pulse, when the voltage returns to its steady-state value, the defects start to emit trapped carriers due to the thermal emission process. The technique observes the device space charge region capacitance where the defect charge state recovery causes the capacitance transient. The voltage pulse followed by the defect charge state recovery are cycled allowing an application of different signal processing methods for defect recharging process analysis.

The DLTS technique has a higher sensitivity than almost any other semiconductor diagnostic technique. For example, in silicon it can detect impurities and defects at a concentration of one part in 10^{12} of the material host atoms. This feature together with a technical simplicity of its design made it very popular in research labs and semiconductor material production factories.

The DLTS technique was pioneered by D. V. Lang (David Vern Lang of Bell Laboratories) in 1974.US Patent was awarded to Lang in 1975.

DLTS Methods

Conventional DLTS

In conventional DLTS the capacitance transients are investigated by using a lock-in amplifier or double box-car averaging technique when the sample temperature is slowly varied (usually in a range from liquid nitrogen temperature to room temperature 300K or above). The equipment reference frequency is the voltage pulse repetition rate. In the conventional DLTS method this frequency multiplied by some constant (depending on the hardware used) is called the "rate window". When during the sample temperature variation the emission rate of carriers from some defect equals to the rate window one obtains in the spectrum a peak. By setting up different rate windows in subsequent DLTS spectra measurements one obtains different temperatures at which some particular peak appears. Having a set of the emission rate and corresponding temperature pairs one can make an Arrhenius plot which allows for the deduction of defect activation energy for the thermal emission process. Usually this energy (sometimes called the defect energy level) together with the plot intersept value are defect parameters used for its identification or analysis.

Recently, DLTS has been used to study quantum dots.

MCTS and Minority-carrier DLTS

For the Schottky diodes, majority carrier traps are observed by the application of a reverse bias pulse, while minority carrier traps can be observed when the reverse bias voltage pulses are replaced with light pulses with the photon energy from the above semiconductor bandgap spectral range. This method is called Minority Carrier Transient Spectroscopy (MCTS). The minority carrier traps can be also observed for the p-n junctions by application of forward bias pulses which inject minority carriers into the space charge region. In DLTS plots the minority carrier spectra usually are depicted with an opposite sign of amplitude in respect to the majority carrier trap spectra.

Laplace DLTS

There is an extension to DLTS known as a high resolution Laplace transform DLTS (LDLTS). Laplace DLTS is an isothermal technique in which the capacitance transients are digitised and averaged at a fixed temperature. Then the defect emission rates are obtained with a use of numerical methods being equivalent to the inverse Laplace transformation. The obtained emission rates are presented as a spectral plot. The main advantage of Laplace DLTS in comparison to conventional DLTS is the substantial increase in energy resolution understood here as an ability to distinguish very similar signals.

Laplace DLTS in combination with uniaxial stress results in a splitting of the defect energy level. Assuming a random distribution of defects in non-equivalent orientations, the number of split lines and their intensity ratios reflect the symmetry class of the given defect.

Application of LDLTS to MOS capacitors needs device polarisation voltages in a range where the Fermi level extrapolated from semiconductor to the semiconductor-oxide interface intersects this interface within the semiconductor bandgap range. The electronic interface states present at this interface can trap carriers similarly to defects described above. If their occupancy with electrons or holes is disturbed by a small voltage pulse then the device capacitance recovers after the pulse to its initial value as the interface states start to emit carriers. This recovery process can be analysed with the LDLTS method for different device polarisation voltages. Such a procedure allows to obtain the energy state distribution of the interface electronic states at the semiconductor-oxide (or dielectric) interfaces.

Constant-Capacitance DLTS

In general, the analysis of the capacitance transients in the DLTS measurements assumes that the concentration of investigated traps is much smaller than the material doping concentration. In cases when this assumption is not fulfilled then the constant capacitance DLTS (CCDLTS) method is used for more accurate determination of the trap concentration. When the defects recharge and their concentration is high then the width of the device space region varies making the analysis of the capacitance transient inaccurate. The additional electronic circuitry maintaining the total device capacitance constant by varying the device bias voltage helps to keep the depletion region width constant. As a result, the varying device voltage reflects the defect recharge process. An analysis of the CCDLTS system using feedback theory was provided by Lau and Lam in 1982.

I-DLTS and PITS

There is an important shortcoming for DLTS: it cannot be used for insulating materials. (Note: an insulator can be considered as a very large bandgap semiconductor.) For insulating materials it is difficult or impossible to produce a device having a space region which width could be changed by the external voltage bias and thus the capacitance measurement-based DLTS methods cannot be applied for the defect analysis. Basing on experiences of the thermally stimulated current (TSC) spectroscopy, the current transients are analysed with the DLTS methods (I-DLTS), where the light pulses are used for the defect occupancy disturbance. This method in the literature is sometimes called the Photoinduced Transient Spectroscopy (PITS). I-DLTS or PITS are also used for studying defects in the i-region of a p-i-n diode.

Focused Ion Beam

Focused ion beam, also known as FIB, is a technique used particularly in the semiconductor industry, materials science and increasingly in the biological field for site-specific analysis, deposition, and ablation of materials. An FIB setup is a scientific instrument that resembles a scanning electron microscope (SEM). However, while the SEM uses a focused beam of electrons to image the sample in the chamber, an FIB setup uses a focused beam of ionsinstead. FIB can also be incorporated in a system with both electron and ion beam columns, allowing the same feature to be investigated using either of the beams. FIB should not be confused with using a beam of focused ions for direct write lithography (such as in proton beam writing). These are generally quite different systems where the material is modified by other mechanisms.

Ion beam source

Most widespread are instruments using liquid-metal ion sources (LMIS), especially gallium ion sources. Ion sources based on elemental gold and iridium are also available. In a gallium LMIS, gallium metal is placed in contact with a tungsten needle and heated gallium wets the tungsten and flows to the tip of the needle where the opposing forces of surface tension and electric field form the gallium into a cusp shaped tip called a Taylor cone. The tip radius of this cone is extremely small (~2 nm). The huge electric field at this small tip (greater than 1×10^8 volts per centimetre) causes ionization and field emission of the gallium atoms.

Source ions are then generally accelerated to an energy of 1–50 keV (kiloelectronvolts), and focused onto the sample by electrostatic lenses. LMIS produce high current density ion beams with very small energy spread. A modern FIB can deliver tens of nanoamperes of current to a sample, or can image the sample with a spot size on the order of a few nanometres.

Principle

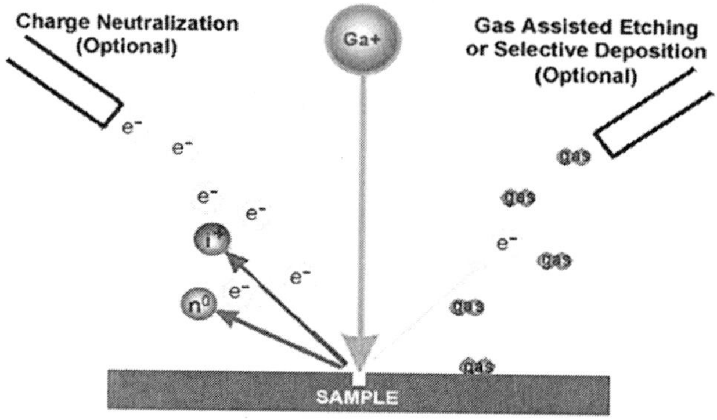

Figure: The principle of FIB

Focused ion beam (FIB) systems have been produced commercially for approximately twenty years, primarily for large semiconductor manufacturers. FIB systems operate in a similar fashion to a scanning electron microscope (SEM) except, rather than a beam of electrons and as the name implies, FIB systems use a finely focused beam of ions (usually gallium) that can be operated at low beam currents for imaging or high beam currents for site specific sputtering or milling.

As the diagram on the right shows, the gallium (Ga+) primary ion beam hits the sample surface and sputters a small amount of material, which leaves the surface as either secondary ions (i+ or i-) or neutral atoms (n^0). The primary beam also produces secondary electrons (e^-). As the primary beam rasters on the sample surface, the signal from the sputtered ions or secondary electrons is collected to form an image.

At low primary beam currents, very little material is sputtered and modern FIB systems can easily achieve 5 nm imaging resolution (imaging resolution with Ga ions is limited to ~5 nm by sputtering and detector efficiency). At higher primary currents, a great deal of material can be removed by sputtering, allowing precision milling of the specimen down to a sub micrometre or even a nano scale.

If the sample is non-conductive, a low energy electron flood gun can be used to provide charge neutralineutralis. In this manner, by imaging with positive secondary ions using the positive primary ion beam, even highly insulating samples may be imaged and milled without a conducting surface coating, as would be required in a SEM.

Until recently, the overwhelming usage of FIB has been in the semiconductor industry. Such applications as defect analysis, circuit modification, mask repair and transmission electron microscope sample preparation of site specific locations on integrated circuits have become commonplace procedures. The latest FIB systems have high resolution imaging capability; this capability coupled with in situ sectioning has eliminated the need, in many cases, to examine FIB sectioned specimens in a separate SEM instrument. SEM imaging is still required for the highest resolution imaging and to prevent damage to sensitive samples. However, the combination of SEM and FIB columns onto the same chamber enables the benefits of both to be utilised.

Why Ions?

• The most fundamental difference between FIB and focused electron beam techniques such as SEM, STEM or EBID is the use of ions instead of electrons, and this has major consequences for the interactions that occur at the sample surface. The most important characteristics and the consequences for the sample interaction are:

Ions are larger than electrons

- Because ions are much larger than electrons, they cannot easily penetrate within individual atoms of the sample. Interaction mainly involves outer shell interaction resulting in atomic ionization and breaking of chemical bonds of the substrate atoms.
- The penetration depth of the ions is much lower than the penetration of electrons of the same energy.
- When the ion has come to a stop within the material, it is caught in the matrix of the material.

ions are heavier than electrons

- Because ions are far heavier than electrons, ions can gain a high momentum. For the same energy, the momentum of the gallium ion is about 370 times larger.
- For the same energy ions move a lot slower than electrons. However, they are still fast compared to the image collection mode and usually this has no real consequences except when the ion beam is scanning very quickly.
- Magnetic lenses are less effective on ions than they would be on electrons with the same energy. As a consequence the focused ion beam system is equipped with electrostatic lenses and not with magnetic lenses.

Ions are positive and electrons are negative

- This difference has negligible consequences and is taken care of by the polarity of fields to control the beam and accelerate the ions.

In summary, ions are positive, large, heavy and slow, whereas electrons are negative, small, light and fast. The most important consequence of the properties listed above is that ion beams will remove atoms from the substrate and because the beam position, dwell time and size are so well controlled it can be applied to remove material locally in a highly controlled manner, down to the nanometre scale.

Usage

Unlike an electron microscope, FIB is inherently destructive to the specimen. When the high-energy gallium ions strike the sample, they will sputter atoms from the surface. Gallium atoms will also be implanted into the top few nanometres of the surface, and the surface will be made amorphous.

Because of the sputtering capability, the FIB is used as a micro- and nano-machining tool, to modify or machine materials at the micro- and nanoscale. FIB micro machining has become a broad field of its own, but nano machining with FIB is a field that is still developing. Commonly the smallest beam size for imaging is 2.5–6 nm. The smallest milled features are somewhat larger (10–15 nm) as this is dependent on the total beam size and interactions with the sample being milled.

FIB tools are designed to etch or machine surfaces, an ideal FIB might machine away one atom layer without any disruption of the atoms in the next layer, or any residual disruptions above the surface. Yet currently because of the sputter the machining typically roughens surfaces at the submicrometre length scales. An FIB can also be used to deposit material via ion beam induced deposition. FIB-assisted chemical vapor deposition occurs when a gas, such as tungsten hexacarbonyl ($W(CO)_6$) is introduced to the vacuum chamber and allowed to chemisorb onto the sample. By scanning an area with the beam, the precursor gas will be decomposed into volatile and non-volatile components; the non-volatile component, such as tungsten, remains on the surface as a deposition. This is useful, as the deposited metal can be used as a sacrificial layer, to protect the underlying sample from the destructive sputtering of the beam. From nanometres to hundred of micrometres in length, tungsten metal deposition allows to put metal lines right where needed. Other materials such as platinum, cobalt, carbon, gold, etc., can also be locally deposited. Gas assisted deposition and FIB etching process are shown below.

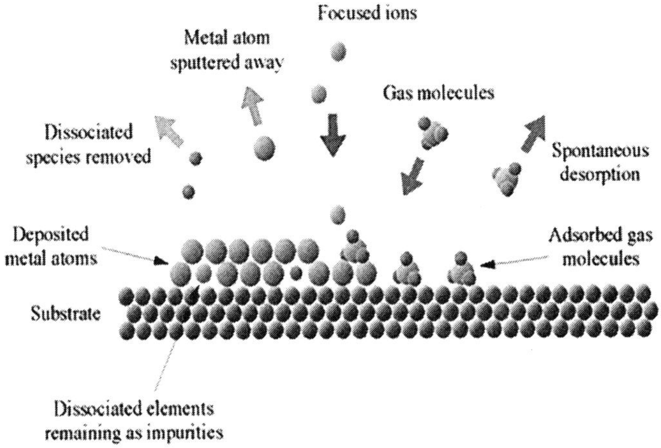

1. Adsorption of the precursor molecules on the substrate
2. Ion beam induced dissociation of the gas molecules
3. Deposition of the material atoms and removal of the organic ligands

Figure: Gas-assisted deposition

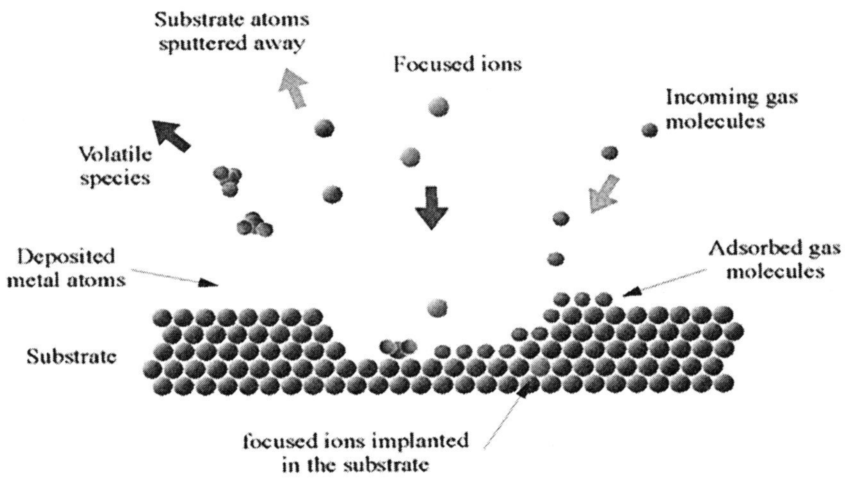

1. Adsorption of the gas molecules on the substrate
2. Interaction of the gas molecules with the substrate
   Formation of volatile and non volatile species
3. Evaporation of volatile species and sputtering of non volatile species

Figure: Gas-assisted FIB etching

Figure: Orsay Physics Canion 31 Plus UHV FIB on a TOF-SIMS 6600 from Physical Electronics

FIB is often used in the semiconductor industry to patch or modify an existing semiconductor device. For example, in an integrated circuit, the gallium beam could be used to cut unwanted electrical connections, and/or to deposit conductive material in order to make a connection. The high level of surface interaction is exploited in patterned doping of semiconductors. FIB is also used for maskless implantation.

The FIB is also commonly used to prepare samples for the transmission electron microscope. The TEM requires very thin samples, typically ~100 nanometres. Other techniques, such as ion milling or electropolishing can be used to prepare such thin samples. However, the nanometre-scale resolution of the FIB allows the exact thin region to be chosen. This is vital, for example, in integrated circuit failure analysis. If a particular transistor out of several million on a chip is bad, the only tool capable of preparing an electron microscope sample of that single transistor is the FIB.

The drawbacks to FIB sample preparation are the above-mentioned surface damage and implantation, which produce noticeable effects when using techniques such as high-resolution "lattice imaging" TEM or electron energy loss spectroscopy. This damaged layer can be minimised by FIB milling with lower beam voltages, or by further milling with a low-voltage argon ion beam after completion of the FIB process.

FIB preparation can be used with cryogenically frozen samples in a suitably equipped instrument, allowing cross sectional analysis of samples containing liquids or fats, such as biological samples, pharmaceuticals, foams, inks, and food products

FIB is also used for Secondary ion mass spectrometry (SIMS). The ejected secondary ions are collected and analysed after the surface of the specimen has been sputtered with a primary focused ion beam.

FIB Imaging

At lower beam currents, FIB imaging resolution begins to rival the more familiar scanning electron microscope (SEM) in terms of imaging topography, however the FIB's two imaging modes, using secondary electrons and secondary ions, both produced by the primary ion beam, offer many advantages over SEM.

FIB secondary electron images show intense grain orientation contrast. As a result, grain morphology can be readily imaged without resorting to chemical etching. Grain boundary contrast can also be enhanced through careful selection of imaging parameters. FIB secondary ion images also reveal chemical differences, and are especially useful in corrosion studies, as secondary ion yields of metals can increase by three orders of magnitude in the presence of oxygen, clearly revealing the presence of corrosion

History

History of FIB technology

- 1975: The first FIB systems based on field emission technology were developed by Levi-Settiand by Orloff and Swanson and used gas field ionization sources (GFISs).
- 1978: The first FIB based on an LMIS was built by Seliger et al.

Physics of LMIS

- 1600: Gilbert documented that fluid under high tension forms a cone.
- 1914: Zeleny observed and filmed cones and jets
- 1959: Feynman suggested the use of ion beams.
- 1964: Taylor produced exactly conical solution to equations of electro hydrodynamics (EHD)
- 1975: Krohn and Ringo produced first high brightness ion source: LMIS

Some pioneers of LMIS & FIB

- Mahoney (1969)
- Sudraud et al. Paris XI Orsay (1974)
- University of Oxford Mair (1980)

- Culham UK, Roy Clampitt Prewett (1980)
- Oregon Graduate Centre L.Swanson (1980)
- Oregon Graduate Centre J. Orloff (1974)
- MIT, J. Melngailis (1980).

Helium Ion Microscope (HeIM)

Another ion source seen in commercially available instruments is a helium ion source, which is inherently less damaging to the sample than Ga ions although it will still sputter small amounts of material especially at high magnifications and long scan times. As helium ions can be focused into a small probe size and provide a much smaller sample interaction than high energy (>1 kV) electrons in the SEM, the He ion microscope can generate equal or higher resolution images with good material contrast and a higher depth of focus. Commercial instruments are capable of sub 1 nm resolution.

Wien Filter in Focused Ion Beam Setup

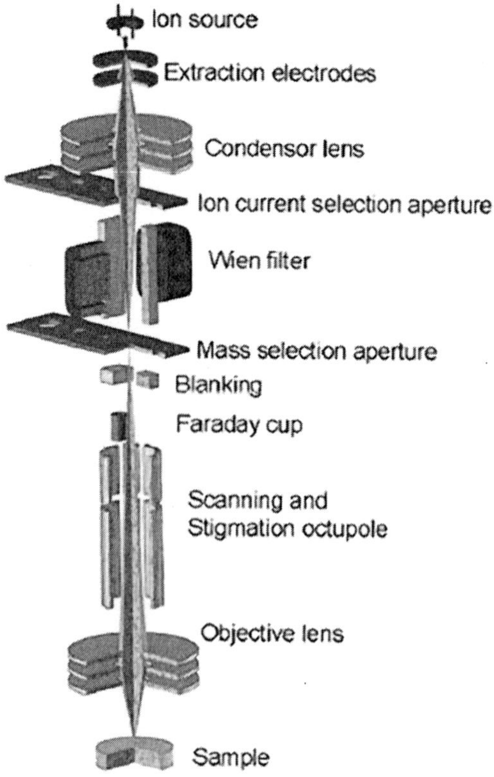

Figure: ExB Column from Orsay Physics

Imaging and milling with Ga ions always result in Ga incorporation near the sample surface. As the sample surface is sputtered away at a rate proportional to the sputtering yield and the ion flux (ions per area per time), the Ga is implanted further into the sample, and a steady-state profile of Ga is reached. This implantation is often a problem in the range of the semiconductor where silicon can be amorphised by the gallium. In order to get an alternative solution to Ga LMI sources, mass-filtered columns have been developed, based on a Wien filter technology. Such sources include Au-Si, Au-Ge and Au-Si-Ge sources providing Si, Cr, Fe, Co, Ni, Ge, In, Sn, Au, Pb and other elements.

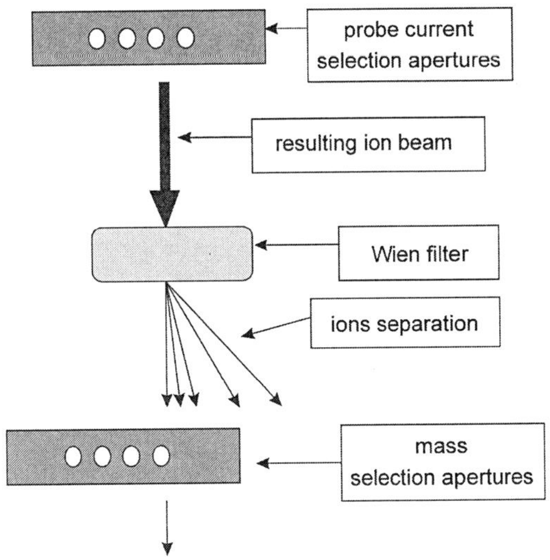

Figure: Mass selection in the FIB column

The principle of a Wien filter is based on the equilibrium of the opposite forces induced by perpendicular electrostatic and a magnetic fields acting on accelerated particles. The proper mass trajectory remains straight and passes through the mass selection aperture while the other masses are stopped.

Besides allowing the use of sources others than gallium, these columns can switch from different species simply by adjusting the properties of the Wien filter. Larger ions can be used to make rapid milling before refining the contours with smaller ones. The user also benefits from the possibility to dope his sample with elements of suitable alloy sources. The latter property has found great interests in the investigation of magnetic materials and devices. Khizroev and

Litvinov have shown, with the help of magnetic force microscopy (MFM), that there is a critical dose of ions that a magnetic material can be exposed to without experiencing a change in the magnetic properties. Exploiting FIB from such an unconventional perspective is especially favourable today when the future of so many novel technologies depends on the ability to rapidly fabricate prototype nanoscale magnetic devices.

Dye Penetrant Inspection

Dye penetrant inspection (DPI), also called liquid penetrant inspection (LPI) or penetrant testing (PT), is a widely applied and low-cost inspection method used to locate surface-breaking defects in all non-porous materials (metals, plastics, or ceramics). The penetrant may be applied to all non-ferrous materials and ferrous materials, although for ferrous components magnetic-particle inspection is often used instead for its subsurface detection capability. LPI is used to detect casting, forging and welding surface defects such as hairline cracks, surface porosity, leaks in new products, and fatigue crackson in-service components.

History

The oil and whiting method used in the railroad industry in the early 1900's was the first recognised use of the principles of penetrants to detect cracks. The oil and whiting method used an oil solvent for cleaning followed by the application of a whiting or chalk coating, which absorbed oil from the cracks revealing their locations. Soon a dye was added to the liquid. By the 1940's, fluorescent or visible dye was added to the oil used to penetrate test objects.

Experience showed that temperature and soak time were important. This started the practice of written instructions to provide standard, uniform results. The use of written procedures has evolved, giving the ability for design engineers and manufacturers to get the high standard results from any properly trained and certified liquid penetrant testing technician.

Principles

DPI is based upon capillary action, where low surface tension fluid penetrates into clean and dry surface-breaking discontinuities. Penetrant may be applied to the test component by dipping, spraying, or brushing. After adequate penetration time has been allowed, the excess penetrant is removed, a developer is applied. The developer helps to draw penetrant out of the flaw where a invisible indication becomes visible to the inspector. Inspection is performed under ultraviolet or white light, depending upon the type of dye used - fluorescent or nonfluorescent (visible).

Materials

Penetrants are classified into sensitivity levels. Visible penetrants are typically red in colour, and represent the lowest sensitivity. Fluorescent penetrants contain two or more dyes that fluoresce when excited by ultraviolet (UV-A) radiation (also known as black light). Since Fluorescent penetrant inspection is performed in a darkened environment, and the excited dyes emit brilliant yellow-green light that contrasts strongly against the dark background, this material is more sensitive to small defects.

When selecting a sensitivity level one must consider many factors, including the environment under which the test will be performed, the surface finish of the specimen, and the size of defects sought. One must also assure that the test chemicals are compatible with the sample so that the examination will not cause permanent staining, or degradation. This technique can be quite portable, because in its simplest form the inspection requires only 3 aerosol spray cans, some lint free clothes, and adequate visible light. Stationary systems with dedicated application, wash, and development stations, are more costly and complicated, but result in better sensitivity and higher samples through-put.

Inspection Steps

Below are the main steps of Liquid Penetrant Inspection:

1. Pre-cleaning: The test surface is cleaned to remove any dirt, paint, oil, grease or any loose scale that could either keep penetrant out of a defect, or cause irrelevant or false indications. Cleaning methods may include solvents, alkaline cleaning steps, vapor degreasing, or media blasting. The end goal of this step is a clean surface where any defects present are open to the surface, dry, and free of contamination. Note that if media blasting is used, it may "work over" small discontinuities in the part, and an etching bath is recommended as a post-bath treatment.
2. Application of Penetrant: The penetrant is then applied to the surface of the item being tested. The penetrant is allowed "dwell time" to soak into any flaws (generally 5 to 30 minutes). The dwell time mainly depends upon the penetrant being used, material being testing and the size of flaws sought. As expected,

smaller flaws require a longer penetration time. Due to their incompatible nature one must be careful not to apply solvent-based penetrant to a surface which is to be inspected with a water-washable penetrant.

3. Excess Penetrant Removal: The excess penetrant is then removed from the surface. The removal method is controlled by the type of penetrant used. Water-washable, solvent-removable, lipophilic post-emulsifiable, or hydrophilic post-emulsifiable are the common choices. Emulsifiers represent the highest sensitivity level, and chemically interact with the oily penetrant to make it removable with a water spray. When using solvent remover and lint-free cloth it is important to not spray the solvent on the test surface directly, because this can remove the penetrant from the flaws. If excess penetrant is not properly removed, once the developer is applied, it may leave a background in the developed area that can mask indications or defects. In addition, this may also produce false indications severely hindering your ability to do a proper inspection.

4. Application of Developer: After excess penetrant has been removed a white developer is applied to the sample. Several developer types are available, including: non-aqueous wet developer, dry powder, water suspendable, and water soluble. Choice of developer is governed by penetrant compatibility (one can't use water-soluble or suspendable developer with water-washable penetrant), and by inspection conditions. When using non-aqueous wet developer (NAWD) or dry powder, the sample must be dried prior to application, while soluble and suspendable developers are applied with the part still wet from the previous step. NAWD is commercially available in aerosol spray cans, and may employ acetone,isopropyl alcohol, or a propellant that is a combination of the two. Developer should form a semi-transparent, even coating on the surface.

   The developer draws penetrant from defects out onto the surface to form a visible indication, commonly known as bleed-out. Any areas that bleed-out can indicate the location, orientation and possible types of defects on the surface. Interpreting the results and characterising defects from the indications found may require some training and/or experience [the indication size is not the actual size of the defect]

5. Inspection: The inspector will use visible light with adequate intensity (100 foot-candles or 1100 lux is typical) for visible dye penetrant. Ultraviolet (UV-A) radiation of adequate intensity (1,000 micro-watts per centimetre squared is common), along with low ambient light levels (less than 2 foot-candles) for fluorescent penetrant examinations. Inspection of the test surface should take place after a 10 minute development time. This time delay allows the blotting action to occur. The inspector may observe the sample for indication formation when using visible dye. It is also good practice to observe indications as they form because the characteristics of the bleed out are a significant part of interpretation characterisation of flaws.
6. Post Cleaning: The test surface is often cleaned after inspection and recording of defects, especially if post-inspection coating processes are scheduled.

Advantages and Disadvantages

The main advantages of DPI are the speed of the test and the low cost. The main disadvantages are that it only detects surface flaws and it does not work on very rough surfaces. Also, on certain surfaces a great enough colour contrast cannot be achieved or the dye will stain the workpiece.

Limited training is required for the operator — although experience is quite valuable. Proper cleaning is necessary to assure that surface contaminants have been removed and any defects present are clean and dry. Some cleaning methods have been shown to be detrimental to test sensitivity, so acid etching to remove metal smearing and re-open the defect may be necessary.

Scanning Electron Microscope

A scanning electron microscope (SEM) is a type of electron microscope that images a sample by scanning it with a high-energy beam of electrons in a raster scanpattern. The electrons interact with the atoms that make up the sample producing signals that contain information about the sample's surface topography, composition, and other properties such as electrical conductivity.

The types of signals produced by an SEM include secondary electrons, back-scattered electrons (BSE), characteristic X-rays, light (cathodoluminescence), specimen current and transmitted electrons. Secondary electron detectors are common in all SEMs, but it is rare that a single machine would have detectors for all possible signals.

The signals result from interactions of the electron beam with atoms at or near the surface of the sample. In the most common or standard detection mode, secondary electron imaging or SEI, the SEM can produce very high-resolution images of a sample surface, revealing details less than 1 nm in size. Due to the very narrow electron beam, SEM micrographs have a large depth of field yielding a characteristic three-dimensional appearance useful for understanding the surface structure of a sample. This is exemplified by the micrograph of pollen shown to the right. A wide range of magnifications is possible, from about 10 times (about equivalent to that of a powerful hand-lens) to more than 500,000 times, about 250 times the magnification limit of the best light microscopes. Back-scattered electrons (BSE) are beam electrons that are reflected from the sample by elastic scattering. BSE are often used in analytical SEM along with the spectra made from the characteristic X-rays. Because the intensity of the BSE signal is strongly related to the atomic number (Z) of the specimen, BSE images can provide information about the distribution of different elements in the sample. For the same reason, BSE imaging can image colloidal gold immuno-labels of 5 or 10 nm diametre, which would otherwise be difficult or impossible to detect in secondary electron images in biological specimens. Characteristic X-rays are emitted when the electron beam removes an inner shell electron from the sample, causing a higher energy electron to fill the shell and release energy. These characteristic X-rays are used to identify the composition and measure the abundance of elements in the sample.

History

The first SEM image was obtained by Max Knoll, who in 1935 obtained an image of silicon steelshowing electron channelling contrast. Further pioneering work on the physical principles of the SEM and beam specimen interactions was performed by Manfred von Ardenne in 1937, who produced a British patent but never made a practical instrument. The SEM was further developed by Professor Sir Charles Oatley and his postgraduate student Gary Stewart and was first marketed in 1965 by the Cambridge Scientific Instrument Company as the "Stereoscan". The first instrument was delivered to DuPont.

Scanning Process and Image Formation

In a typical SEM, an electron beam is thermionicallyemitted from an electron gun fitted with a tungsten filamentcathode. Tungsten is normally used in thermionic electron guns because it has the highest melting point and lowest vapour pressure of all metals, thereby allowing it to be heated for electron emission, and because of its low cost. Other types of electron emitters include lanthanum hexaboride (LaB_6) cathodes, which can be used in a standard tungsten filament SEM if the vacuum system is upgraded and field emission guns (FEG), which may be of the cold-cathode type using tungsten single crystal emitters or the thermally assisted Schottky type, using emitters of zirconium oxide.

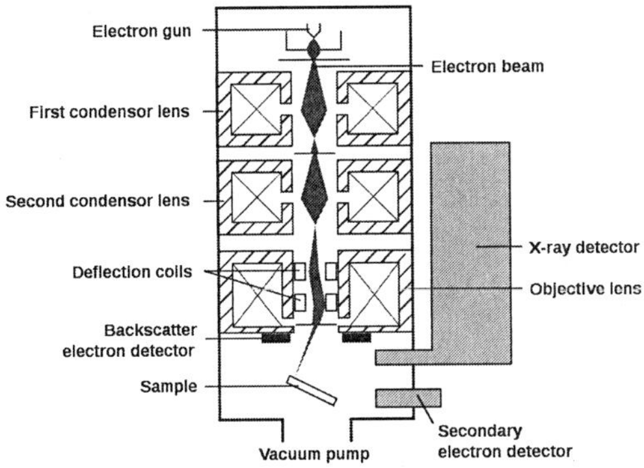

Figure: Schematic diagram of an SEM.

The electron beam, which typically has an energy ranging from 0.2 keV to 40 keV, is focused by one or two condenser lenses to a spot about 0.4 nm to 5 nm in diametre. The beam passes through pairs of scanning coils or pairs of deflector plates in the electron column, typically in the final lens, which deflect the beam in the x and y axes so that it scans in a raster fashion over a rectangular area of the sample surface.

When the primary electron beam interacts with the sample, the electrons lose energy by repeated random scattering and absorption within a teardrop-shaped volume of the specimen known as the interaction volume, which extends from less than 100 nm to around 5 μm into the surface. The size of the interaction volume depends on the electron's landing energy, the atomic number of the specimen and the specimen's density.

The energy exchange between the electron beam and the sample results in the reflection of high-energy electrons by elastic scattering, emission of secondary electrons by inelastic scattering and the emission of electromagnetic radiation, each of which can be detected by specialised detectors. The beam current absorbed by the specimen can also be detected and used to create images of the distribution of specimen current. Electronic amplifiers of various types are used to amplify the signals, which are displayed as variations in brightness on a cathode ray tube. The raster scanning of the CRT display is synchronised with that of the beam on the specimen in the microscope, and the resulting image is therefore a distribution map of the intensity of the signal being emitted from the scanned area of the specimen. The image may be captured by photography from a high-resolution cathode ray tube, but in modern machines is digitally captured and displayed on a computer monitor and saved to a computer's hard disk.

Magnification

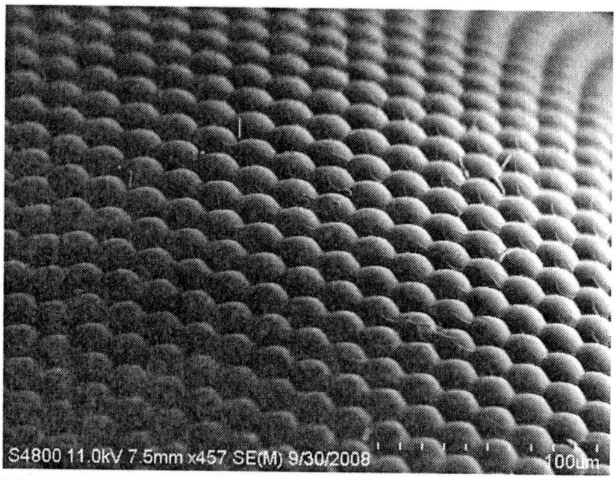

Figure: An SEM micrograph of a house fly compound eye surface at 450× magnification.

Magnification in a SEM can be controlled over a range of up to 6 orders of magnitude from about 10 to 500,000 times. Unlike optical and transmission electron microscopes, image magnification in the SEM is not a function of the power of the objective lens. SEMs may have condenser and objective lenses, but their function is to focus the beam to a spot, and not to image the specimen.

Provided the electron gun can generate a beam with sufficiently small diametre, a SEM could in principle work entirely without condenser or objective lenses, although it might not be very versatile or achieve very high resolution.

In a SEM, as in scanning probe microscopy, magnification results from the ratio of the dimensions of the raster on the specimen and the raster on the display device. Assuming that the display screen has a fixed size, higher magnification results from reducing the size of the raster on the specimen, and vice versa. Magnification is therefore controlled by the current supplied to the x, y scanning coils, or the voltage supplied to the x, y deflector plates, and not by objective lens power.

Sample Preparation

Figure: 13mm radius aluminium stubs (the base to support and prepare a sample on) for sample preparation and gold coating

All samples must also be of an appropriate size to fit in the specimen chamber and are generally mounted rigidly on a specimen holder called a specimen stub. Several models of SEM can examine any part of a 6-inch (15 cm) semiconductor wafer, and some can tilt an object of that size to 45°.

For conventional imaging in the SEM, specimens must be electrically conductive, at least at the surface, and electrically grounded to prevent the accumulation of electrostatic charge at the surface. Metal objects require little special preparation for SEM except for cleaning and mounting on a specimen stub. Nonconductive specimens tend to charge when scanned by the electron beam, and especially in secondary electron imaging mode, this causes scanning faults and other image artifacts. They are therefore usually coated with an ultrathin coating of electrically conducting material, deposited on the sample either by low-vacuum sputter coating or by high-vacuum evaporation. Conductive materials in current use for specimen coating include gold, gold/palladium alloy, platinum, osmium, iridium, tungsten, chromium, and graphite. Additionally, coating may increase signal/noise ratio for samples of low atomic number (Z). The improvement arises because secondary electron emission for high-Z materials is enhanced.

An alternative to coating for some biological samples is to increase the bulk conductivity of the material by impregnation with osmium using variants of the OTO staining method (O-osmium, T-thiocarbohydrazide, O-osmium). Nonconducting specimens may be imaged uncoated using specialised SEM instrumentation such as the "Environmental SEM" (ESEM) or field emission gun (FEG) SEMs operated at low voltage. Environmental SEM instruments place the specimen in a relatively high-pressure chamber where the working distance is short and the electron optical column is differentially pumped to keep vacuum adequately low at the electron gun. The high-pressure region around the sample in the ESEM neutralises charge and provides an amplification of the secondary electron signal. Low-voltage SEM of non-conducting specimens can be operationally difficult to accomplish in a conventional SEM and is typically a research application for specimens that are sensitive to the process of applying conductive coatings. Low-voltage SEM of non-conducting specimens can be operationally difficult to accomplish in a conventional SEM and is typically a research application for specimens that are sensitive to the process of applying conductive coatings. Low-voltage SEM is typically conducted in an FEG-SEM because the FEG is capable of producing high primary electron brightness even at low accelerating potentials. Operating conditions to prevent charging of non-conductive specimens must be adjusted such that the incoming beam current was equal to sum of outcoming secondary and backscattered electrons currents. It usually occurs at accelerating voltages of 0.5-4 kV.

Embedding in a resin with further polishing to a mirror-like finish can be used for both biological and materials specimens when imaging in backscattered electrons or when doing quantitative X-ray microanalysis.

Biological Samples

For SEM, a specimen is normally required to be completely dry, since the specimen chamber is at high vacuum. Hard, dry materials such as wood, bone, feathers, dried insects, or shells can be examined with little further treatment, but living cells and tissues and whole, soft-bodied organisms usually require chemical fixation to preserve and stabilise their structure. Fixation is usually performed by incubation in a solution of a buffered chemical fixative, such as glutaraldehyde, sometimes in combination with formaldehyde and other fixatives, and optionally followed by postfixation with osmium tetroxide. The fixed tissue is then dehydrated. Because air-drying causes collapse and shrinkage, this is commonly achieved by replacement of water in the cells with organic solvents such as ethanol or acetone, and replacement of these solvents in turn with a transitional fluid such as liquid carbon dioxide by critical point drying. The carbon dioxide is finally removed while in a supercritical state, so that no gas-liquid interface is present within the sample during drying. The dry specimen is usually mounted on a specimen stub using an adhesive such as epoxy resin or electrically conductive double-sided adhesive tape, and sputter-coated with gold or gold/palladium alloy before examination in the microscope.

If the SEM is equipped with a cold stage for cryo-microscopy, cryofixation may be used and low-temperature scanning electron microscopy performed on the cryogenically fixed specimens. Cryo-fixed specimens may be cryo-fractured under vacuum in a special apparatus to reveal internal structure, sputter-coated, and transferred onto the SEM cryo-stage while still frozen. Low-temperature scanning electron microscopy is also applicable to the imaging of temperature-sensitive materials such as ice (see e.g. illustration at left) and fats.

Freeze-fracturing, freeze-etch or freeze-and-break is a preparation method particularly useful for examining lipid membranes and their incorporated proteins in "face on" view. The preparation method reveals the proteins embedded in the lipid bilayer.

Gold has a high atomic number and sputter-coating with gold produces high topographic contrast and resolution. However, the coating has a thickness of a few nanometres, and can obscure the underlying fine detail of the specimen at very high magnification. Low-vacuum SEMs with differential pumping apertures allow samples to be imaged without such coating and without the loss of natural contrast caused by the coating, but are unable to achieve the resolution attainable by conventional SEMs with coated specimens.

Materials

Back scattered electron imaging, quantitative X-ray analysis, and X-ray mapping of specimens often requires that the surfaces be ground and polished to an ultra smooth surface. Specimens that undergo WDS or EDS analysis are often carbon coated. In general, metals are not coated prior to imaging in the SEM because they are conductive and provide their own pathway to ground.

Fractography is the study of fractured surfaces that can be done on a light microscope or commonly, on an SEM. The fractured surface is cut to a suitable size, cleaned of any organic residues, and mounted on a specimen holder for viewing in the SEM.

Integrated circuits may be cut with a focused ion beam (FIB) or other ion beam milling instrument for viewing in the SEM. The SEM in the first case may be incorporated into the FIB. Metals, geological specimens, and integrated circuits all may also be chemically polished for viewing in the SEM. Special high-resolution coating techniques are required for high-magnification imaging of inorganic thin films.

ESEM

Environmental SEM allows use of hydrated specimens, omitting dehydration procedure. The accumulation of electric charge on the surfaces of non-metallic specimens can be avoided by using environmental SEM in which the specimen is placed in an internal chamber at higher pressure, rather than the vacuum in the electron optical column. Positively charged ions generated by beam interactions with the gas help to neutralise the negative charge on the specimen surface. The pressure of gas in the chamber can be controlled, and the type of gas used can be varied according to need. Coating is thus unnecessary. In ESEM mode X-ray analysis is prone to artifacts arose from electron beam scattering on gases of specimen chamber.

Detection of Secondary Electrons

The most common imaging mode collects low-energy (<50 eV) secondary electrons that are ejected from the k-orbitals of the specimen atoms by inelastic scattering interactions with beam electrons. Due to their low energy, these electrons originate within a few nanometres from the sample surface. The electrons are detected by an Everhart-Thornley detector, which is a type of scintillator-photomultiplier system. The secondary electrons are first collected by attracting them towards an electrically biased grid at about +400 V, and then further accelerated towards a phosphor or scintillator positively biased to about +2,000 V. The accelerated secondary electrons are now sufficiently energetic to cause the scintillator to emit flashes of light (cathodoluminescence), which are conducted to a photomultiplier outside the SEM column via a light pipe and a window in the wall of the specimen chamber. The amplified electrical signal output by the photomultiplier is displayed as a two-dimensional intensity distribution that can be viewed and photographed on an analogue video display, or subjected to analog-to-digital conversion and displayed and saved as a digital image. This process relies on a raster-scanned primary beam. The brightness of the signal depends on the number of secondary electrons reaching the detector. If the beam enters the sample perpendicular to the surface, then the activated region is uniform about the axis of the beam and a certain number of electrons "escape" from within the sample. As the angle of incidence increases, the "escape" distance of one side of the beam will decrease, and more secondary electrons will be emitted. Thus steep surfaces and edges tend to be brighter than flat surfaces, which results in images with a well-defined, three-dimensional appearance. Using the signal of secondary electrons image resolution less than 0.5 nm is possible.

Detection of Backscattered Electrons

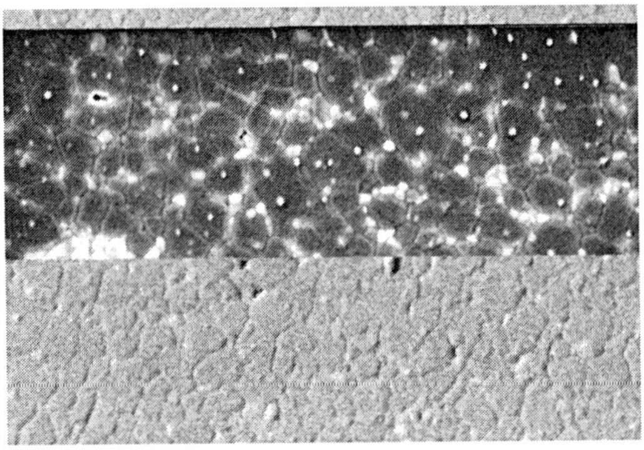

Figure: Comparison of SEM techniques:

Top: backscattered electron analysis - composition

Bottom: secondary electron analysis - topography

Backscattered electrons (BSE) consist of high-energy electrons originating in the electron beam, that are reflected or backscattered out of the specimen interaction volume by elastic scattering interactions with specimen atoms. Since heavy elements (high atomic number) backscatter electrons more strongly than light elements (low atomic number), and thus appear brighter in the image, BSE are used to detect contrast between areas with different chemical compositions. The Everhart-Thornley detector, which is normally positioned to one side of the specimen, is inefficient for the detection of backscattered electrons because few such electrons are emitted in the solid angle subtended by the detector, and because the positively biased detection grid has little ability to attract the higher energy BSE electrons. Dedicated backscattered electron detectors are positioned above the sample in a "doughnut" type arrangement, concentric with the electron beam, maximising the solid angle of collection.

BSE detectors are usually either of scintillator or of semiconductor types. When all parts of the detector are used to collect electrons symmetrically about the beam, atomic number contrast is produced. However, strong topographic contrast is produced by collecting backscattered electrons from one side above the specimen using an asymmetrical, directional BSE detector; the resulting contrast appears as illumination of the topography from that side. Semiconductor detectors can be made in radial segments that can be switched in or out to control the type of contrast produced and its directionality.

Backscattered electrons can also be used to form an electron backscatter diffraction (EBSD) image that can be used to determine the crystallographic structure of the specimen.

Beam-injection Analysis of Semiconductors

The nature of the SEM's probe, energetic electrons, makes it uniquely suited to examining the optical and electronic properties of semiconductor materials. The high-energy electrons from the SEM beam will inject charge carriers into the semiconductor. Thus, beam electrons lose energy by promoting electrons from the valence band into the conduction band, leaving behind holes.

In a direct bandgap material, recombination of these electron-hole pairs will result in cathodoluminescence; if the sample contains an internal electric field, such as is present at a p-n junction, the SEM beam injection of carriers will cause electron beam induced current (EBIC) to flow.

Cathodoluminescence and EBIC are referred to as "beam-injection" techniques, and are very powerful probes of the optoelectronic behaviour of semiconductors, in particular for studying nanoscale features and defects.

Cathodoluminescence

Cathodoluminescence, the emission of light when atoms excited by high-energy electrons return to their ground state, is analogous to UV-induced fluorescence, and some materials such as zinc sulfide and some fluorescent dyes, exhibit both phenomena. Cathodoluminescence is most commonly experienced in everyday life as the light emission from the inner surface of the cathode ray tube in television sets and computer CRT monitors. In the SEM, CL detectors either collect all light emitted by the specimen or can analyse the wavelengths emitted by the specimen and display an emission spectrum or an image of the distribution of cathodoluminescence emitted by the specimen in real colour.

X-ray Microanalysis

X-rays, which are also produced by the interaction of electrons with the sample, may also be detected in an SEM equipped for energy-dispersive X-ray spectroscopy or wavelength dispersive X-ray spectroscopy.

Resolution of the SEM

A video illustrating a typical practical magnification range of a scanning electron microscope designed for biological specimens. The video starts at 25x, about 6 mm across the whole field of view, and zooms in to 12000x, about 12 ìm across the whole field of view. The spherical objects are glass beads with a diametre of 10 ìm, similar in diametre to a red blood cell.

The spatial resolution of the SEM depends on the size of the electron spot, which in turn depends on both the wavelength of the electrons and the electron-optical system that produces the scanning beam. The resolution is also limited by the size of the interaction volume, or the extent to which the material interacts with the electron beam. The spot size and the interaction volume are both large compared to the distances between atoms, so the resolution of the SEM is not high enough to image individual atoms, as is possible in the shorter wavelength (i.e. higher energy) transmission electron microscope (TEM). The SEM has compensating advantages, though, including the ability to image a comparatively large area of the specimen; the ability to image bulk materials (not just thin films or foils); and the variety of analytical modes available for measuring the composition and properties of the specimen. Depending on the instrument, the resolution can fall somewhere between less than 1 nm and 20 nm. By 2009, The world's highest SEM resolution at high-beam energies (0.4 nm at 30 kV) is obtained with the Hitachi S-5500. At low-beam energies, the best resolution (by 2009) is achieved by the Magellan XHR system from FEI Company (0.9 nm at 1 kV).

Environmental SEM

Main article: Environmental scanning electron microscope

Conventional SEM requires samples to be imaged under vacuum, because a gas atmosphere rapidly spreads and attenuates electron beams. As a consequence, samples that produce a significant amount of vapour, e.g. wet biological samples or oil-bearing rock, must be either dried or cryogenically frozen. Processes involving phase transitions, such as the drying of adhesives or melting of alloys, liquid transport, chemical reactions, and solid-air-gas systems, in general cannot be observed. Some observations of living insects have been possible, however.

The first commercial development of the Environmental SEM (ESEM) in the late 1980s allowed samples to be observed in low-pressure gaseous environments (e.g. 1-50 Torr) and high relative humidity (up to 100%). This was made possible by the development of a secondary-electron detector capable of operating in the presence of water vapour and by the use of pressure-limiting apertures with differential pumping in the path of the electron beam to separate the vacuum region (around the gun and lenses) from the sample chamber.

The first commercial ESEMs were produced by the ElectroScan Corporation in USA in 1988. ElectroScan was taken over by Philips (who later sold their electron-optics division to FEI Company) in 1996.

ESEM is especially useful for non-metallic and biological materials because coating with carbon or gold is unnecessary. Uncoated Plastics and Elastomers can be routinely examined, as can uncoated biological samples. Coating can be difficult to reverse, may conceal small features on the surface of the sample and may reduce the value of the results obtained. X-ray analysis is difficult with a coating of a heavy metal, so carbon coatings are routinely used in conventional SEMs, but ESEM makes it possible to perform X-ray microanalysis on uncoated non-conductive specimens. ESEM may be the preferred for electron microscopy of unique samples from criminal or civil actions, where forensic analysis may need to be repeated by several different experts.

3D in SEM

3D data can be measured in the SEM with different methods such as:

- photogrammetry (2 or 3 images from tilted specimen)
- photometric stereo (use of 4 images from BSE detector)
- inverse reconstruction using electron-material interactive models

Possible applications are roughness measurement, measurement of fractal dimension, corrosion measurement and height step measurement.

Electron Beam Induced Current

Electron beam induced current (EBIC) is a semiconductor analysis technique performed in a scanning electron microscope (SEM) or scanning transmission electron microscope (STEM). It is used to identify buried junctions or defects in semiconductors, or to examine minority carrier properties. EBIC is similar to cathodoluminescence in that it depends on the creation ofelectron–hole pairs in the semiconductor sample by the microscope's electron beam. This technique is used in semiconductor failure analysis and solid-state physics.

Physics of the Technique

If the semiconductor sample contains an internal electric field, as will be present in the depletion region at a p-n junction or schottky junction, the electron–hole pairs will be separated by drift due to the electric field. If the p- and n-sides (or semiconductor and schottky contact, in the case of a schottky device) are connected through a picoammetre, a current will flow.

EBIC is best understood by analogy: in a solar cell, photons of light fall on the entire cell, thus delivering energy and creating electron hole pairs, and cause a current to flow. In EBIC, energetic electrons take the role of the photons, causing the EBIC current to flow. However, because the electron beam of an SEM or STEM is very small, it is scanned across the sample and variations in the induced EBIC are used to map the electronic activity of the sample.

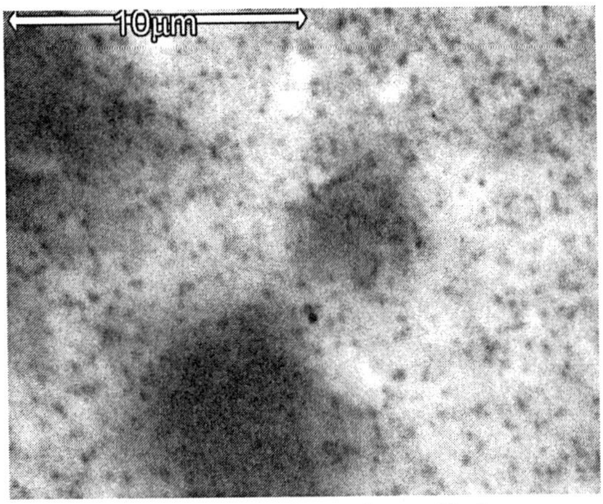

Figure: Plan-view EBIC showing defects in a diode

By using the signal from the picoammetre as the imaging signal, an EBIC image is formed on the screen of the SEM or STEM. When a semiconductor device is imaged in cross-section, the depletion region will show bright EBIC contrast. The shape of the contrast can be treated mathematically to determine the minority carrier properties of the semiconductor, such as diffusion length and surface recombination velocity. In plan-view, areas with good crystal quality will show bright contrast, and areas containing defects will show dark EBIC contrast.

As such, EBIC is a semiconductor analysis technique useful for evaluating minority carrier properties and defect populations.

EBIC has also been extended to the study of local defects in insulators. For example, W.S. Lau (Lau Wai Shing) developed "true oxide electron beam induced current" in the 1990s. Thus, besides p-n junction or Schottky junction, EBIC can also be applied to MOS diodes. Local defects in semiconductor and local defects in the insulator could be distinguished. There exists a kind of defect which originates in the silicon substrate and extends into the insulator on top of the silicon substrate. (Please see references below.)

Recently, EBIC has been applied to high-k dielectric used in advanced CMOS technology. (Please see references below.)

Quantitative EBIC

Most EBIC images are qualitative and only show the EBIC signal as contrast image. Use of a external scan control generator on the SEM and a dedicated data acquisition system allow for sub-picoamp measurements and can give quantitative results. Some systems that are commercially available that do this , and provide the ability to provide functional imaging by biasing and applying gate voltages to semiconductor devices.

Charge Induced Voltage Alteration

Charge induced voltage alteration (CIVA) is a technique which uses a scanning electron microscope to locate open conductors on CMOS integrated circuits. This technique is used in semiconductor failure analysis.

Theory of Operation

The scanning of an electron beam across the surface of the device may result in additional charge buildup on conductors which are disconnected from the remainder of the circuit (floating conductors). If a CMOS device is under active bias, the presence of open conductors may not prevent the circuit from operating at low clock frequencies as the result of quantum tunnelling effects. By injecting charge into floating conductors that are operating in this tunnelling mode, it is possible to produce additional loading which can be detected by monitoring the power supply current. These changes in supply current may be associated with the visual image of the device at the coordinates at which the change was detected. The result is a scanning electron microscope image which has an overlay of the floating conductors superimposed on it.

Electron Backscatter Diffraction

Electron backscatter diffraction (EBSD), also known as backscatter Kikuchi diffraction (BKD) is a microstructural-crystallographic technique used to examine the crystallographic orientation of many materials, which can be used to elucidate texture or preferred orientation of any crystalline or polycrystalline material. EBSD can be used to index and identify the seven crystal systems, and as such it is applied to crystal orientation mapping, defect studies, phase identification, grain boundary and morphology studies, regional heterogeneity investigations, material discrimination, microstrain mapping, and using complementary techniques, physico-chemical identification. Traditionally these types of studies have been carried out using X-ray diffraction (XRD), neutron diffraction and/or electron diffraction in a TEM.

Experimentally EBSD is conducted using a Scanning Electron Microscope (SEM) equipped with an EBSD detector containing at least a phosphor screen, compact lens and low light CCD camera chip. Commercially available EBSD systems typically come with one of two different CCD cameras: for fast measurements the CCD chip has a native resolution of 640x480 pixels; for slower, and more sensitive measurements, the CCD chip resolution can go up to 1600x1200 pixels. However, with higher resolutions the readouts are more time-consuming. The biggest advantage of the high-resolution detectors is their higher sensitivity. For texture and orientation measurements, the images are binned in order to reduce their size and reduce computational times. Thus, the transfer and interpretation of up to 880 images/s is possible if the diffraction signal is sufficient .

For an EBSD measurement a flat/polished crystalline specimen is placed in the SEM chamber at a highly tilted angle (~70° from horizontal) towards the diffraction camera (to increase the contrast in the resultant electron backscatter diffraction pattern). The phosphor screen is located within the specimen chamber of the SEM at an angle off approximately 90° to the pole piece and is coupled to a compact lens which focusses the image from the phosphor screen onto the CCD camera. Electrons can backscatter within the material as these backscattering electrons exit the crystal they may exit at the Bragg condition related to the spacing of the periodic atomic lattice planes of the crystalline structure. These diffracted electrons can escape the material and some will collide and excite the phosphor causing it to fluoresce.

An electron backscatter diffraction pattern (EBSP) is formed when many different planes diffract different electrons to form kikuchi bands which correspond to each of the lattice diffracting planes. If the system geometry is well described, it is possible to relate the bands present in the EBSP to the underlying crystal phase and orientation of the material within the electron interaction volume. Each band can be indexed individually by the Miller indices of the diffracting plane which formed it. In most materials, only three bands/planes which intercept are required to describe a unique solution to the crystal orientation (based upon their interplanar angles) and most commercial systems use look up tables with international crystal data bases to perform indexing.

While this 'geometric' description related to the kinematic solution (using the Bragg condition) is very powerful and useful for orientation and texture analysis, it only describes the geometry of the crystalline lattice and ignores many physical processes involved within the diffracting material. To adequately describe finer features within the EBSP, one must use a many beam dynamical model (e.g. the variation in band intensities in an experimental pattern does not fit the kinematic solution related to the structure factor).

Indexing

Often, the first step in the EBSD process after pattern collection is indexing. This allows for identification of the crystal orientation at the single volume of the sample from where the pattern was collected. With EBSD software, pattern bands are typically detected via a mathematical routine called the Hough transform, in which every pixel in Hough space denotes a unique line/band in the EBSP. The Hough transform is used to enable band detection, which are difficult to locate by computer in the original EBSP. Once the band locations have been detected it is possible to relate these locations to the underlying crystal orientation, as angles between bands represent angles between lattice planes. Thus when the position / angles between three bands are known an orientation solution can be determined. In highly symmetric materials typically more than three bands are utilised to obtain and verify the orientation measurement.

There are two leading methods of indexing performed by most commercial EBSD software: triplet voting; and minimising the 'fit' between the experimental pattern and a computationally determined orientation.

Triplet voting involves identify multiple 'triplets' associated with different solutions to the crystal orientation; each crystal orientation determined from each triplet receives one vote. Should four bands identify the same crystal orientation then four (four choose three) votes will be cast for that particular solution. Thus the candidate orientation with the highest number of votes will be the most likely solution to the underlying crystal orientation present. The ratio of votes for the solution chosen as compared to the total number of votes describes the confidence in the underlying solution. Care must be taken in interpreting this 'confidence index' as some pseudo-symmetric orientations may result in low confidence for one candidate solution vs. another.

Minimising the fit involves starting with all possible orientations for a triplet. More bands are included that reduces the number of candidate orientations. As the number of bands increases, the number of possible orientations converge ultimately to one solution. The 'fit' between the measured orientation and the captured pattern can be determined.

Pattern Centre

In order to relate the orientation of a crystal, much like in X-ray diffraction, the geometry of the system must be known. In particular the pattern centre, which describes both the distance of the interaction volume to the detector and the location of the nearest point between the phosphor and the sample on the phosphor screen. Early work utilised a single crystal of known orientation being inserted into the SEM chamber and a particular feature of the EBSP was known to correspond to the pattern centre. Later developments involved exploiting various geometric relationships between the generation of an EBSP and the chamber geometry (shadow casting and phosphor movement).

Unfortunately each of these methods are cumbersome, prone to some systematic errors for a general operator and can not be easily utilised in modern SEMs with multiple designated uses. Thus most commercial EBSD systems utilise the indexing algorithm combined with an iterative movement of both crystal orientation and suggested pattern centre location. Minimising the fit between bands located within experimental patterns and those in look up tables tends to converge on the pattern centre location to an accuracy of ~0.5–1% of the pattern width.

Orientation Mapping

EBSD can be used to find the crystal orientation of the material located within the incident electron beam's interaction volume. Thus by scanning the electron beam in a prescribed fashion (typically in a square or hexagonal grid, correcting for the image foreshortening due to the sample tilt) results in many maps. These maps can spatially describe the crystal orientation of the material being interrogated and can be used to examine microtexture and sample morphology. Some of these maps describe grain orientation, grain boundary, image quality. Various statistical tools can be used to measure the average misorientation, grain size, and crystallographic texture. From this dataset numerous maps, charts and plots can be generated.

From orientation data, a wealth of information can be devised that aids in the understanding of the samples microstructure and processing history. Recent developments include understanding: the prior texture of parent phases at elevated temperature; the storage and residual deformation after mechanical testing; the population of various microstructural features, including precipitates and grain boundary character.

Integrated EBSD/EDS Mapping

When simultaneous EDS/EBSD collection can be achieved, the capabilities of both techniques can be enhanced. There are applications where sample chemistry or phase cannot be differentiated via EDS alone because of similar composition; and structure cannot be solved with EBSD alone because of ambiguous structure solutions. To accomplish integrated mapping, analysis area is scanned and at each point Hough peaks and EDS region-of-interest counts are stored. Positions of phases are determined in X-ray maps and measured EDS intensities are given in charts for each element. For each phase the chemical intensity ranges are set to select the grains. All patterns are then re-indexed off-line. The recorded chemistry determines which phase / crystal structure file is used for indexing of each point. Each pattern is indexed by only one phase and maps displaying clearly distinguished phases are generated.

EBSD when used together with other in-SEM techniques such as cathodoluminescence (CL), wavelength dispersive X-ray spectroscopy (WDS) and/or energy dispersive X-ray spectroscopy (EDS) can provide a deeper insight into the specimen's properties. For example, the minerals calcite (limestone) and aragonite (shell) have the same chemical composition – calcium carbonate ($CaCO_3$) therefore EDS/WDS cannot tell them apart, but they have different microcrystalline structures so EBSD can differentiate between them.

Energy-dispersive X-ray Spectroscopy

Energy-dispersive X-ray spectroscopy (EDS or EDX or EDAX) is an analytical technique used for the elemental analysis or chemical characterisation of a sample. It relies on the investigation of an interaction of a some source of X-ray excitation and a sample. Its characterisation capabilities are due in large part to the fundamental principle that each element has a unique atomic structure allowing X-rays that are characteristic of an element's atomic structure to be identified uniquely from one another.

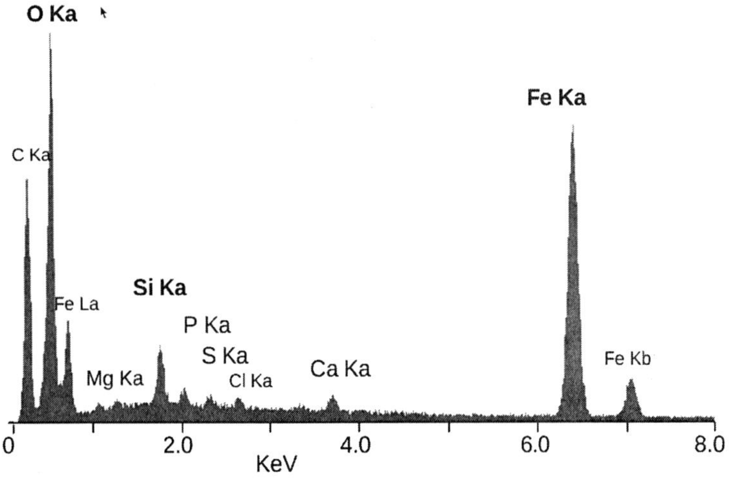

Figure: EDS spectrum of the mineral crust of Rimicaris exoculata

To stimulate the emission of characteristic X-rays from a specimen, a high-energy beam of charged particles such as electrons or protons (see PIXE), or a beam of X-rays, is focused into the sample being studied. At rest, an atom within the sample contains ground state (or unexcited) electrons in discrete energy levels or electron shells bound to the nucleus. The incident beam may excite an electron in an inner shell, ejecting it from the shell while creating an electron hole where the electron was. An electron from an outer, higher-energy shell then fills the hole, and the difference in energy between the higher-energy shell and the lower energy shell may be released in the form of an X-ray. The number and energy of the X-rays emitted from a specimen can be measured by an energy-dispersive spectrometre. As the energy of the X-rays are characteristic of the difference in energy between the two shells, and of the atomic structure of the element from which they were emitted, this allows the elemental composition of the specimen to be measured.

X-ray Generation

The equipment measures the number of emitted X-rays. Higher accelerating voltages penetrate deeper into the sample; the beam also penetrates deeper into lighter elements, which thus generate more counts.

Equipment: the Connection with SEM

There are four primary components of the EDS setup: the beam source; the X-ray detector; the pulse processor; and the analyser.

A number of free-standing EDS systems exist. However, EDS systems are most commonly found on scanning electron microscopes (SEM-EDS) and electron microprobes. Scanning electron microscopes are equipped with a cathode and magnetic lenses to create and focus a beam of electrons, and since the 1960s they have been equipped with elemental analysis capabilities. A detector is used to convert X-ray energy into voltage signals; this information is sent to a pulse processor, which measures the signals and passes them onto an analyser for data display and analysis.

Technological Variants

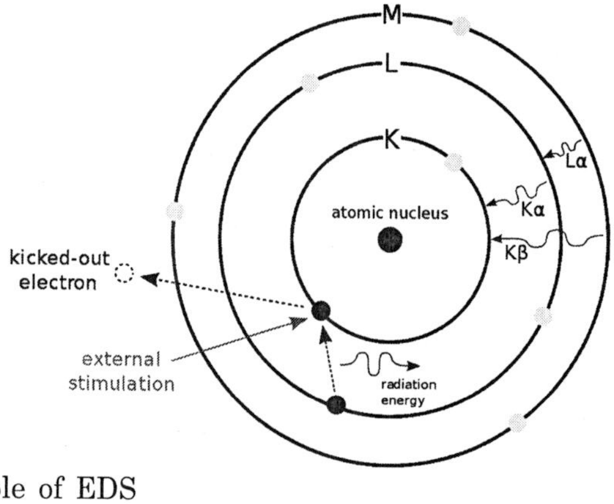

Figure: Principle of EDS

The excess energy of the electron that migrates to an inner shell to fill the newly-created hole can do more than emit an X-ray. Often, instead of X-ray emission, the excess energy is transferred to a third electron from a further outer shell, prompting its ejection. This ejected species is called an Auger electron, and the method for its analysis is known as Auger electron spectroscopy (AES).

X-ray photoelectron spectroscopy (XPS) is another close relative of EDS, utilising ejected electrons in a manner similar to that of AES. Information on the quantity and kinetic energy of ejected electrons is used to determine the binding energy of these now-liberated electrons, which is element-specific and allows chemical characterisation of a sample.

EDS is often contrasted with its spectroscopic counterpart, WDS (wavelength dispersive X-ray spectroscopy). WDS differs from EDS in that it uses the diffraction patterns created by light-matter interaction as its raw data. WDS has a much finer spectral resolution than EDS. WDS also avoids the problems associated with artifacts in EDS (false peaks, noise from the amplifiers, and microphonics). In WDS, only one element can be analysed at a time, while EDS gathers a spectrum of all elements, within limits, of a sample.

Accuracy of EDS

Accuracy of EDS spectrum can be affected by many factors. The first three elements of the Periodic Table (H, He, and Li) do not have enough electrons to produce characteristic X-ray, so they are not detectable by X-ray spectroscopy. Also, many elements will have overlapping peaks (e.g., Ti $K_{\hat{a}}$ and V $K_{\acute{a}}$, Mn $K_{\hat{a}}$ and Fe $K_{\acute{a}}$). The accuracy of the spectrum can also be affected by the nature of the sample. X-rays can be generated by any atom in the sample that is sufficiently excited by the incoming beam. These X-rays are emitted in any direction, and so they may not all escape the sample. The likelihood of an X-ray escaping the specimen, and thus being available to detect and measure, depends on the energy of the X-ray and the amount and density of material it has to pass through. This can result in reduced accuracy in inhomogeneous and rough samples.

Emerging Technology

There is a trend towards a newer EDS detector, called the silicon drift detector (SDD). The SDD consists of a high-resistivity silicon chip where electrons are driven to a small collecting anode. The advantage lies in the extremely low capacitance of this anode, thereby utilising shorter processing times and allowing very high throughput. Benefits of the SDD include:

1. High count rates and processing,
2. Better resolution than traditional Si(Li) detectors at high count rates,
3. Lower dead time (time spent on processing X-ray event),
4. Faster analytical capabilities and more precise X-ray maps or particle data collected in seconds,
5. Ability to be stored and operated at relatively high temperatures, eliminating the need for liquid nitrogen cooling.

Because the capacitance of the SDD chip is independent of the active area of the detector, much larger SDD chips can be utilised (40

mm^2 or more). This allows for even higher count rate collection. Further benefits of large area chips include:

1. Minimising SEM beam current allowing for optimisation of imaging under analytical conditions,
2. Reduced sample damage and
3. Smaller beam interaction and improved spatial resolution for high speed maps.

In recent years, a different type of EDS detector, based upon a microcalorimetre, has become commercially available. This new model allegedly has the simultaneous detection capabilities of EDS as well as the high spectral resolution of WDS. The EDS microcalorimetre relies highly on two components: an absorber, and a thermistor. The former absorbs X-rays emitted from the sample and converts this energy into heat; the latter measures the subsequent change in temperature due to the influx of heat (in essence, a thermometre). The EDS microcalorimetre has suffered from a number of drawbacks; including low count rates, poor collection efficiencies and small detector areas. The count rate is hampered by its reliance on the time constant of the calorimetre's electrical circuit. The collection efficiency is a function of the absorber material and remains to be optimised. The detector area must be small in order to keep the heat capacity as small as possible and maximise thermal sensitivity (resolution). Innovative engineering solutions are necessary for further improvement of spectroscopic microanalysis.

Transmission Electron Microscopy

Transmission electron microscopy (TEM) is a microscopy technique whereby a beam of electrons is transmitted through an ultra thin specimen, interacting with the specimen as it passes through. An image is formed from the interaction of the electrons transmitted through the specimen; the image is magnified and focused onto an imaging device, such as a fluorescent screen, on a layer of photographic film, or to be detected by a sensor such as a CCD camera.

TEMs are capable of imaging at a significantly higher resolution than light microscopes, owing to the small de Broglie wavelength of electrons. This enables the instrument's user to examine fine detail— even as small as a single column of atoms, which is tens of thousands times smaller than the smallest resolvable object in a light microscope. TEM forms a major analysis method in a range of scientific fields, in both physical and biological sciences. TEMs find application in cancer research, virology, materials science as well as pollution, nanotechnology, and semiconductor research.

At smaller magnifications TEM image contrast is due to absorption of electrons in the material, due to the thickness and composition of the material. At higher magnifications complex wave interactions modulate the intensity of the image, requiring expert analysis of observed images. Alternate modes of use allow for the TEM to observe modulations in chemical identity, crystal orientation, electronic structure and sample induced electron phase shift as well as the regular absorption based imaging.

The first TEM was built by Max Knoll and Ernst Ruska in 1931, with this group developing the first TEM with resolving power greater than that of light in 1933 and the first commercial TEM in 1939.

History of Initial Development

Figure: The first practical TEM, Originally installed at I. G Farben-Werke and now on display at the Deutsches Museum in Munich, Germany

Ernst Abbe originally proposed that the ability to resolve detail in an object was limited by the wavelength of the light used in imaging, thus limiting the useful obtainable magnification from an optical microscope to a few micrometres. Developments into ultraviolet(UV) microscopes, led by Köhler and Rohr, allowed for an increase in resolving power of about a factor of two. However this required more expensive quartz optical components, due to the absorption of UV by glass. At this point it was believed that obtaining an image with sub-micrometre information was simply impossible due to this wavelength constraint.

It had earlier been recognised by Plücker in 1858 that the deflection of "cathode rays" (electrons) was possible by the use of magnetic fields. This effect had been utilised to build primitive cathode ray oscilloscopes (CROs) as early as 1897 by Ferdinand Braun, intended as a measurement device. Indeed in 1891 it was recognised by Riecke that the cathode rays could be focused by these magnetic fields, allowing for simple lens designs. Later this theory was extended by Hans Busch in his work published in 1926, who showed that the lens maker's equation, could under appropriate assumptions, be applicable to electrons.

In 1928, at the Technological University of Berlin Adolf Matthias, Professor of High voltage Technology and Electrical Installations, appointed Max Knoll to lead a team of researchers to advance the CRO design. The team consisted of several PhD students including Ernst Ruska and Bodo von Borries. This team of researchers concerned themselves with lens design and CRO column placement, which they attempted to obtain the parameters that could be optimised to allow for construction of better CROs, as well as the development of electron optical components which could be used to generate low magnification (nearly 1:1) images. In 1931 the group successfully generated magnified images of mesh grids placed over the anode aperture. The device used two magnetic lenses to achieve higher magnifications, arguably the first electron microscope. In that same year, Reinhold Rudenberg, the scientific director of the Siemens company, had patented an electrostatic lens electron microscope.

Improving Resolution

At this time the wave nature of electrons, which were considered charged matter particles, had not been fully realised until the publication of the De Broglie hypothesis in 1927. The group was unaware of this publication until 1932, where it was quickly realised that the De Broglie wavelength of electrons was many orders of magnitude smaller than that for light, theoretically allowing for imaging at atomic scales. In April 1932, Ruska suggested the construction of a new electron microscope for direct imaging of specimens inserted into the microscope, rather than simple mesh grids or images of apertures. With this device successful diffraction and normal imaging of aluminium sheet was achieved, however exceeding the magnification achievable with light microscopy had still not been successfully demonstrated. This goal was achieved in September 1933, using images of cotton fibres, which were quickly acquired before being damaged by the electron beam.

At this time, interest in the electron microscope had increased, with other groups, such as Albert Prebus and James Hillier at the University of Toronto who constructed the first TEM in North America in 1938, continually advancing TEM design.

Research continued on the electron microscope at Siemens in 1936, the aim of the research was the development improvement of TEM imaging properties, particularly with regard to biological specimens. At this time electron microscopes were being fabricated for specific groups, such as the "EM1" device used at the UK National Physical Laboratory. In 1939 the first commercial electron microscope, pictured, was installed in the Physics department of I. G Farben-Werke. Further work on the electron microscope was hampered by the destruction of a new laboratory constructed at Siemens by an air-raid, as well as the death of two of the researchers, Heinz Müller and Friedrick Krause during World War II.

Further Research

After World War II, Ruska resumed work at Siemens, where he continued to develop the electron microscope, producing the first microscope with 100k magnification. The fundamental structure of this microscope design, with multi-stage beam preparation optics, is still used in modern microscopes. The worldwide electron microscopy community advanced with electron microscopes being manufactured in Manchester UK, the USA (RCA), Germany (Siemens) and Japan. The first international conference in electron microscopy was in Delft in 1942, with more than one hundred attendees. Later conferences included the "First" international conference in Paris, 1950 and then in London in 1954.

With the development of TEM, the associated technique of scanning transmission electron microscopy (STEM) was re-investigated and did not become developed until the 1970s, with Albert Crewe at the University of Chicago developing the field emission gun and adding a high quality objective lens to create the modern STEM. Using this design, Crewe demonstrated the ability to image atoms using annular dark-field imaging. Crewe and coworkers at the University of Chicago developed the cold field electron emission source and built a STEM able to visualize single heavy atoms on thin carbon substrates.

Electrons

Theoretically, the maximum resolution, d, that one can obtain with a light microscope has been limited by the wavelength of the photons that are being used to probe the sample, λ and the numerical aperture of the system, NA.

$$d = \frac{\lambda}{2n\sin\alpha} \approx \frac{\lambda}{2NA}$$

Early twentieth century scientists theorised ways of getting around the limitations of the relatively large wavelength of visible light (wavelengths of 400–700 nanometres) by using electrons. Like all matter, electrons have both wave and particle properties (as theorised by Louis-Victor de Broglie), and their wavelike properties mean that a beam of electrons can be made to behave like a beam of electromagnetic radiation. The wavelength of electrons is found by equating the de Broglie equation to the kinetic energy of an electron. An additional correction must be made to account for relativistic effects, as in a TEM an electron's velocity approaches the speed of light, c.

$$\lambda_e \approx \frac{h}{\sqrt{2m_0 E\left(1 + \frac{E}{2m_0 e^2}\right)}}$$

Where, h is Planck's constant, m_0 is the rest mass of an electron and E is the energy of the accelerated electron. Electrons are usually generated in an electron microscope by a process known as thermionic emission from a filament, usually tungsten, in the same manner as a light bulb, or alternatively by field electron emission. The electrons are then accelerated by an electric potential (measured in volts) and focused by electrostatic and electromagnetic lenses onto the sample. The transmitted beam contains information about electron density, phase and periodicity; this beam is used to form an image.

Source Formation

From the top down, the TEM consists of an emission source, which may be a tungsten filament, or a lanthanum hexaboride (LaB_6)

source. For tungsten, this will be of the form of either a hairpin-style filament, or a small spike-shaped filament. LaB_6sources utilise small single crystals. By connecting this gun to a high voltage source (typically ~100-300 kV) the gun will, given sufficient current, begin to emit electrons either by thermionic or field electron emission into the vacuum. This extraction is usually aided by the use of a Wehnelt cylinder. Once extracted, the upper lenses of the TEM allow for the formation of the electron probe to the desired size and location for later interaction with the sample.

Manipulation of the electron beam is performed using two physical effects. The interaction of electrons with a magnetic field will cause electrons to move according to the right hand rule, thus allowing for electromagnets to manipulate the electron beam. The use of magnetic fields allows for the formation of a magnetic lens of variable focusing power, the lens shape originating due to the distribution of magnetic flux. Additionally, electrostatic fields can cause the electrons to be deflected through a constant angle. Coupling of two deflections in opposing directions with a small intermediate gap allows for the formation of a shift in the beam path, this being used in TEM for beam shifting, subsequently this is extremely important to STEM. From these two effects, as well as the use of an electron imaging system, sufficient control over the beam path is possible for TEM operation. The optical configuration of a TEM can be rapidly changed, unlike that for an optical microscope, as lenses in the beam path can be enabled, have their strength changed, or be disabled entirely simply via rapid electrical switching, the speed of which is limited by effects such as the magnetic hysteresis of the lenses.

Optics

The lenses of a TEM allow for beam convergence, with the angle of convergence as a variable parameter, giving the TEM the ability to change magnification simply by modifying the amount of current that flows through the coil, quadrupole or hexapole lenses. The quadrupole lens is an arrangement of electromagnetic coils at the vertices of the square, enabling the generation of a lensing magnetic fields, the hexapole configuration simply enhances the lens symmetry by using six, rather than four coils. Typically a TEM consists of three stages of lensing. The stages are the condensor lenses, the objective lenses, and the projector lenses. The condensor lenses are responsible for primary beam formation, whilst the objective lenses focus the beam that comes through the sample itself (in STEM scanning mode, there are also objective lenses above the sample to make the incident electron beam convergent). The projector lenses are used to expand the beam onto the phosphor screen or other imaging device, such as film. The magnification of the TEM is due to the ratio of the distances between the specimen and the objective lens' image plane. Additional quad or hexapole lenses allow for the correction of asymmetrical beam distortions, known as astigmatism. It is noted that TEM optical configurations differ significantly with implementation, with manufacturers using custom lens configurations, such as in spherical aberration corrected instruments, or TEMs utilising energy filtering to correct electron chromatic aberration.

Display

Imaging systems in a TEM consist of a phosphor screen, which may be made of fine (10-100 ìm) particulate zinc sulphide, for direct observation by the operator. Optionally, an image recording system such as film based or doped YAG screen coupled CCDs. Typically these devices can be removed or inserted into the beam path by the operator as required.

Components

Figure: The electron source of the TEM is at the top, where the lensing system (4,7 and 8) focuses the beam on the specimen and then projects it onto the viewing screen (10). The beam control is on the right (13 and 14)

A TEM is composed of several components, which include a vacuum system in which the electrons travel, an electron emission source for generation of the electron stream, a series of electromagnetic lenses, as well as electrostatic plates. The latter two allow the operator to guide and manipulate the beam as required. Also required is a device to allow the insertion into, motion within, and removal of specimens from the beam path. Imaging devices are subsequently used to create an image from the electrons that exit the system.

Vacuum System

To increase the mean free path of the electron gas interaction, a standard TEM is evacuated to low pressures, typically on the order of 10^{-4} Pa. The need for this is twofold: first the allowance for the voltage difference between the cathode and the ground without generating an arc, and secondly to reduce the collision frequency of electrons with gas atoms to negligible levels—this effect is characterised by the mean free path. TEM components such as specimen holders and film cartridges must be routinely inserted or replaced requiring a system with the ability to re-evacuate on a regular basis. As such, TEMs are equipped with multiple pumping systems and airlocks and are not permanently vacuum sealed.

The vacuum system for evacuating a TEM to an operating pressure level consists of several stages. Initially a low or roughing vacuum is achieved with either a rotary vane pump or diaphragm pumps bringing the TEM to a sufficiently low pressure to allow the operation of a turbomolecular or diffusion pump which brings the TEM to its high vacuum level necessary for operations. To allow for the low vacuum pump to not require continuous operation, while continually operating the turbomolecular pumps, the vacuum side of a low-pressure pump may be connected to chambers which accommodate the exhaust gases from the turbomolecular pump. Sections of the TEM may be isolated by the use of gate valves, to allow for different vacuum levels in specific areas, such as a higher vacuum of 10^{-4} to 10^{-7} Pa or higher in the electron gun in high-resolution or field-emission TEMs.

High-voltage TEMs require ultra-high vacuums on the range of 10^{-7} to 10^{-9} Pa to prevent generation of an electrical arc, particularly at the TEM cathode. As such for higher voltage TEMs a third vacuum system may operate, with the gun isolated from the main chamber either by use of gate valves or by the use of a differential pumping aperture. The differential pumping aperture is a small hole that prevents diffusion of gas molecules into the higher vacuum gun area faster than they can be pumped out. For these very low pressures either an ion pump or a getter material is used.

Poor vacuum in a TEM can cause several problems, from deposition of gas inside the TEM onto the specimen as it is being viewed through a process known as electron beam induced deposition, or in more severe cases damage to the cathode from an electrical discharge. Vacuum problems due to specimen sublimation are limited by the use of a cold trap to adsorb sublimated gases in the vicinity of the specimen.

Specimen Stage

TEM specimen stage designs include airlocks to allow for insertion of the specimen holder into the vacuum with minimal increase in pressure in other areas of the microscope. The specimen holders are adapted to hold a standard size of grid upon which the sample is placed or a standard size of self-supporting specimen. Standard TEM grid sizes is a 3.05 mm diametre ring, with a thickness and mesh size ranging from a few to 100 ìm. The sample is placed onto the inner meshed area having diametre of approximately 2.5 mm. Usual grid materials are copper, molybdenum, gold or platinum. This grid is placed into the sample holder which is paired with the specimen stage. A wide variety of designs of stages and holders exist, depending upon the type of experiment being performed. In addition to 3.05 mm grids, 2.3 mm grids are sometimes, if rarely, used. These grids were particularly used in the mineral sciences where a large degree of tilt can be required and where specimen material may be extremely rare. Electron transparent specimens have a thickness around 100 nm, but this value depends on the accelerating voltage.

Once inserted into a TEM, the sample often has to be manipulated to present the region of interest to the beam, such as in single grain diffraction, in a specific orientation. To accommodate this, the TEM stage includes mechanisms for the translation of the sample in the XY plane of the sample, for Z height adjustment of the sample holder, and usually for at least one rotation degree of freedom for the sample. Thus a TEM stage may provide four degrees of freedom for the motion of the specimen. Most modern TEMs provide the ability for two orthogonal rotation angles of movement with specialised holder designs called double-tilt sample holders. Of note however is that some stage designs, such as top-entry or vertical insertion stages once common for high resolution TEM studies, may simply only have X-Y translation available. The design criteria of TEM stages are complex, owing to the simultaneous requirements of mechanical and electron-optical constraints and have thus generated many unique implementations. A TEM stage is required to have the ability to hold a specimen and be manipulated to bring the region of interest into the path of the electron beam. As the TEM can operate over a wide range of magnifications, the stage must simultaneously be highly resistant to mechanical drift, with drift requirements as low as a few nm/minute while being able to move several um/minute, with repositioning accuracy on the order of nanometres. Earlier designs of TEM accomplished this with a complex set of mechanical downgearing devices, allowing the operator to finely control the motion of the stage by several rotating rods. Modern devices may use electrical stage designs, using screw gearing in concert with stepper motors, providing the operator with a computer-based stage input, such as a joystick or trackball.

Two main designs for stages in a TEM exist, the side-entry and top entry version. Each design must accommodate the matching holder to allow for specimen insertion without either damaging delicate TEM optics or allowing gas into TEM systems under vacuum.

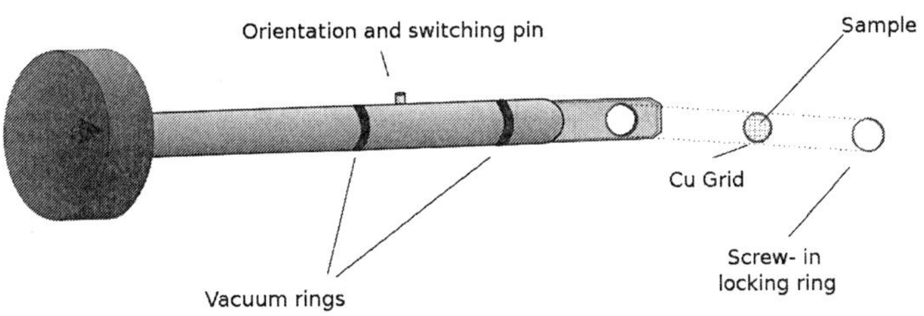

Figure: A diagram of a single axis tilt sample holder for insertion into a TEM goniometre. Titling of the holder is achieved by rotation of the entire goniometre

The most common is the side entry holder, where the specimen is placed near the tip of a long metal (brass or stainless steel) rod, with the specimen placed flat in a small bore. Along the rod are several polymer vacuum rings to allow for the formation of a vacuum seal of sufficient quality, when inserted into the stage. The stage is thus designed to accommodate the rod, placing the sample either in between or near the objective lens, dependent upon the objective design. When inserted into the stage, the side entry holder has its tip contained within the TEM vacuum, and the base is presented to atmosphere, the airlock formed by the vacuum rings.

Insertion procedures for side entry TEM holders typically involve the rotation of the sample to triggermicro switches that initiate evacuation of the airlock before the sample is inserted into the TEM column.

The second design is the top-entry holder consists of a cartridge that is several cm long with a bore drilled down the cartridge axis. The specimen is loaded into the bore, possibly utilising a small screw ring to hold the sample in place. This cartridge is inserted into an airlock with the bore perpendicular to the TEM optic axis. When sealed, the airlock is manipulated to push the cartridge such that the cartridge falls into place, where the bore hole becomes aligned with the beam axis, such that the beam travels down the cartridge bore and into the specimen. Such designs are typically unable to be tilted without blocking the beam path or interfering with the objective lens.

Electron Gun

The electron gun is formed from several components: the filament, a biasing circuit, a Wehnelt cap, and an extraction anode. By connecting the filament to the negative component power supply, electrons can be "pumped" from the electron gun to the anode plate, and TEM column, thus completing the circuit. The gun is designed to create a beam of electrons exiting from the assembly at some given angle, known as the gun divergence semiangle, á. By constructing the Wehnelt cylinder such that it has a higher negative charge than the filament itself, electrons that exit the filament in a diverging manner are, under proper operation, forced into a converging pattern the minimum size of which is the gun crossover diametre.

The thermionic emission current density, J, can be related to the work function of the emitting material and is a Boltzmann distribution given below, where A is a constant, Ö is the work function and T is the temperature of the material.

$$J = AT^2 \exp\left(-\frac{\Phi}{kT}\right)$$

This equation shows that in order to achieve sufficient current density it is necessary to heat the emitter, taking care not to cause damage by application of excessive heat, for this reason materials with either a high melting point, such as tungsten, or those with a low work function (LaB_6) are required for the gun filament. Furthermore both lanthanum hexaboride and tungsten thermionic sources must be heated in order to achieve thermionic emission, this can be achieved by the use of a small resistive strip. To prevent thermal shock, there is often a delay enforced in the application of current to the tip, to prevent thermal gradients from damaging the filament, the delay is usually a few seconds for LaB_6, and significantly lower for tungsten.

Electron Lens

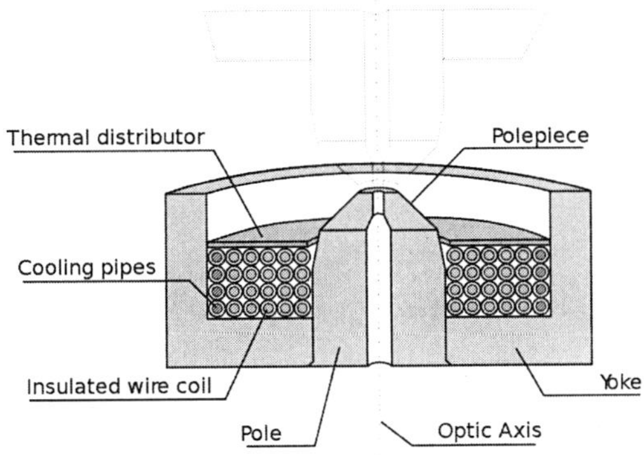

Figure: Diagram of a TEM split polepiece design lens

Electron lenses are designed to act in a manner emulating that of an optical lens, by focusing parallel rays at some constant focal length. Lenses may operate electrostatically or magnetically. The majority of electron lenses for TEM utilise electromagnetic coils to generate a convex lens. For these lenses the field produced for the lens must be radially symmetric, as deviation from the radial symmetry of the magnetic lens causes aberrations such as astigmatism, and worsens spherical and chromatic aberration. Electron lenses are manufactured from iron, iron-cobalt or nickel cobalt alloys, such as permalloy. These are selected for their magnetic properties, such as magnetic saturation, hysteresis and permeability.

The components include the yoke, the magnetic coil, the poles, the polepiece, and the external control circuitry. The polepiece must be manufactured in a very symmetrical manner, as this provides the boundary conditions for the magnetic field that forms the lens. Imperfections in the manufacture of the polepiece can induce severe distortions in the magnetic field symmetry, which induce distortions that will ultimately limit the lenses' ability to reproduce the object plane. The exact dimensions of the gap, pole piece internal diametre and taper, as well as the overall design of the lens is often performed by finite element analysis of the magnetic field, whilst considering the thermal and electrical constraints of the design. The coils which produce the magnetic field are located within the lens yoke. The coils can contain a variable current, but typically utilise high voltages, and therefore require significant insulation in order to prevent short-circuiting the lens components. Thermal distributors are placed to ensure the extraction of the heat generated by the energy lost to resistance of the coil windings. The windings may be water-cooled, using a chilled water supply in order to facilitate the removal of the high thermal duty.

Apertures

Apertures are annular metallic plates, through which electrons that are further than a fixed distance from the optic axis may be excluded. These consist of a small metallic disc that is sufficiently thick to prevent electrons from passing through the disc, whilst permitting axial electrons. This permission of central electrons in a TEM causes two effects simultaneously: firstly, apertures decrease the beam intensity as electrons are filtered from the beam, which may be desired in the case of beam sensitive samples. Secondly, this filtering removes electrons that are scattered to high angles, which may be due to unwanted processes such as spherical or chromatic aberration, or due to diffraction from interaction within the sample.

Apertures are either a fixed aperture within the column, such as at the condensor lens, or are a movable aperture, which can be inserted or withdrawn from the beam path, or moved in the plane perpendicular to the beam path. Aperture assemblies are mechanical devices which allow for the selection of different aperture sizes, which may be used by the operator to trade off intensity and the filtering effect of the aperture. Aperture assemblies are often equipped with micrometres to move the aperture, required during optical calibration.

Imaging Methods

Imaging methods in TEM utilise the information contained in the electron waves exiting from the sample to form an image. The projector lenses allow for the correct positioning of this electron wave distribution onto the viewing system. The observed intensity of the image, I, assuming sufficiently high quality of imaging device, can be approximated as proportional to the time-average amplitude of the electron wavefunctions, where the wave which form the exit beam is denoted by ψ .

$$I(x) = \frac{k}{t_1 - t_0} \int_{t_0}^{t_1} \Psi\Psi^* dt$$

Different imaging methods therefore attempt to modify the electron waves exiting the sample in a form that is useful to obtain information with regards to the sample, or beam itself. From the previous equation, it can be deduced that the observed image depends not only on the amplitude of beam, but also on the phase of the electrons, although phase effects may often be ignored at lower magnifications. Higher resolution imaging requires thinner samples and higher energies of incident electrons. Therefore the sample can no longer be considered to be absorbing electrons, via a Beer's law effect, rather the sample can be modelled as an object that does not change the amplitude of the incoming electron wavefunction. Rather the sample modifies the phase of the incoming wave; this model is known as a pure phase object, for sufficiently thin specimens phase effects dominate the image, complicating analysis of the observed intensities. For example, to improve the contrast in the image the TEM may be operated at a slight defocus to enhance contrast, owing to convolution by the contrast transfer function of the TEM, which would normally decrease contrast if the sample was not a weak phase object.

Contrast Formation

Contrast formation in the TEM depends greatly on the mode of operation. Complex imaging techniques, which utilise the unique ability to change lens strength or to deactivate a lens, allow for many operating modes. These modes may be used to discern information that is of particular interest to the investigator.

Bright Field

The most common mode of operation for a TEM is the bright field imaging mode. In this mode the contrast formation, when considered classically, is formed directly by occlusion and absorption of electrons in the sample. Thicker regions of the sample, or regions with a higher atomic number will appear dark, whilst regions with no sample in the beam path will appear bright – hence the term "bright field". The image is in effect assumed to be a simple two dimensional projection of the sample down the optic axis, and to a first approximation may be modelled via Beer's law, more complex analyses require the modelling of the sample to include phase information.

Diffraction Contrast

Samples can exhibit diffraction contrast, whereby the electron beam undergoes Bragg scattering, which in the case of a crystalline sample, disperses electrons into discrete locations in the back focal plane. By the placement of apertures in the back focal plane, i.e. the objective aperture, the desired Bragg reflections can be selected (or excluded), thus only parts of the sample that are causing the electrons to scatter to the selected reflections will end up projected onto the imaging apparatus.

If the reflections that are selected do not include the unscattered beam (which will appear up at the focal point of the lens), then the image will appear dark wherever no sample scattering to the selected peak is present, as such a region without a specimen will appear dark. This is known as a dark-field image.

Modern TEMs are often equipped with specimen holders that allow the user to tilt the specimen to a range of angles in order to obtain specific diffraction conditions, and apertures placed above the specimen allow the user to select electrons that would otherwise be diffracted in a particular direction from entering the specimen.

Applications for this method include the identification of lattice defects in crystals. By carefully selecting the orientation of the sample, it is possible not just to determine the position of defects but also to determine the type of defect present. If the sample is oriented so that one particular plane is only slightly tilted away from the strongest diffracting angle (known as the Bragg Angle), any distortion of the crystal plane that locally tilts the plane to the Bragg angle will produce particularly strong contrast variations. However, defects that produce only displacement of atoms that do not tilt the crystal to the Bragg angle (i. e. displacements parallel to the crystal plane) will not produce strong contrast.

Electron Energy Loss

Utilising the advanced technique of EELS, for TEMs appropriately equipped electrons can be rejected based upon their voltage (which, due to constant charge is their energy), using magnetic sector based devices known as EELS spectrometres. These devices allow for the selection of particular energy values, which can be associated with the way the electron has interacted with the sample. For example different elements in a sample result in different electron energies in the beam after the sample. This normally results in chromatic aberration – however this effect can, for example, be used to generate an image which provides information on elemental composition, based upon the atomic transition during electron-electron interaction.

EELS spectrometres can often be operated in both spectroscopic and imaging modes, allowing for isolation or rejection of elastically scattered beams. As for many images inelastic scattering will include information that may not be of interest to the investigator thus reducing observable signals of interest, EELS imaging can be used to enhance contrast in observed images, including both bright field and diffraction, by rejecting unwanted components.

Phase Contrast

Crystal structure can also be investigated by high-resolution transmission electron microscopy (HRTEM), also known as phase contrast. When utilising a Field emission source, of uniform thickness, the images are formed due to differences in phase of electron waves, which is caused by specimen interaction. Image formation is given by the complex modulus of the incoming electron beams. As such, the image is not only dependent on the number of electrons hitting the screen, making direct interpretation of phase contrast images more complex. However this effect can be used to an advantage, as it can be manipulated to provide more information about the sample, such as in complex phase retrieval techniques.

Diffraction

As previously stated, by adjusting the magnetic lenses such that the back focal plane of the lens rather than the imaging plane is placed on the imaging apparatus a diffraction pattern can be generated. For thin crystalline samples, this produces an image that consists of a pattern of dots in the case of a single crystal, or a series of rings in the case of a polycrystalline or amorphous solid material. For the single crystal case the diffraction pattern is dependent upon the orientation of the specimen and the structure of the sample illuminated by the electron beam. This image provides the investigator with information about the space group symmetries in the crystal and the crystal's orientation to the beam path. This is typically done without utilising any information but the position at which the diffraction spots appear and the observed image symmetries.

Diffraction patterns can have a large dynamic range, and for crystalline samples, may have intensities greater than those recordable by CCD. As such, TEMs may still be equipped with film cartridges for the purpose of obtaining these images, as the film is a single use detector.

Analysis of diffraction patterns beyond point-position can be complex, as the image is sensitive to a number of factors such as specimen thickness and orientation, objective lens defocus, spherical and chromatic aberration. Although quantitative interpretation of the contrast shown in lattice images is possible, it is inherently complicated and can require extensive computer simulation and analysis, such as electron multislice analysis.

More complex behaviour in the diffraction plane is also possible, with phenomena such as Kikuchi lines arising from multiple diffraction within the crystalline lattice. In convergent beam electron diffraction (CBED) where a non-parallel, i.e. converging, electron wavefront is produced by concentrating the electron beam into a fine probe at the sample surface, the interaction of the convergent beam can provide information beyond structural data such as sample thickness.

Three-dimensional Imaging

As TEM specimen holders typically allow for the rotation of a sample by a desired angle, multiple views of the same specimen can be obtained by rotating the angle of the sample along an axis perpendicular to the beam. By taking multiple images of a single TEM sample at differing angles, typically in 1° increments, a set of images known as a "tilt series" can be collected. This methodology was proposed in the 1970s by Walter Hoppe. Under purely absorption contrast conditions, this set of images can be used to construct a three-dimensional representation of the sample.

The reconstruction is accomplished by a two-step process, first images are aligned to account for errors in the positioning of a sample; such errors can occur due to vibration or mechanical drift. Alignment methods use image registration algorithms, such as autocorrelation methods to correct these errors. Secondly, using a technique known as filtered back projection, the aligned image slices can be transformed from a set of two-dimensional images, $I_j(x, y)$, to a single three-dimensional image, $I'_j(x, y, z)$. This three-dimensional image is of particular interest when morphological information is required, further study can be undertaken using computer algorithms, such as isosurfaces and data slicing to analyse the data.

As TEM samples cannot typically be viewed at a full 180° rotation, the observed images typically suffer from a "missing wedge" of data, which when using Fourier-based back projection methods decreases the range of resolvable frequencies in the three-dimensional reconstruction. Mechanical refinements, such as multi-axis tilting (two tilt series of the same specimen made at orthogonal directions) and conical tomography (where the specimen is first tilted to a given fixed angle and then imaged at equal angular rotational increments through one complete rotation in the plane of the specimen grid) can be used to limit the impact of the missing data on the observed specimen morphology. In addition, numerical techniques exist which can improve the collected data.

All the above-mentioned methods involve recording tilt series of a given specimen field. This inevitably results in the summation of a high dose of reactive electrons through the sample and the accompanying destruction of fine detail during recording. The technique of low-dose (minimal-dose) imaging is therefore regularly applied to mitigate this effect. Low-dose imaging is performed by deflecting illumination and imaging regions simultaneously away from the optical axis to image an adjacent region to the area to be recorded (the high-dose region). This area is maintained centred during tilting and refocused before recording. During recording the deflections are removed so that the area of interest is exposed to the electron beam only for the duration required for imaging. An improvement of this technique (for objects resting on a sloping substrate film) is to have two symmetrical off-axis regions for focusing followed by setting focus to the average of the two high-dose focus values before recording the low-dose area of interest.

Non-tomographic variants on this method, referred to as single particle analysis, use images of multiple (hopefully) identical objects at different orientations to produce the image data required for three-dimensional reconstruction. If the objects do not have significant preferred orientations, this method does not suffer from the missing data wedge (or cone) which accompany tomographic methods nor does it incur excessive radiation dosage, however it assumes that the different objects imaged can be treated as if the 3D data generated from them arose from a single stable object.

Sample Preparation

Sample preparation in TEM can be a complex procedure. TEM specimens are required to be at most hundreds of nanometres thick, as unlike neutron or X-Ray radiation the electron beam interacts readily with the sample, an effect that increases roughly with atomic number squared (z^2). High quality samples will have a thickness that is comparable to the mean free path of the electrons that travel through the samples, which may be only a few tens of nanometres. Preparation of TEM specimens is specific to the material under analysis and the desired information to obtain from the specimen. As such, many generic techniques have been used for the preparation of the required thin sections. Materials that have dimensions small enough to be electron transparent, such as powders or nanotubes, can be quickly prepared by the deposition of a dilute sample containing the specimen onto support grids or films. In the biological sciences in order to withstand the instrument vacuum and facilitate handling, biological specimens can be fixated using either a negative staining material such as uranyl acetate or by plastic embedding. Alternately samples may be held at liquid nitrogen temperatures after embedding in vitreous ice. In material science and metallurgy the specimens tend to be naturally resistant to vacuum, but still must be prepared as a thin foil, or etched so some portion of the specimen is thin enough for the beam to penetrate. Constraints on the thickness of the material may be limited by the scattering cross-section of the atoms from which the material is comprised.

Tissue Sectioning

By passing samples over a glass or diamond edge, small, thin sections can be readily obtained using a semi-automated method. This method is used to obtain thin, minimally deformed samples that allow for the observation of tissue samples. Additionally inorganic samples have been studied, such as aluminium, although this usage is limited owing to the heavy damage induced in the less soft samples. To prevent charge build-up at the sample surface, tissue samples need to be coated with a thin layer of conducting material, such as carbon, where the coating thickness is several nanometres. This may be achieved via an electric arc deposition process using a sputter coating device.

Sample Staining

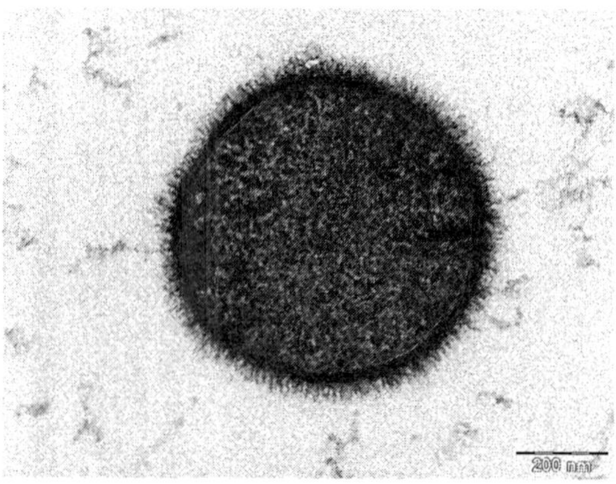

Figure: A section of a cell of Bacillus subtilis, taken with a Tecnai T-12 TEM. The scale bar is 200 nm.

Details in light microscope samples can be enhanced bystains that absorb light; similarly TEM samples of biological tissues can utilise high atomic number stains to enhance contrast. The stain absorbs electrons or scatters part of the electron beam which otherwise is projected onto the imaging system. Compounds of heavy metals such asosmium, lead, uranium or gold (in immunogold labelling) may be used prior to TEM observation to selectively deposit electron dense atoms in or on the sample in desired cellular or protein regions, requiring an understanding of how heavy metals bind to biological tissues.

Mechanical Milling

Mechanical polishing may be used to prepare samples. Polishing needs to be done to a high quality, to ensure constant sample thickness across the region of interest. A diamond, or cubic boron nitride polishing compound may be used in the final stages of polishing to remove any scratches that may cause contrast fluctuations due to varying sample thickness. Even after careful mechanical milling, additional fine methods such as ion etching may be required to perform final stage thinning.

Chemical Etching

Certain samples may be prepared by chemical etching, particularly metallic specimens. These samples are thinned using a chemical etchant, such as an acid, to prepare the sample for TEM observation. Devices to control the thinning process may allow the operator to control either the voltage or current passing through the specimen, and may include systems to detect when the sample has been thinned to a sufficient level of optical transparency.

Ion Etching

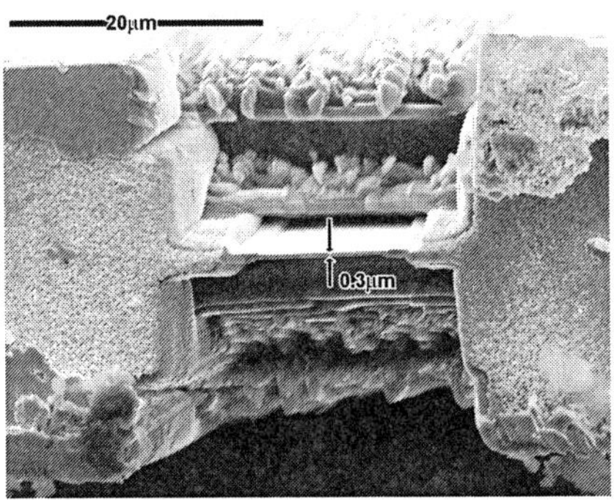

Figure: SEM image of a thin TEM sample milled by FIB. The thin membrane shown here is suitable for TEM examination; however, at ~300-nm thick, it would not be suitable for high-resolution TEM without further milling.

Ion etching is a sputtering process that can remove very fine quantities of material. This is used to perform a finishing polish of specimens polished by other means. Ion etching uses an inert gas passed through an electric field to generate a plasma stream that is directed to the sample surface. Acceleration energies for gases such as argon are typically a few kilovolts. The sample may be rotated to promote even polishing of the sample surface. The sputtering rate of such methods is on the order of tens of micrometres per hour, limiting the method to only extremely fine polishing.

More recently focused ion beam methods have been used to prepare samples. FIB is a relatively new technique to prepare thin samples for TEM examination from larger specimens. Because FIB can be used to micro-machine samples very precisely, it is possible to mill very thin membranes from a specific area of interest in a sample, such as a semiconductor or metal. Unlike inert gas ion sputtering, FIB makes use of significantly more energetic gallium ions and may alter the composition or structure of the material through gallium implantation.

Replication

Samples may also be replicated using cellulose acetate film, the film subsequently coated with a heavy metal, the original film melted away, and the replica imaged on the TEM. This technique is used for both materials and biological samples.

Modifications

The capabilities of the TEM can be further extended by additional stages and detectors, sometimes incorporated on the same microscope. An electron cryomicroscope(CryoTEM) is a TEM with a specimen holder capable of maintaining the specimen at liquid nitrogen or liquid helium temperatures. This allows imaging specimens prepared in vitreous ice, the preferred preparation technique for imaging individual molecules or macromolecular assemblies.

A TEM can be modified into a scanning transmission electron microscope (STEM) by the addition of a system that rasters the beam across the sample to form the image, combined with suitable detectors. Scanning coils are used to deflect the beam, such as by an electrostatic shift of the beam, where the beam is then collected using a current detector such as a faraday cup, which acts as a direct electron counter. By correlating the electron count to the position of the scanning beam (known as the "probe"), the transmitted component of the beam may be measured. The non-transmitted components may be obtained either by beam tilting or by the use of annular dark field detectors.

In-situ experiments may also be conducted with experiments such as in-situ reactions or material deformation testing.

Modern research TEMs may include aberration correctors, to reduce the amount of distortion in the image. Incident beam Monochromators may also be used which reduce the energy spread of the incident electron beam to less than 0.15 eV. Major TEM makers include JEOL, Hitachi High-technologies, FEI Company (from merging with Philips Electron Optics), Carl Zeiss and NION.

Low-voltage Electron Microscope (LVEM)

The low-voltage electron microscope (LVEM) is a combination of SEM, TEM and STEM in one instrument, which operated at relatively low electron accelerating voltage of 5 kV. Low voltage increases image contrast which is especially important for biological specimens. This increase in contrast significantly reduces, or even eliminates the need to stain. Sectioned samples generally need to be thinner than they would be for conventional TEM (20-65 nm). Resolutions of a few nm are possible in TEM, SEM and STEM modes.

Cryo-Electron Microscopy

This technique allows TEM's to be used to see molecular structure of proteins and large molecules. Cryoelectron microscopy involves viewing unaltered macromolecular assemblies by vitrifying them, placing them on a grid and obtaining images by detecting electrons that transmit through the specimen.

Limitations

There are a number of drawbacks to the TEM technique. Many materials require extensive sample preparation to produce a sample thin enough to be electron transparent, which makes TEM analysis a relatively time consuming process with a low throughput of samples. The structure of the sample may also be changed during the preparation process. Also the field of view is relatively small, raising the possibility that the region analysed may not be characteristic of the whole sample. There is potential that the sample may be damaged by the electron beam, particularly in the case of biological materials.

Resolution Limits

The limit of resolution obtainable in a TEM may be described in several ways, and is typically referred to as the information limit of the microscope. One commonly used value is a cut-off value of the contrast transfer function, a function that is usually quoted in the frequency domain to define the reproduction of spatial frequencies of objects in the object plane by the microscope optics.

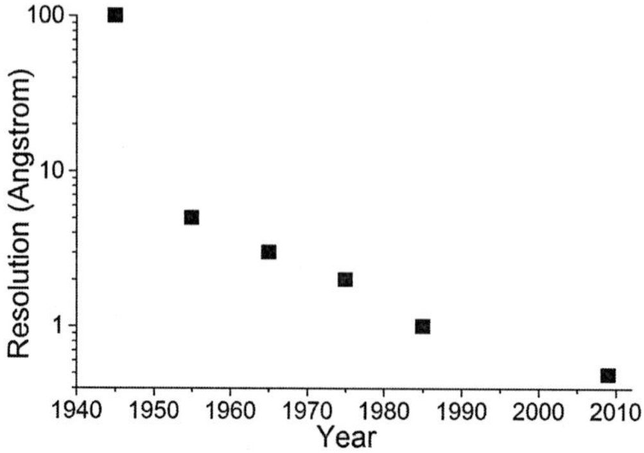

Figure: Approximate tendency of spatial resolution achieved with TEM.

A cut-off frequency, q_{max}, for the transfer function may be approximated with the following equation, where C_s is the spherical aberration coefficient and λ is the electron wavelength:

$$q_{\max} = \frac{1}{0.67\left(C_s\lambda^3\right)^{1/4}}.$$

For a 200 kV microscope, with partly corrected spherical aberrations ("to the third order") and a C_s value of 1 µm, a theoretical cut-off value might be $1/q_{max}$ = 42 pm. The same microscope without a corrector would have C_s = 0.5 mm and thus a 200-pm cut-off. The spherical aberrations are suppressed to the third or fifth order in the "aberration-corrected" microscopes. Their resolution is however limited by electron source geometry and brightness and chromatic aberrations in the objective lens system.

The frequency domain representation of the contrast transfer function may often have an oscillatory nature, which can be tuned by adjusting the focal value of the objective lens. This oscillatory nature implies that some spatial frequencies are faithfully imaged by the microscope, whilst others are suppressed. By combining multiple images with different spatial frequencies, the use of techniques such as focal series reconstruction can be used to improve the resolution of the TEM in a limited manner. The contrast transfer function can, to some extent, be experimentally approximated through techniques such as Fourier transforming images of amorphous material, such as amorphous carbon. More recently, advances in aberration corrector design have been able to reduce spherical aberrations and to achieve resolution below 0.5 Ångströms (50 pm) at magnifications above 50 million times. Improved resolution allows for the imaging of lighter atoms that scatter electrons less efficiently, such as lithium atoms in lithium battery materials. The ability to determine the position of atoms within materials has made the HRTEM an indispensable tool for nanotechnology research and development in many fields, including heterogeneous catalysis and the development of semiconductor devices for electronics and photonics.

Optical Beam Induced Current

Optical beam induced current (OBIC) is a semiconductor analysis technique performed using laser signal injection. The technique uses a scanning laser beam to create electron–hole pairs in a semiconductor sample. This induces a current which may be analysed to determine the sample's properties, especially defects or anomalies.

Conventional OBIC scans an ultrafast laser beam over the surface of the sample, exciting someelectrons into the conduction band through what is known as 'single-photon absorption'. As its name implies, single-photon absorption involves just a single photon to excite the electron into conduction. This can only occur if that single photon carries enough energy to overcome the band gap of the semiconductor (1.2 eV for Si) and provide the electron with enough energy to make it jump into the conduction band.

Uses

The technique is used in semiconductor failure analysis to locate buried diffusion regions, damaged junctions and gate oxide shorts.

The OBIC technique may be used to detect the point at which a focused ion beam (FIB) milling operation in bulk silicon of an IC must be terminated (also known as endpoint). This is accomplished by using a laser to induce a photocurrent in the silicon while simultaneously monitoring the magnitude of the photocurrent by connecting an ammetre to the device's power and ground. As the bulk silicon is thinned, the photocurrent is increased and reaches a peak as the depletion region of the well to substrate junction is reached. This way, endpoint can be achieved to just below the well depth and the device remains operational.

Light Induced Voltage Alteration

Light induced voltage alteration (LIVA) is a semiconductor analysis technique that uses a laser or infrared light source to induce voltage changes in a device while scanning the beam of light across its surface. The technique relies upon the generation of electron-hole pairs in the semiconductor material when exposed to photons.

Theory of Operation

The device to be analysed is biased using a constant current power supply. As the light source is scanned across the surface of the silicon, electron-hole pairs are generated. This causes subtle alterations of the operating characteristics of the device, which may result in slight changes to the power supply voltage. Any changes that are detected in the power supply voltage are noted and correlated with the position of the light source on the device. This allows the physical locations corresponding to power supply fluctuations to be mapped onto an image of the device. This provides the device analyst with specific locations at which the device may be examined for defects.

Laser Assisted Device Alteration

Laser Assisted Device Alteration (LADA) is a laser-based timing analysis technique used in the failure analysis of semiconductor devices. The laser is used to temporarily alter the operating characteristics of transistors on the device.

Theory of Operation

The LADA technique targets a variable power continuous wave (CW) laser at specific device transistors. The laser is typically of a short wavelength variety on the order of 1064 nm. This allows the laser to generate photo carriers in the silicon without resulting in localised heating of the device. The LADA technique is somewhat similar in execution to the Soft Defect Localisation (SDL) technique, except that SDL utilises a longer wavelength laser (1340 nm) in order to induce localised heating rather than generate photo carriers. Both techniques require the device to be scanned with a laser while it is under active stimulation by the tester.

The device being tested is electrically stimulated and the device output is monitored. This technique is applied to the back side of the semiconductor device, thereby allowing direct access of the laser to the device active diffusion regions. The effect of the laser on the active transistor region is to generate a localised photocurrent. This photocurrent is a temporary effect and only occurs during the time that the laser is stimulating the target region. The creation of this photocurrent alters the transistor operating parameters, which may be observed as a change in function of the device. The effect of this change in parameters may be to speed up or slow down the operation of the device. This makes LADA a suitable technique for determining critical timing paths within a semiconductor circuit.

The laser has differing effects on NMOS and PMOS transistors. In the case of NMOS, the transistor will turn on. For PMOS, however, the effect is to lower the transistor threshold voltage. The effect on the PMOS transistor becomes proportionately stronger as the laser power is increased. The effect is to either increase or decrease the speed of the device being tested.

Setup for a LADA analysis involves connecting the device to a test stimulus. The test parameters for operating voltage and device speed are then adjusted in order to place the device into a state which borders on a Pass/Fail or Fail/Pass transition. It is useful to use a tester Shmoo plot in order to select the appropriate operating conditions. The effect of scanning the laser over sensitive regions is to trip the device from a Pass into a Fail condition, or from a Fail into a Pass condition.

Applications

LADA is useful for confirming or disproving an existing theory for the cause of failure. It may be used to confirm suspected transistor leakage or bus noise. It has also found wide use in localising process defects as the LADA effect easily modulates transistor characteristics in the same path as the process defect.

LADA has been used to analyse failures in Domino logic, state elements in memories and leakage.

Thermal Laser Stimulation

Thermal laser stimulation represents a class of defect imaging techniques which employ a laser to produce a thermal variation in a semiconductor device. This technique may be used for semiconductor failure analysis. There are four techniques associated with thermal laser stimulation: optical beam induced resistance change (OBIRCH), thermally induced voltage alteration (TIVA)), external induced voltage alteration (XIVA) and Seebeck effect imaging (SEI)

Optical Beam Induced Resistance Change

Optical beam induced resistance change (OBIRCH) is an imaging technique which uses a laser beam to induce a thermal change in the device. Laser stimulation highlights differences in thermal characteristics between areas containing defects and areas which are defect-free. As the laser locally heats a defective area on a metal line which is carrying a current, the resulting resistance changes can be detected by monitoring the input current to the device. OBIRCH is useful for detecting electromigration effects resulting in open metal lines.

A constant voltage is applied to the device-under-test (DUT). An area of interest is selected on the device, and a laser beam is used to scan the area. The input current being drawn by the device is monitored for changes during this process. When a change in current is noted, the position of the laser at the time that the change occurred is marked on the image of the device.

When the laser beam strikes a location which does not contain a void, good thermal transmission exists and the change in electrical resistance is small. In areas containing voids, however, thermal transmission is impeded, resulting in a larger change in resistance. The degree of resistance change is displayed visually on an image of the device, with areas of higher resistance being displayed as bright spots.

Thermally Induced Voltage Alteration

Thermally induced voltage alteration (TIVA) is an imaging technique which uses a laser beam to pinpoint the location of electrical shorts on a device. The laser induces local thermal gradients in the device, which result in changes to the amount of power that the device uses.

A laser is scanned over the surface of the device while it is under electrical bias. The device is biased using a constant current source, and the power supply pin voltage is monitored for changes. When the laser strikes an area containing a short circuit, localised heating occurs. This heating changes the resistance of the short, resulting in a change in power consumption of the device. These changes in power consumption are plotted onto an image of the device in locations corresponding to the position of the laser at the time that the change was detected.

External Induced Voltage Alteration

External induced voltage alteration (XIVA) maintains a constant voltage bias and constant current sensing on the device under test. When the scanning laser passes over a defective location, a sudden change in impedance is created. This would normally result in a change in current, however, the constant current choke prevents this from happening. The detection of these events allows the position of the defect to be determined.

Seebeck Effect Imaging

Seebeck effect imaging (SEI) uses a laser to generate thermal gradients in conductors. The thermal gradients induced generate corresponding electric potential gradients. This correlation of thermal and electric gradients is known as the Seebeck effect. The SEI technique is used to locate electrically floating conductors.

When the laser changes the thermal gradient of a floating conductor, its electrical potential changes. This change in potential will change the bias of any transistors connected to the floating conductor, which affects the heat dissipation of the device. These changes are mapped to a visual image of the device in order to physically locate the floating conductors.

Mechanical Probe Station

A Mechanical probe station is used to physically acquire signals from the internal nodes of a semiconductor device. The probe station utilises manipulators which allow the precise positioning of thin needles on the surface of a semiconductor device. If the device is being electrically stimulated, the signal is acquired by the mechanical probe and is displayed on an oscilloscope. The mechanical probe station is often used in the failure analysis of semiconductor devices.

Figure: Close-up of passive probes on the same probe station

There are two types of mechanical probes: active and passive. Passive probes usually consist of a thin tungsten needle. Active probes utilise a FET device on the probe tip in order to significantly reduce loading on the circuit.

Research

Mechanical probe stations are often used in academic research on electronics and materials science. It is often faster and more flexible to test a new electronic device or sample with a probe station than to wire bond and package the device before testing.

Figure: This Desert Cryogenics probe station encloses the sample and probes in a vacuum chamber and permits the introduction of cryogens to cool the sample to low temperatures.

Electron Beam Prober

The electron beam prober (e-beam prober) is a specialised adaption of a standard scanning electron microscope (SEM) that is used for semiconductor failure analysis. While a standard SEM may be operated in a voltage range of 25–30 KeV, the e-beam Prober typically operates at 1 KeV. The e-beam prober is capable of measuring voltage and timing waveforms on internal semiconductor signal structures. Waveforms may be measured on metal line, polysilicon and diffusion structures that have an electrically active, changing signal. The operation of the prober is similar to that of a sampling oscilloscope. A continuously looping, repeating test pattern must be applied to the device-under-test (DUT). E-beam probers are used primarily for front side semiconductor analysis. With the advent offlip-chip technology, many e-beam probers have been replaced with back side analysis instruments.

Theory of Operation

The e-beam prober generates a SEM image by raster-scanning a focused electron beam over a selected region of the semiconductor surface. The high energy electrons in the primary beam strike the surface of the silicon, producing a number of low energy secondary electrons. The secondary electrons are guided back up through the SEM column to a detector. The varying numbers of secondary electrons reaching the detector are interpreted to produce the SEM image.

During waveform acquisition mode, the primary electron beam is focused on a single point on the device surface. As the DUT cycles through its test pattern, the signal at the point being probed changes. The signal changes produce a corresponding change in the local electric field surrounding the point being probed. This affects the number of secondary electrons that escape the device surface and reach the detector. Since electrons are negatively charged, a conductor at a +5-volt potential inhibits the escape of electrons, while a 0-volt potential allows a greater number of electrons to reach the detector. By monitoring these changes in potential, a voltage and timing waveform may be produced for the signal at the point being probed.

Laser Voltage Prober

The laser voltage probe (LVP) is a laser-based voltage and timing waveform acquisition system which is used to perform failure analysis on flip-chip integrated circuits. The device to be analysed is de-encapsulated in order to expose the silicon surface. The silicon substrate is thinned mechanically using a back side mechanical thinning tool. The thinned device is then mounted on a movable stage and connected to an electrical stimulus source. Signal measurements are performed through the back side of the device after substrate thinning has been performed. The device being probed must be electrically stimulated using a repeating test pattern, with a trigger pulse provided to the LVP as reference. The operation of the LVP is similar to that of a sampling oscilloscope.

Theory of Operation

The LVP instrument measures voltage waveform signals in the device diffusion regions. Device imaging is accomplished through the use of a laser scanning microscope (LSM). The LVP uses dual infra-red(IR) lasers to perform both device imaging and waveform acquisition. One laser is used to acquire images or waveforms from the device, while the second laser provides a reference which may be used to subtract unwanted noise from the signal data being acquired. On an electrically active device, the instrument monitors the changes in the phase of the electromagnetic field surrounding a signal being applied to a junction are monitored. The instrument obtains voltage waveform and timing information by monitoring the interaction of laserlight with the changes in the electric field across a p-n junction. As the laser reaches the silicon surface, a certain amount of that light is reflected back. The amount of reflected laser light from the junction is sampled at various points in time. The changing electromagnetic field at the junction affects the amount of laser light that is reflected back. By plotting the variations in reflected laser light versus time, it is possible to construct a timing waveform of the signal at the junction. As the test pattern continues to loop, additional measurements are acquired and averaged into the previous measurements. Over a period of time, this averaging of measurements produces a more refined waveform. The end result is a waveform that is representative of the electrical signal present at the junction.

Time Resolved Photon Emission Prober

The Time-Resolved Photon Emission Prober (TRPE) is an instrument which is used to measure timing waveforms on semiconductor devices. TRPE measurements are performed on the back side of the semiconductor device. The substrate of the device-under-test (DUT) must first be thinned mechanically. The device is mounted on a movable X-Y stage in an enclosure which shields it from all sources of light. The DUT is connected to an active electrical stimulus. The stimulus pattern is continuously looped and a trigger signal is sent to the TRPE instrument in order to tell it when the pattern repeats. The TRPE prober operates in a manner similar to a sampling oscilloscope, and is used to perform semiconductor failure analysis.

Theory of Operation

As the electrical stimulus pattern is repetitively applied to the DUT, internal transistors switch on and off. As pMOS and nMOS transistors switch on or off, they emit photons. These photons emissions are recorded by a sensitive photon detector. By counting the number of photons emitted for a specific transistor across a period of time, a photon histogram may be constructed. The photon histogram records an increase in photon emissions during times that the transistor switches on or off. By detecting the combined photon emissions of pairs p- and n-channel transistors contained in logic gates, it is possible to use the resulting histogram to determine the locations in time of the rising and falling edges of the signal at that node. The waveform produced is not representative of a true voltage waveform, but more accurately represents the derivative of the waveform, with photon spikes being seen only at rising or falling edges.

CAD Navigation

CAD Navigation refers to software tools which are used for the correlation of electronic semiconductor design data with a physical semiconductor device. CAD Navigation tools consist of software that is capable of reading and displaying the physical layout and logical schematic for the device. The logical design consists of a netlist and/or a schematic. The physical design consists of a set of polygons which precisely represent the location of all electrical conductors, diffusions and interconnections in the physical semiconductor device. CAD navigation tools are often used to provide a cross-correlation between the logical design and the physical design. CAD navigation tools are used extensively with E-beam probers, focused-ion beam systems and photon probers for the purpose of semiconductor failure analysis.

Automatic Test Pattern Generation

ATPG (acronym for both Automatic Test Pattern Generation and Automatic Test Pattern Generator) is an electronic design automation method/technology used to find an input (or test) sequence that, when applied to a digital circuit, enables automatic test equipment to distinguish between the correct circuit behaviour and the faulty circuit behaviour caused by defects. The generated patterns are used to test semiconductor devices after manufacture, and in some cases to assist with determining the cause of failure (failure analysis.) The effectiveness of ATPG is measured by the amount of modelled defects, or fault models, that are detected and the number of generated patterns. These metrics generally indicate test quality (higher with more fault detections) and test application time (higher with more patterns). ATPG efficiency is another important consideration. It is influenced by the fault model under consideration, the type of circuit under test (full scan, synchronous sequential, or asynchronous sequential), the level of abstraction used to represent the circuit under test (gate, register-transistor, switch), and the required test quality.

Basics of ATPG

A defect is an error introduced into a device during the manufacturing process. A fault model is a mathematical description of how a defect alters design behaviour. The logic values observed at the device's primary outputs, while applying a test pattern to some device under test (DUT), are called the output of that test pattern. The output of a test pattern, when testing a fault-free device that works exactly as designed, is called the expected output of that test pattern. A fault is said to be detected by a test pattern if the output of that test pattern, when testing a device that has only that one fault, is different than the expected output. The ATPG process for a targeted fault consists of two phases: fault activation and fault propagation. Fault activation establishes a signal value at the fault model site that is opposite of the value produced by the fault model. Fault propagation moves the resulting signal value, or fault effect, forward by sensitizing a path from the fault site to a primary output.

ATPG can fail to find a test for a particular fault in at least two cases. First, the fault may be intrinsically undetectable, such that no patterns exist that can detect that particular fault. The classic example of this is a redundant circuit, designed so that no single fault causes the output to change. In such a circuit, any single fault will be inherently undetectable.

Second, it is possible that a pattern(s) exist, but the algorithm cannot find it. Since the ATPG problem is NP-complete (by reduction from the Boolean satisfiability problem) there will be cases where patterns exist, but ATPG gives up since it will take an incredibly long time to find them (assumingP“‘NP, of course).

The Stuck-at Fault Model

In the past several decades, the most popular fault model used in practice is the single stuck-at faultmodel. In this model, one of the signal lines in a circuit is assumed to be stuck at a fixed logic value, regardless of what inputs are supplied to the circuit. Hence, if a circuit has n signal lines, there are potentially 2n stuck-at faults defined on the circuit, of which some can be viewed as being equivalent to others. The stuck-at fault model is a logical fault model because no delay information is associated with the fault definition. It is also called a permanent fault model because the faulty effect is assumed to be permanent, in contrast to intermittent faults which occur (seemingly) at random and transient faults which occur sporadically, perhaps depending on operating conditions (e.g. temperature, power supply voltage) or on the data values (high or low voltage states) on surrounding signal lines. The single stuck-at fault model is structural because it is defined based on a structural gate-level circuit model.

A pattern set with 100% stuck-at fault coverage consists of tests to detect every possible stuck-at fault in a circuit. 100% stuck-at fault coverage does not necessarily guarantee high quality, since faults of many other kinds—such as bridging faults, opens faults, and transition (aka delay) faults—often occur.

Combinational ATPG

The combinational ATPG method allows testing the individual nodes (or Flip-Flops) of the logic circuit without being concerned with the operation of the overall circuit. During test, a so-called scan-mode is enabled forcing all FF's to be connected in a simplified fashion, effectively bypassing their interconnections as intended during normal operation. This allows using a relatively simple vector matrix to quickly test all the comprising FF's, as well as to trace failures to specific FF's.

Sequential ATPG

Sequential-circuit ATPG searches for a sequence of vectors to detect a particular fault through the space of all possible vector sequences. Various search strategies and heuristics have been devised to find a shorter sequence and/or to find a sequence faster. However, according to reported results, no single strategy/heuristic outperforms others for all applications/circuits. This observation implies that a test generator should include a comprehensive set of heuristics.

Even a simple stuck-at fault requires a sequence of vectors for detection in a sequential circuit. Also, due to the presence of memory elements, the controllability and observability of the internal signals in a sequential circuit are in general much more difficult than those in a combinational logic circuit. These factors make the complexity of sequential ATPG much higher than that of combinational ATPG, where the a scan-chain (i.e. switchable, for-test-only signal chain) is added to allow simple access to the individual nodes.

Due to the high complexity of the sequential ATPG, it remains a challenging task for large, highly sequential circuits that do not incorporate any Design For Testability (DFT) scheme. However, these test generators, combined with low-overhead DFT techniques such as partial scan, have shown a certain degree of success in testing large designs. For designs that are sensitive to area and/or performance overhead, the solution of using sequential-circuit ATPG and partial scan offers an attractive alternative to the popular full-scan solution, which is based on combinational-circuit ATPG.

ATPG and Nanometre Technologies

Historically, ATPG has focused on a set of faults derived from a gate-level fault model. As design trends move toward nanometre technology, new manufacture testing problems are emerging. During design validation, engineers can no longer ignore the effects of crosstalk and power supply noise on reliability and performance. Current fault modelling and vector-generation techniques are giving way to new models and techniques that consider timing information during test generation, that are scalable to larger designs, and that can capture extreme design conditions. For nanometre technology, many current design validation problems are becoming manufacturing test problems as well, so new fault-modelling and ATPG techniques will be needed.

Algorithmic Methods

Testing very-large-scale integrated circuits with a high fault coverage is a difficult task because of complexity. Therefore many different ATPG methods have been developed to address combinational and sequential circuits.

- Early test generation algorithms such as boolean difference and literal proposition were not practical to implement on a computer.
- The D Algorithm was the first practical test generation algorithm in terms of memory requirements. The D Algorithm introduced D Notation which continues to be used in most ATPG algorithms.
- Path-Oriented Decision Making (PODEM) is an improvement over the D Algorithm. PODEM was created in 1981 when shortcomings in D Algorithm became evident when design innovations resulted in circuits that D Algorithm could not realise.
- Fan-Out Oriented (FAN Algorithm) is an improvement over PODEM. It limits the ATPG search space to reduce computation time and accelerates backtracing.
- Methods based on Boolean satisfiability are sometimes used to generate test vectors.
- Pseudorandom test generation is the simplest method of creating tests. It uses a pseudorandom number generator to generate test vectors, and relies on logic simulation to compute good machine results, and fault simulation to calculate the fault coverage of the generated vectors.

Relevant Conferences

ATPG is a topic that is covered by several conferences throughout the year. The primary US conferences are the International Test Conference and The VLSI Test Symposium, while in Europe the topic is covered by DATE and ETS.

Accident Analysis

Accident analysis is carried out in order to determine the cause or causes of an accident or series of accidents so as to prevent further incidents of a similar kind. It is also known as accident investigation. It may be performed by a range of experts, including forensic scientists, forensic engineers or health and safety advisers. Accident investigators, particularly those in the aircraft industry, are colloquially known as "tin-kickers".

Sequence

Accident analysis is performed in four steps:

1. Fact gathering: After an accident happened a forensic process starts to gather all possibly relevant facts that may contribute to understanding the accident.
2. Fact Analysis: After the forensic process has been completed or at least delivered some results, the facts are put together to give a "big picture." The history of the accident is reconstructed and checked for consistency and plausibility.
3. Conclusion Drawing: If the accident history is sufficiently informative, conclusions can be drawn about causation and contributing factors.
4. Counter-measures: In some cases the development of counter-measures is desired or recommendations have to be issued to prevent further accidents of the same kind.

Methods

There exist numerous forms of Accident Analysis methods. These can be divided into three categories:

1. Causal Analysis uses the principle of causality to determine the course of events. Though people casually speak of a "chain of events", results from Causal Analysis usually have the form of directed a-cyclic graphs-the nodes being events and the edges the cause-effect relations. Methods of Causal Analysis differ in their respective notion of causation.
2. Expert Analysis relies on the knowledge and experience of field experts. This form of analysis usually lacks a rigorous (formal/semiformal) methodological approach. This usually affects falsify-ability and objectivity of analyses. This is of importance when conclusions are heavily disputed among experts.

3. Organisational Analysis relies on systemic theories of organisation. Most theories imply that if a system's behaviour stayed within the bounds of the ideal organisation then no accidents can occur. Organisational Analysis can be falsified and results from analyses can be checked for objectivity. Choosing an organisational theory for accident analysis comes from the assumption that the system to be analysed conforms to that theory.

Stress Intensity Factor

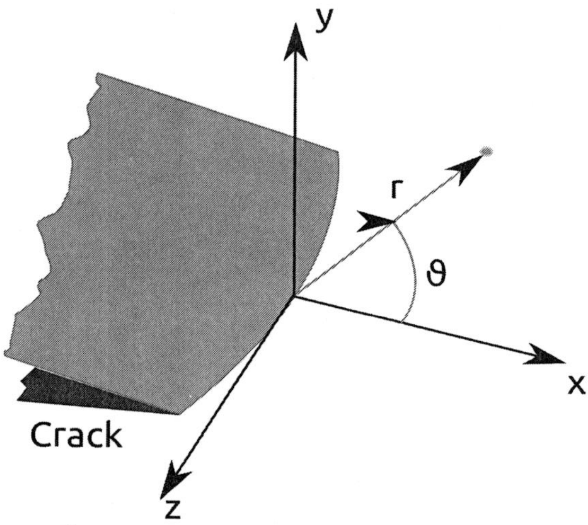

Figure: Polar coordinates at the crack tip.

The stress intensity factor, K, is used infrastructure mechanics to predict the stress state ("stress intensity") near the tip of a crack caused by a remote load or residual stresses. It is a theoretical construct usually applied to a homogeneous, linear elastic material and is useful for providing a failure criterion for brittle materials. The concept can also be applied to materials that exhibit small-scale yielding at a crack tip.

The magnitude of K depends on sample geometry, the size and location of the crack, and the magnitude and the modal distribution of loads on the material.

Linear elastic theory predicts that the stress distribution ($ó_{ij}$) near the crack tip, in polar coordinates (r,è) with origin at the crack tip, has the form

$$\sigma_{ij}(r,\theta) = \frac{K}{\sqrt{2\pi r}} f_{ij}(\theta) + \textit{higher order terms}$$

where K is the stress intensity factor (with units of stress length$^{1/2}$) and f_{ij} is a dimensionless quantity that depends on the load and geometry. This relation breaks down very close to the tip (smallr) because as r goes to 0, the stress ó$_{ij}$ goes to . Plastic distortion typically occurs at high stresses and the linear elastic solution is no longer applicable close to the crack tip. However, if the crack-tip plastic zone is small, it can be assumed that the stress distribution near the crack is still given by the above relation.

Stress Intensity Factors for Various Modes

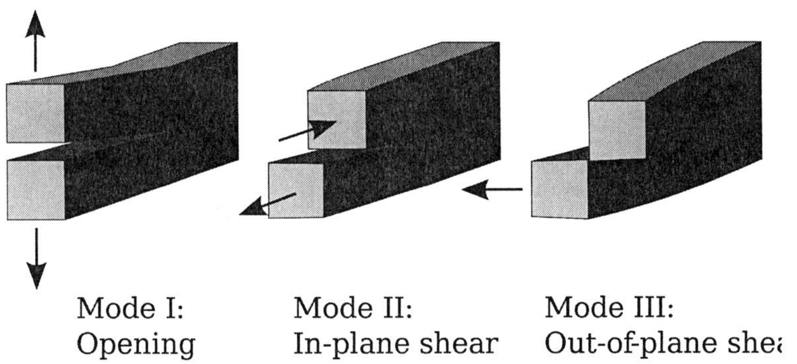

Figure: Mode I, Mode II, and Mode III crack loading.

Three linearly independent cracking modes are used in fracture mechanics. These load types are categorised as Mode I, II, or III as shown in the figure.

Mode I, shown to the right, is an opening (tensile) mode where the crack surfaces move directly apart. Mode II is a sliding (in-plane shear) mode where the crack surfaces slide over one another in a direction perpendicular to the leading edge of the crack. Mode III is a tearing (antiplane shear) mode where the crack surfaces move relative to one another and parallel to the leading edge of the crack. Mode I is the most common load type encountered in engineering design.

Different subscripts are used to designate the stress intensity factor for the three different modes. The stress intensity factor for mode I is designated K_I and applied to the crack opening mode.

The mode II stress intensity factor, K_{II}, applies to the crack sliding mode and the mode III stress intensity factor, K_{III}, applies to the tearing mode. These factors are formally defined as

$$K_I = \lim_{r \to 0} \sqrt{2\pi r}\,\sigma_{yy}(r,0)$$

$$K_{II} = \lim_{r \to 0} \sqrt{2\pi r}\,\sigma_{yx}(r,0)$$

$$K_{III} = \lim_{r \to 0} \sqrt{2\pi r}\,\sigma_{yz}(r,0).$$

Relationship to energy release rate and J-integral

The strain energy release rate (G) for a crack under mode I loading is related to the stress intensity factor by

$$G = K_1^2 \left(\frac{1-v^2}{E} \right)$$

where E is the Young's modulus and í is the Poisson's ratio of the material. The material is assumed to be an isotropic, homogeneous, and linear elastic. Plane strain has been assumed and the crack has been assumed to extend along the direction of the initial crack. For plane stress conditions, the above relation simplifies to

$$G = K_1^2 \left(\frac{1}{E} \right).$$

For pure mode II loading, we have similar relations

$$G = K_{II}^2 \left(\frac{1-v^2}{E} \right) \quad or \quad G = K_{II}^2 \left(\frac{1}{E} \right).$$

For pure mode III loading,

$$G = K_{III}^2 \left(\frac{1}{2\mu} \right)$$

where ì is the shear modulus. For general loading in plane strain, the relationship between the strain energy and the stress intensity factors for the three modes is

$$G = K_1^2 \left(\frac{1-v^2}{E} \right) + K_{II}^2 \left(\frac{1-v^2}{E} \right) + K_{III}^2 \left(\frac{1}{2\mu} \right).$$

A similar relation is obtained for plane stress by adding the contributions for the three modes.

The above relations can also be used to connect the J integral to the stress intensity factor because

$$G = J = \int_{\Gamma} \left(W\, dx_2 - t \cdot \frac{\partial u}{\partial x_1} ds \right).$$

Critical Stress Intensity Factor

The stress intensity factor, K, is a parameter that amplifies the magnitude of the applied stress that includes the geometrical parameter Y (load type). Stress intensity in any mode situation is directly proportional to the applied load on the material. If a very sharp crack can be made in a material, the minimum value of K_I can be empirically determined, which is the critical value of stress intensity required to propagate the crack. This critical value determined for mode I loading in plane strain is referred to as the critical fracture toughness (K_{Ic}) of the material. K_{Ic} has units of stress times the root of a distance. The units of K_{Ic} imply that the fracture stress of the material must be reached over some critical distance in order for K_{Ic} to be reached and crack propagation to occur. The Mode I critical stress intensity factor, K_{Ic}, is the most often used engineering design parameter in fracture mechanics and hence must be understood if we are to design fracture tolerant materials used in bridges, buildings, aircraft, or even bells. Polishing cannot detect a crack. Typically, if a crack can be seen it is very close to the critical stress state predicted by the stress intensity factor.

G–criterion

The G-criterion is a fracture criterion that relates the critical stress intensity factor (or fracture toughness) to the stress intensity factors for the three modes. This failure criterion is written as

$$K_{Ic}^2 = K_I^2 + K_{II}^2 + \frac{E'}{2\mu} K_{III}^2$$

where K_{Ic} is the mode I fracture toughness, E' = E / (1 " í²) for plane strain and E' = E for plane stress. The critical stress intensity factor for plane stress is often written as K_c.

Puncture Resistance

Puncture resistance denotes the relative ability of a material to inhibit the progression of a tear once it has been pierced by a cut or a nick. Tests devised to measure puncture resistance are generally application-specific, covering items such as roofing and packaging materials, protective gloves, needle disposal facilities, etc. Puncture resistance in fabrics can be obtained through very tight woven fabrics, small ceramic plates in fabric coating or tight woven fabrics with a coating of hard crystals. All described methods significantly reduce the softness and flexibility of the fabric.

The puncture resistance will depend on the nature of puncture attempt, with the two most important features being point sharpness and force. A fine sharp point such as a hypodermic needle will require a high ability to absorb and distribute the force to avoid penetration, but the total forces applied are still limited. The EN388 glove standard use a more pencil-like object with a flat tip of 1mm diametre. The EN388 test is highly dependent on the materials ability to withstand high forces through high tenacity and to a lesser extent to avoid cut or separation of the material.

There is no or limited correlation between the protections provided in the low force/needle protection and the high force/pencil like EN388 test.

Needle-resistant materials as described above are generally pierced by a force between 2-10N by a 25 gauge needle perpendicular to the fabric. The forces in the EN 388 test results are rated according to a score from 0-4 (0-<20N, 1-20N, 2-60N 3-100N 4-150N+). Three layers of SuperFabric have a higher needle protection than single layers of woven para aramid (Kevlar) and SuperFabric has a significantly higher level of protection against the pencil-like high force EN388 test compared to the coated woven para aramid.

Brittle-ductile Transition Zone

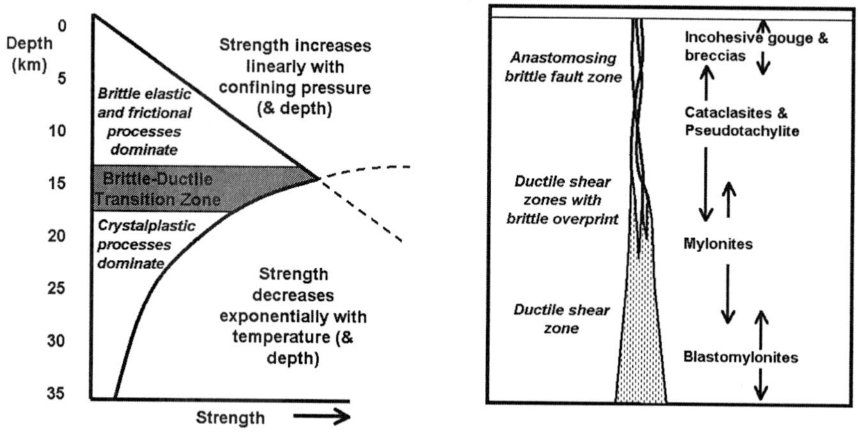

Figure: Variation of strength with depth in continental crust and changes in dominant deformation mechanisms and fault rocks in a conceptual vertical fault zone.

The brittle-ductile transition zone is the strongest part of the Earth's crust. For quartz and feldspar rich rocks in continental crust this occurs at an approximate depth of 13–18 km (roughly equivalent to temperatures in the range 250-400°C). At this depth rock becomes less likely to fracture, and more likely to deform ductilely by creep. This happens because the brittle strength of a material is increased by the confining pressure, whilst the ductile strength of a material decreases with increasing temperature. The transition zone occurs at the level in the crust where the downwards increasing brittle strength equals the upwards increasing ductile strength, giving a characteristic "saw-tooth" crustal strength profile. This zone is, therefore, the strongest part of the crust and the depth at which many earthquakes occur. The level of the transition zone depends on both strain rate and temperature gradient, being shallower for slow deformation and/or high heat flow and deeper for fast deformation and/or low heat flow. Crustal composition will also affect the depth at which this zone occurs.

Sections of fault zones once active in the transition zone, and now exposed at the surface, typically have a complex overprinting of brittle and ductile rock types. Cataclasites or pseudotachylite breccias with mylonite clasts are common, as are ductily deformed cataclasites and pseudotachylites.

Charpy Impact Test

The Charpy impact test, also known as the Charpy v-notch test, is a standardised high strain-rate test which determines the amount of energy absorbed by a material during fracture. This absorbed energy is a measure of a given material's toughness and acts as a tool to study temperature-dependent brittle-ductile transition. It is widely applied in industry, since it is easy to prepare and conduct and results can be obtained quickly and cheaply. But a major disadvantage is that all results are only comparative.

The test was developed in 1905 by the French scientist Georges Charpy. It was pivotal in understanding the fracture problems of ships during the Second World War. Today it is used in many industries for testing building and construction materials used in the construction of pressure vessels, bridges and to see how storms will affect materials used in building.

Definition

The apparatus consists of a pendulum axe swinging at a notched sample of material. The energy transferred to the material can be inferred by comparing the difference in the height of the hammer before and after a big fracture.

The notch in the sample affects the results of the impact test, thus it is necessary for the notch to be of regular dimensions and geometry. The size of the sample can also affect results, since the dimensions determine whether or not the material is in plane strain. This difference can greatly affect conclusions made.

Figure: A vintage impact test machine.

The "Standard methods for Notched Bar Impact Testing of Metallic Materials" can be found in ASTM E23, ISO 148-1or EN 10045-1, where all the aspects of the test and equipment used.

Quantitative Results

The quantitative result of the impact tests the energy needed to fracture a material and can be used to measure the toughness of the material and the yield strength. Also, the strain rate may be studied and analysed for its effect on fracture. The ductile-brittle transition temperature (DBTT) may be derived from the temperature where the energy needed to fracture the material drastically changes. However, in practice there is no sharp transition and so it is difficult to obtain a precise transition temperature. An exact DBTT may be empirically derived in many ways: a specific absorbed energy, change in aspect of fracture (such as 50% of the area is cleavage), etc.

Qualitative Results

The qualitative results of the impact test can be used to determine the ductility of a material. If the material breaks on a flat plane, the fracture was brittle, and if the material breaks with jagged edges or shear lips, then the fracture was ductile. Usually a material does not break in just one way or the other, and thus comparing the jagged to flat surface areas of the fracture will give an estimate of the percentage of ductile and brittle fracture.

Sample sizes

According to ASTM A370, the standard specimen size for Charpy impact testing is 10mm×10mm×55mm. Subsize specimen sizes are: 10mm×7.5mm×55mm, 10mm×6.7mm×55mm, 10mm×5mm×55mm, 10mm×3.3mm×55mm, 10mm×2.5mm×55mm. Details of specimens as per ASTM A370 (Standard Test Method and Definitions for Mechanical Testing of Steel Products). According to EN 10045-1, standard specimen sizes are 10mmx10mmx55mm. Subsize specimens are: 10mmx7.5mmx55mm and 10mmx5mmx55mm.

Izod Impact Strength Test

Izod impact strength testing is an ASTM standard method of determining impact strength. A notched sample is generally used to determine impact strength. The test is named after the English engineer Edwin Gilbert Izod (1876–1946), who described it in his 1903 address to the British Association, subsequently published in Engineering. Impact is a very important phenomenon in governing the life of a structure. For example, in the case of an aircraft, impact can take place by a bird hitting a plane while it is cruising, or during take off and landing the aircraft may be struck by debris present on the runway, and as well as other causes.

An arm held at a specific height (constant potential energy) is released. The arm hits the sample and breaks it. From the energy absorbed by the sample, its impact strength is determined.

The North American standard for Izod Impact testing is ASTM D256. The results are expressed in energy lost per unit of thickness (such as ft-lb/in or J/cm) at the notch. Alternatively, the results may be reported as energy lost per unit cross-sectional area at the notch (J/m^2 or $ft\text{-}lb/in^2$). In Europe, ISO 180 methods are used and results are based only on the cross-sectional area at the notch (J/m^2).

The dimensions of a standard specimen for ASTM D256 are 63.5 x 12.7 x 3.2 mm (2.5" x 0.5" x 1/8"). The most common specimen thickness is 3.2 mm (0.125"), but the width can vary between 3.0 and 12.7 mm (0.118" and 0.500"). The Izod impact test differs from the Charpy impact test in that the sample is held in a cantilevered beam configuration as opposed to a three-point bending configuration. This test can also be used to determine the notch sensitivity. Impact tests are used in studying the toughness of material. A material's toughness is a factor of its ability to absorb energy during plastic deformation. Brittle materials have low toughness as a result of the small amount of plastic deformation that they can endure. The impact value of a material can also change with temperature. Generally, at lower temperatures, the impact strength of a material is decreased. The size of the specimen may also affect the value of the Izod impact test because it may allow a different number of imperfections in the material, which can act as stress risers and lower the impact strength.

Ceramography

Ceramography is the art and science of preparation, examination and evaluation of ceramic microstructures. Ceramography can be thought of as the metallography of ceramics. The microstructure is the structure level of approximately 0.1 to 100 μm, between the minimum wavelength of visible light and the resolution limit of the naked eye. The microstructure includes most grains, secondary phases, grain boundaries, pores, micro-cracks and hardness microindentions. Most bulk mechanical, optical, thermal, electrical and magnetic properties are significantly affected by the microstructure. The fabrication method and process conditions are generally indicated by the microstructure. The root cause of many ceramic failures is evident in the microstructure. Ceramography is part of the broader field of materialography, which includes all the microscopic techniques of material analysis, such as metallography, petrography and plastography. Ceramography is usually reserved for high-performance ceramics for industrial applications, such as 85–99.9%alumina (Al_2O_3) in Fig. below, zirconia (ZrO_2), silicon carbide (SiC), silicon nitride (Si_3N_4), and ceramic-matrix composites. It is seldom used on whiteware ceramics such as sanitaryware, wall tiles and dishware.

Ceramographic Microstructures

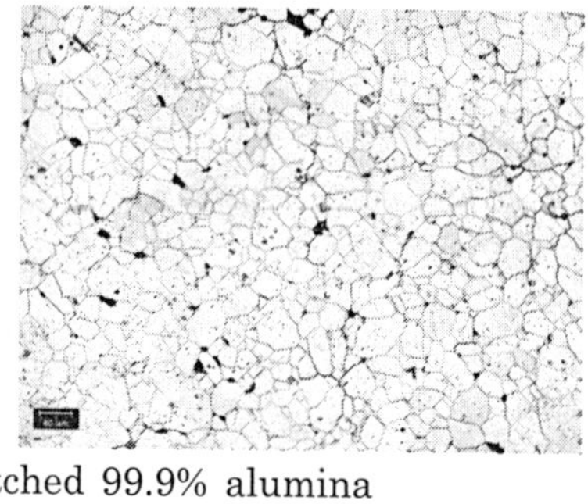

Figure: Thermally etched 99.9% alumina

Figure: Thin section of 99.9% alumina

A Brief History of Ceramography

Ceramography evolved along with other branches of materialography and ceramic engineering. Alois de Widmanstätten of Austria etched a meteorite in 1808 to reveal proeutectoid ferrite bands that grew on prior austenite grain boundaries. Geologist Henry Clifton Sorby, the "father of metallography," applied petrographic techniques to the steel industry in the 1860s in Sheffield, England. French geologist Auguste Michel-Lévy devised a chart that correlated the optical properties of minerals to their transmitted colour and thickness in the 1880s. Swedish metallurgist J.A. Brinell invented the first quantitative hardness scale in 1900. Smith and Sandland developed the first microindention hardness test at Vickers Ltd. in London in 1922. Swiss-born microscopist A.I. Buehler started the first metallographic equipment manufacturer near Chicago in 1936. Frederick Knoop and colleagues at the National Bureau of Standards developed a less-penetrating (than Vickers) microindention test in 1939. Struers A/S of Copenhagen introduced the electrolytic polisher to metallography in 1943. George Kehl of Columbia University wrote a book that was considered the bible of materialography until the 1980s. Kehl co-founded a group within the Atomic Energy Commission that became the International Metallographic Society in 1967.

Preparation of Ceramographic Specimens

The preparation of ceramic specimens for microstructural analysis consists of five broad steps: sawing, embedding, grinding, polishing and etching. The tools and consumables for ceramographic preparation are available worldwide from metallography equipment vendors and laboratory supply companies.

- Sawing: most ceramics are extremely hard and must be wet-sawed with a circular blade embedded with diamond particles. A metallography or lapidary saw equipped with a low-density diamond blade is usually suitable. The blade must be cooled by a continuous liquid spray.

- Embedding: to facilitate further preparation, the sawed specimen is usually embedded (or mounted or encapsulated) in a plastic disc, 25, 30 or 35 mm in diametre. A thermosetting solidresin, activated by heat and compression, e.g. mineral-filled epoxy, is best for most applications. A castable (liquid) resin such as unfilled epoxy, acrylic or polyester may be used for porous refractoryceramics or microelectronic devices. The castable resins are also available with fluorescent dyes that aid in fluorescence microscopy. The left and right specimens in Fig. 3 were embedded in mineral-filled epoxy. The centre refractory in Fig. 3 was embedded in castable, transparent acrylic.
- Grinding is abrasion of the surface of interest by abrasive particles, usually diamond, that are bonded to paper or a metal disc. Grinding erases saw marks, coarsely smooths the surface, and removes stock to a desired depth. A typical grinding sequence for ceramics is one minute on a 240-grit metal-bonded diamond wheel rotating at 240 rpm and lubricated by flowing water, followed by a similar treatment on a 400-grit wheel. The specimen is washed in an ultrasonic bath after each step.
- Polishing is abrasion by free abrasives that are suspended in a lubricant and can roll or slide between the specimen and paper. Polishing erases grinding marks and smooths the specimen to a mirror-like finish. Polishing on a bare metallic platen is called lapping. A typical polishing sequence for ceramics is 5–10 minutes each on 15-, 6- and 1-μm diamond paste or slurry on napless paper rotating at 240 rpm. The specimen is again washed in an ultrasonic bath after each step. The three sets of specimens in Fig. 3 have been sawed, embedded, ground and polished.
- Etching reveals and delineates grain boundaries and other microstructural features that are not apparent on the as-polished surface. The two most common types of etching in ceramography are selective chemical corrosion, and a thermal treatment that causes relief. As an example, alumina can be chemically etched by immersion in boiling concentrated phosphoric acid for 30–60 s, or thermally etched in a furnace for 20–40 min at 1,500 °C (2,730 °F) in air. The plastic

encapsulation must be removed before thermal etching. The alumina was thermally etched.

Alternatively, non-cubic ceramics can be prepared as thin sections, also known as petrography, for examination by polarised transmitted light microscopy. In this technique, the specimen is sawed to ~1 mm thick, glued to a microscope slide, and ground to a thickness (x) approaching 30 μm. A cover slip is glued onto the exposed surface. The adhesives, such as epoxy or Canada balsam resin, must have approximately the same refractive index $(\eta \approx 1.54)$ as glass. Most ceramics have a very small absorption coefficient $\alpha \approx$ (0.5 cm^{-1} for alumina in Fig. 2) in the Beer-Lambert law below, and can be viewed in transmitted light. Cubic ceramics, e.g. yttria-stabilised zirconia and spinel, have the same refractive index in all crystallographic directions and are, therefore, opaque when the microscope's polariser is 90° out of phase with its analyser.

$$I_t = I_0 e^{-\alpha x} \text{ (Beer–Lambert eqn)}$$

Ceramographic specimens are electrical insulators in most cases, and must be coated with a conductive ~10-nm layer of metal or carbon for electron microscopy, after polishing and etching. Gold or Au-Pd alloy from a sputter coater or evaporative coater also improves the reflection of visible light from the polished surface under a microscope, by the Fresnel formula below. Bare alumina $\left(\eta \approx 1.77 k \approx 10^{-5}\right)$ has a negligible extinction coefficient and reflects only 8% of the incident light from the microscope. Gold-coated $(\eta \approx 0.82, k \approx 1.59 @ \lambda = 500 mm)$ alumina reflects 44% in air, 39% in immersion oil.

$$R = \frac{I_r}{I_i} = \frac{(\eta_1 - \eta_2)^2 + k^2}{(\eta_1 + \eta_2)^2 + k^2 (Fresnel\, eqn)} \quad \text{(Fresnel eqn)}$$

Ceramographic Analysis

Ceramic microstructures are most often analysed by reflected visible-light microscopy in brightfield. Darkfield is used in limited circumstances, e.g., to reveal cracks. Polarised transmitted light is used with thin sections, where the contrast between grains comes from birefringence. Very fine microstructures may require the higher magnification and resolution of a scanning electron microscope(SEM) or confocal laser scanning microscope (CLSM). The cathodoluminescence microscope (CLM) is useful for distinguishing phases of refractories. The transmission electron microscope (TEM) and scanning acoustic microscope (SAM) have speciality applications in ceramography.

Ceramography is often done qualitatively, for comparison of the microstructure of a component to a standard for quality control or failure analysis purposes. Three common quantitative analyses of microstructures are grain size, second-phase content and porosity. Microstructures are measured by the principles of stereology, in which three-dimensional objects are evaluated in 2-D by projections or cross-sections.

Grain size can be measured by the line-fraction or area-fraction methods of ASTM E112. In the line-fraction methods, a statistical grain size is calculated from the number of grains or grain boundaries intersecting a line of known length or circle of known circumference. In the area-fraction method, the grain size is calculated from the number of grains inside a known area. In each case, the measurement is affected by secondary phases, porosity, preferred orientation, exponential distribution of sizes, and non-equiaxed grains. Image analysis can measure the shape factors of individual grains by ASTM E1382.

Second-phase content and porosity are measured the same way in a microstructure, such as ASTM E562. Procedure E562 is a point-fraction method based on the stereological principle of point fraction = volume fraction, i.e., $P_p = V_v$. Second-phase content in ceramics, such as carbide whiskers in an oxide matrix, is usually expressed as a mass fraction. Volume fractions can be converted to mass fractions if the density of each phase is known. Image analysis can measure porosity, pore-size distribution and volume fractions of secondary phases by ASTM E1245. Porosity measurements do not require etching. Multi-phase microstructures do not require etching if the contrast between phases is adequate, as is usually the case.

Grain size, porosity and second-phase content have all been correlated with ceramic properties such as mechanical strength ó by the Hall–Petch equation, hardness, toughness, dielectric constant and many others.

Microindention Hardness and Toughness

The hardness of a material can be measured in many ways. The Knoop hardness test, a method of microindention hardness, is the most reproducible for dense ceramics. The Vickers hardness test and superficial Rockwell scales (e.g., 45N) can also be used, but tend to cause more surface damage than Knoop. The Brinell test is suitable for ductile metals, but not ceramics. In the Knoop test, a diamond indenter in the shape of an elongated pyramid is forced into a polished (but not etched) surface under a predetermined load, typically 500 or 1000 g. The load is held for some amount of time, say 10 s, and the indenter is retracted. The indention long diagonal (d, ìm, in Fig. 4) is measured under a microscope, and the Knoop hardness (HK) is calculated from the load (P, g) and the square of the diagonal length in the equations below. The constants account for the projected area of the indenter and unit conversion factors. Most oxide ceramics have a Knoop hardness in the range of 1000–1500kg_f/mm^2 (10 – 15 GPa), and many carbides are over 2000 (20 GPa). The method is specified in ASTM C849, C1326 & E384. Microindention hardness is also called microindentation hardness or simply microhardness. The hardness of very small particles and thin films of ceramics, on the order of 100 nm, can be measured by nanoindentation methods that use a Berkovich indenter.

$$HK = 14229\frac{P}{d^2} \quad (kg_f/mm^2) \text{ and } HK = 139.54\frac{P}{d^2} \quad (GPa)$$

The toughness of ceramics can be determined from a Vickers test under a load of 10 – 20 kg. Toughness is the ability of a material to resist crack propagation. Several calculations have been formulated from the load (P), elastic modulus (E), microindention hardness (H), crack length (c in Fig. 5) and flexural strength (ó). Modulus of rupture (MOR) bars with a rectangular cross-section are indented in three places on a polished surface. The bars are loaded in 4-point bending with the polished, indented surface in tension, until fracture. The fracture normally originates at one of the indentions. The crack lengths are measured under a microscope. The toughness of most ceramics is 2–4 MPa"m, but toughened zirconia is as much as 13, and cemented carbides are often over 20. The toughness-by-indention methods have been discredited recently and are being replaced by more rigorous methods that measure crack growth in a notched beam in flexure.

$$K_{id} = 0.016\sqrt{\frac{E}{H}}\frac{P}{(C_0)^{1.5}} \text{ initial crack length}$$

$$K_{isb} = 0.59\left(\frac{E}{H}\right)^{1/8}\left[\sigma\left(P^{1/3}\right)\right]^{3/4} \text{ indention strength in bending.}$$

Shock (mechanics)

A mechanical or physical shock is a sudden acceleration or deceleration caused, for example, by impact, drop, kick, earthquake, or explosion. Shock is a transient physical excitation.

Shock is usually measured by an accelerometre. This describes a shock pulse as a plot of acceleration versus time. Acceleration can be reported in units of metre per second squared. Often, for convenience, the magnitude of a shock is stated as a multiple of the standard acceleration due to free fall in the Earth's gravity, a quantity with the symbol g having the value 9.80665 m/s^2.

Thus a shock of "20 g" is equivalent to about 196 m/s^2. A shock can be characterised by the peak acceleration, the duration, and the shape of the shock pulse (half sine, triangular, trapezoidal, etc). The Shock response spectrum is a method for further evaluating a mechanical shock. It is sometimes used as a defence standard for military equipment.

Effects of Shock

Mechanical shock has the potential for damaging an item (e.g., an entire light bulb) or an element of the item (e.g. a filament in an Incandescent light bulb):

- A brittle or fragile item can fracture. For example, two crystal wine glasses may shatter when impacted against each other. A shear pin in an engine is designed to fracture with a specific magnitude of shock. Note that a soft ductile material may sometimes exhibit brittle failure during shock due to time-temperature superposition.
- A ductile item can be bent by a shock. For example, a copper pitcher may bend when dropped on the floor.
- Some items may not be damaged by a single shock but will experience fatigue failure with numerous repeated low-level shocks.
- A shock may result in only minor damage which may not be critical for use. However, cumulative minor damage from several shocks will eventually result in the item being unusable.
- A shock may not produce immediate apparent damage but might cause the service life of the product to be shortened: the reliability is reduced.
- A shock may cause an item to become out of adjustment. For

example, when a precision scientific instrument is subjected to a moderate shock, good metrology practice may be to have it recalibrated before further use.

- Some materials such as primary high explosives may detonate with mechanical shock or impact.
- When glass bottles of liquid are dropped or subjected to shock, the water hammer effect may cause hydrodynamic glass breakage.

Considerations

When laboratory testing, field experience, or engineering judgement indicates that an item could be damaged by mechanical shock, several courses of action might be considered:

- Reduce and control the input shock at the source.
- Modify the item to improve its toughness or support it to better handle shocks.
- Use shock absorbers or cushions to control the shock transmitted to the item. Cushioning reduces the peak acceleration by extending the duration of the shock.
- Plan for failures: accept certain losses. Have redundant systems available, etc.

Chapter 3

Fracture Toughening Mechanisms

In materials science, fracture toughening mechanisms are processes that increase energy absorption during fracture, resulting in higher fracture toughness.

Intrinsic Toughening Mechanisms

Intrinsic toughening mechanisms involve a fundamental change of material properties like ductility and plasticity. These changes can be achieved by creating a more stable microstructure, or by increasing precipitate particle spacing to improve ductility.

Extrinsic Toughening Mechanisms

Extrinsic toughening mechanisms act during crack propagation at the location of the crack, and these mechanisms come in two subcategories based on where they act.

Zone Shielding

Three zone shielding mechanisms act on the crack tip to interfere with crack propagation.

The first is transformation toughening, which occurs when the crack actually changes the crystalline structure of the surrounding material to inhibit crack growth. The second and third mechanisms are microcrack toughening and crack field void formation, where either microcracks or microvoids form around the crack tip and reduce the crack tip stress concentration.

Contact Shielding

Contact shielding mechanisms act behind the propagating crack. Contact shielding mechanisms often induce some sort of crack closure, either by the roughness of the fracture surface, or by unbroken fibres bridging the crack. A sliding crack surface or a wake of plasticity behind the crack are also contact shielding mechanisms.

Pipe (fluid conveyance)

Steel Pipes

Figure: Plastic (PVC) pipes in USA

A pipe is a tubular section or hollow cylinder, usually but not necessarily of circular cross-section, used mainly to convey substances which can flow — liquids and gases (fluids), slurries, powders, masses of small solids. It can also be used for structural applications; hollow pipe is far stiffer per unit weight than solid members.

In common usage the words pipe and tube are usually interchangeable, but in industry and engineering, the terms are uniquely defined. Depending on the applicable standard to which it is manufactured, pipe is generally specified by a nominal diametre with a constant outside diametre (OD) and a schedule that defines the thickness. Tube is most often specified by the OD and wall thickness, but may be specified by any two of OD, inside diametre (ID), and wall thickness. Pipe is generally manufactured to one of several international and national industrial standards. While similar standards exist for specific industry application tubing, tube is often made to custom sizes and a broader range of diametres and tolerances. Many industrial and government standards exist for the production of pipe and tubing. The term "tube" is also commonly applied to non-cylindrical sections, i.e., square or rectangular tubing. In general, "pipe" is the more common term in most of the world, whereas "tube" is more widely used in the United States.

Both "pipe" and "tube" imply a level of rigidity and permanence, whereas a hose (or hosepipe) is usually portable and flexible. Pipe assemblies are almost always constructed with the use of fittings such as elbows, tees, and so on, while tube may be formed or bent into custom configurations. For materials that are inflexible, cannot be formed or where construction is governed by codes or standards, tube assemblies are also constructed with the use of tube fittings.

Uses

A pipe implantation, inside an avenue of Belo Horizonte, Brazil.

- Domestic water systems
- Pipelines transporting gas or liquid over long distances
- Scaffolding
- Structural steel
- As components in mechanical systems such as:
  - Rollers in conveyor belts
  - Compactors (E.g.: steam rollers)
  - Bearing casing
- Casing for concrete pilings used in construction projects
- High temperature or pressure manufacturing processes
- The petroleum industry:
  - Oil well casing
  - Oil refinery equipment
- Delivery of fluids, either gaseous or liquid, in a process plant from one point to another point in the process
- Delivery of bulk solids, in a food or process plant from one point to another point in the process
- The construction of high pressure storage vessels (note that large pressure vessels are constructed from plate, not pipe owing to their wall thickness and size).

Manufacture

There are three processes for metallic pipe manufacture. Centrifugal casting of hot alloyed metal is one of the most prominent process. Ductile iron pipes are generally manufactured in such a fashion. Seamless (SMLS) pipe is formed by drawing a solid billet over a piercing rod to create the hollow shell. Historically seamless pipe was regarded as withstanding pressure better than other types, and was often more easily available than welded pipe. Advances since the 1970's in materials, process control and non-destructive testing allow correctly specified welded pipe to replace seamless in many applications. Welded (also Electric Resistance Welded ("ERW"), and Electric Fusion Welded ("EFW")) pipe is formed by rolling plate and welding the seam.

The weld flash can be removed from the outside or inside surfaces using a scarfing blade. The weld zone can also be heat treated to make the seam less visible. Welded pipe often has tighter dimensional tolerances than seamless, and can be cheaper if manufactured in the same quantities. Large-diametre pipe (25 centimetres (10 in) or greater) may be ERW, EFW or Submerged Arc Welded ("SAW") pipe.

Tubing, either metal or plastic, is generally extruded.

Materials

Figure: Historic water mains from Philadelphia included wooden pipes

Pipe are made in many materials including ceramic, fibreglass, many metals, concrete and plastic. In the past wood and lead (Latin plumbum, from which we get the wordplumbing) were commonly used.

Metallic pipes are commonly made from steel or iron; the finish and metal chemistry are peculiar to the use, fit and form. Typically metallic piping is made of steel or iron, such as unfinished, black (lacquer) steel, carbon steel, stainless steel or galvanized steel, brass, and ductile iron. Aluminium pipe or tubing may be utilised where iron is incompatible with the service fluid or where weight is a concern;

aluminium is also used for heat transfer tubing such as in refrigerant systems. Copper tubing is popular for domestic water (potable) plumbing systems; copper may be used where heat transfer is desirable (i.e. radiators or heat exchangers). Inconel, chrome moly, and titanium steel alloys are used in high temperature and pressure piping in process and power facilities. When specifying alloys for new processes, the known issues of creep and sensitization effect must be taken into account.

Lead piping is still found in old domestic and other water distribution systems, but it is no longer permitted for new potable water piping installations due to its toxicity. Many building codes now require that lead piping in residential or institutional installations be replaced with non-toxic piping or that the tubes' interiors be treated with phosphoric acid. According to a senior researcher and lead expert with the Canadian Environmental Law Association, "...there is no safe level of lead [for human exposure]".

Plastic tubing is widely used for its light weight, chemical resistance, non-corrosive properties, and ease of making connections. Plastic materials include polyvinyl chloride (PVC), chlorinated polyvinyl chloride (CPVC), fibre reinforced plastic (FRP), reinforced polymer mortar (RPMP), polypropylene(PP), polyethylene (PE), cross-linked high-density polyethylene (PEX), polybutylene (PB), and acrylonitrile butadiene styrene (ABS), for example. In many countries, PVC pipes account for most pipe materials used in buried municipal applications for drinking water distribution and wastewater mains.

Pipe may be made from concrete or ceramic, usually for low-pressure applications such as gravity flow or drainage. Pipes for sewage are still predominantly made from concrete or vitrified clay. In other applications plastic pipes provide better properties and processing options.

Reinforced concrete can be used for large-diametre concrete pipes. This pipe material can be used in many types of construction, and is often used in the gravity-flow transport of storm water. Usually such pipe will have a receiving bell or a stepped fitting, with various sealing methods applied at installation.

Sizes

Pipe sizes can be confusing because the terminology may relate to historical dimensions. For example, a half-inch iron pipe does not have any dimension that is a half inch. Initially, a half inch pipe did have an inner diametre of 0.5 inches (13 mm)—but it also had thick walls. As technology improved, thinner walls became possible, but the outside diametre stayed the same so it could mate with existing older pipe, increasing the inner diametre beyond half an inch. The history of copper pipe is similar. In the 1930s, the pipe was designated by its internal diametre and a $^{1}D_{16}$-inch (1.6 mm) wall thickness. Consequently, a 1-inch (25 mm) copper pipe had a 1$^{1}D_{8}$-inch (28.58 mm) outside diametre. The outside diametre was the important dimension for mating with fittings. The wall thickness on modern copper is usually thinner than $^{1}D_{16}$ inches (1.6 mm), so the internal diametre is only "nominal" rather than a controlling dimension. Newer pipe technologies sometimes adopted a sizing system as its own. PVC pipe uses the Nominal Pipe Size.

Pipe sizes are specified by a number of national and international standards, including API 5L,ANSI/ASME B36.10M and B36.19M in the US, BS 1600 and BS EN 10255 in the United Kingdom and Europe.

There are two common methods for designating pipe outside diametre (OD). The North American method is called NPS ("Nominal Pipe Size") and is based on inches (also frequently referred to as NB ("Nominal Bore")). The European version is called DN ("Diametre Nominal" / "Nominal Diametre") and is based on millimetres. Designating the outside diametre allows pipes of the same size to be fit together no matter what the wall thickness.

- For pipe sizes less than NPS 14 inch (DN 350), both methods give a nominal value for the OD that is rounded off and is not the same as the actual OD. For example, NPS 2 inch and DN 50 are the same pipe, but the actual OD is 2.375 inches or 60.33 millimetres. The only way to obtain the actual OD is to look it up in a reference table.
- For pipe sizes of NPS 14 inch (DN 350) and greater the NPS size is the actual diametre in inches and the DN size is equal to NPS times 25 (not 25.4) rounded to a convenient multiple of 50. For example, NPS 14 has an OD of 14 inches or 355.60 millimetres, and is equivalent to DN 350.

Since the outside diametre is fixed for a given pipe size, the inside diametre will vary depending on the wall thickness of the pipe. For example, 2" Schedule 80 pipe has thicker walls and therefore a smaller inside diametre than 2" Schedule 40 pipe.

Steel pipe has been produced for about 150 years. The pipe sizes that are in use today in PVC and galvanized were originally designed years ago for steel pipe. The number system, like Sch 40, 80, 160, were set long ago and seem a little odd. For example, Sch 20 pipe is even thinner than Sch 40, but same OD. And while these pipes are based on old steel pipe sizes, there is other pipe, like cpvc for heated water, that uses pipe sizes, inside and out, based on old copper pipe size standards instead of steel.

Many different standards exist for pipe sizes, and their prevalence varies depending on industry and geographical area. The pipe size designation generally includes two numbers; one that indicates the outside (OD) or nominal diametre, and the other that indicates the wall thickness. In the early twentieth century, American pipe was sized by inside diametre. This practice was abandoned to improve compatibility with pipe fittings that must usually fit the OD of the pipe, but it has had a lasting impact on modern standards around the world.

In North America and the UK, pressure piping is usually specified by Nominal Pipe Size (NPS) and schedule (SCH). Pipe sizes are documented by a number of standards, including API 5L, ANSI/ ASMEB36.10M (Table 1) in the US, and BS 1600 and BS 1387 in the United Kingdom. Typically the pipe wall thickness is the controlled variable, and the Inside Diametre (I.D.) is allowed to vary. The pipe wall thickness has a variance of approximately 12.5 percent.

In the rest of Europe pressure piping uses the same pipe IDs and wall thicknesses as Nominal Pipe Size, but labels them with a metric Diametre Nominal (DN) instead of the imperial NPS. For NPS larger than 14, the DN is equal to the NPS multiplied by 25. (Not 25.4) This is documented by EN10255 (formerly DIN 2448 and BS 1387) and ISO 65, and it is often called DIN or ISO pipe.

Japan has its own set of standard pipe sizes, often called JIS pipe.

The Iron pipe size (IPS) is an older system still used by some manufacturers and legacy drawings and equipment. The IPS number is the same as the NPS number, but the schedules were limited to Standard Wall (STD), Extra Strong (XS), and Double Extra Strong (XXS). STD is identical to SCH 40 for NPS 1/8 to NPS 10, inclusive, and indicates .375" wall thickness for NPS 12 and larger. XS is identical to SCH 80 for NPS 1/8 to NPS 8, inclusive, and indicates .500" wall thickness for NPS 8 and larger. Different definitions exist for XXS, however it is never the same as SCH 160. XXS is in fact thicker than SCH 160 for NPS 1/8" to 6" inclusive, whereas SCH 160 is thicker than XXS for NPS 8" and larger.

Another old system is the Ductile Iron Pipe Size (DIPS), which generally has larger ODs than IPS. Copper plumbing tube for residential plumbing follows an entirely different size system in America, often called Copper Tube Size (CTS); see domestic water system. Its nominal size is neither the inside nor outside diametre. Plastic tubing, such as PVC and CPVC, for plumbing applications also has different sizing standards.

Agricultural applications use PIP sizes, which stands for Plastic Irrigation Pipe. PIP comes in pressure ratings of 22 psi (150 kPa), 50 psi (340 kPa), 80 psi (550 kPa), 100 psi (690 kPa), and 125 psi (860 kPa) and is generally available in diametres of 6", 8", 10", 12", 15", 18", 21", and 24".

Standards

The manufacture and installation of pressure piping is tightly regulated by the ASME "B31" code series such as B31.1 or B31.3 which have their basis in the ASME Boiler and Pressure Vessel Code. This code has the force of law in Canada and the USA. Europe and the rest of the world has an equivalent system of codes. Pressure piping is generally pipe that must carry pressures greater than 10 to 25 atmospheres, although definitions vary. To ensure safe operation of the system, the manufacture, storage, welding, testing, etc. of pressure piping must meet stringent quality standards.

Manufacturing standards for pipes commonly require a test of chemical composition and a series of mechanical strength tests for each heat of pipe. A heat of pipe is all forged from the same cast ingot, and therefore had the same chemical composition. Mechanical tests may be associated to a lot of pipe, which would be all from the same heat and have been through the same heat treatment processes. The manufacturer performs these tests and reports the composition in a mill traceability report and the mechanical tests in a material test report, both of which are referred to by the acronym MTR. Material with these associated test reports is called traceable. For critical applications, third party verification of these tests may be required; in this case an independent lab will produce a certified material test report(CMTR), and the material will be called certified.

Some widely used pipe standards are:

- The API range - now ISO 3183. E.g.: API 5L Grade B - now ISO L245 where the number indicates yield strength in MPa
- ASME SA106 Grade B (Seamless carbon steel pipe for high temperature service)

- ASTM A312 (Seamless and welded austenitic stainless steel pipe)
- ASTM C76 (Concrete Pipe)
- ASTM D3033/3034 (PVC Pipe)
- ASTM D2239 (Polyethylene Pipe)

API 5L was changed in the second half of 2008 to edition 44 from edition 43 to make it identical to ISO 3183. It is important to note that the change has created the requirement that sour service, ERW pipe, pass a hydrogen induced cracking (HIC) test per NACE TM0284 in order to be used for sour service.

- ACPA [American Concrete Pipe Association]
- AWWA [American Water Works Association]
- AWWA M45

See this site for more specification summaries.

Traceability and Positive Material Identification (PMI)

Maintaining the traceability between the material and this paperwork is an important quality assurance issue. QA often requires the heat number to be written on the pipe. Precautions must also be taken to prevent the introduction of counterfeit materials. As a backup to etching/labelling of the material identification on the pipe, Positive Material Identification (PMI) is performed using a handheld device; the device scans the pipe material using an emitted electromagnetic wave (x-ray fluorescence/XRF) and receives a reply that is spectrographically analysed.

Installation

Pipe installation is often more expensive than the material and a variety of specialised tools, techniques, and parts have been developed to assist this. Pipe is usually delivered to a customer or jobsite as either "sticks" or lengths of pipe (typically 20 feet) or they are prefabricated with elbows, tees and valves into a prefabricated pipe spool [A pipe spool is a piece of pre-assembled pipe and fittings, usually prepared in a shop so that installation on the construction site can be more efficient.]. They are usually tagged with a bar code and the ends are capped for protection. The Pipe and Pipe Spools are delivered to a warehouse on a large commercial job and they may be held indoors or in a gridded laydown yard. The Pipe or pipe spool is retrieved, staged, rigged, and then lifted into place. On large process jobs the lift is made using cranes and hoist and other material lifts.

They are typically temporarily supported in the steel structure using beam clamps, straps, and small hoists until the Pipe Supports are attached or otherwise secured.

An example of a tool used for installation for a small plumbing pipe (threaded ends) is the pipe wrench. Small pipe is typically not heavy and can be lifted into place by the installation craft labourer.

Joining

Pipes are commonly joined by welding, using threaded pipe and fittings; sealing the connection with a pipe thread compound, Polytetrafluoroethylene (PTFE) Thread seal tape, oakum, or PTFE string, or by using a mechanical coupling. Process piping is usually joined by welding using a TIG or MIG process. The most common process pipe joint is the butt weld. The ends of pipe to be welded must have a certain weld preparation called an End Weld Prep (EWP) which is typically at an angle of 37.5 degrees to accommodate the filler weld metal. The most common pipe thread in North America is the National Pipe Thread (NPT) or the Dryseal (NPTF) version. Other pipe threads include the British standard pipe thread (BSPT), the garden hose thread (GHT), and the fire hose coupling (NST).

Copper pipes are typically joined by soldering, brazing, compression fittings, flaring, or crimping. Plastic pipes may be joined by solvent welding, heat fusion, or elastomeric sealing.

If frequent disconnection will be required, gasketed pipe flanges or union fittings provide better reliability than threads. Some thin-walled pipes of ductile material, such as the smaller copper or flexible plastic water pipes found in homes for ice makers and humidifiers, for example, may be joined with compression fittings.

Underground pipe typically uses a “push-on” gasket style of pipe that compresses a gasket into a space formed between the two adjoining pieces. Push-on joints are available on most types of pipe. A pipe joint lubricant must be used in the assembly of the pipe. Under buried conditions, gasket-joint pipes allow for lateral movement due to soil shifting as well as expansion/contraction due to temperature differentials. Plastic MDPE and HDPE gas and water pipes are also often joined with Electrofusion fittings. Large above ground pipe typically uses a flanged joint, which is generally available in ductile iron pipe and some others. It is a gasket style where the flanges of the adjoining pipes are bolted together, compressing the gasket into a space between the pipe. Mechanical grooved couplings or Victaulic joints are also frequently used for frequent disassembly & assembly. Developed in the 1920s, these mechanical grooved couplings can operate up to 1,200psi working pressures and available in materials to match the pipe grade. Another type of mechanical coupling is a Swagelok brand fitting; this type of compression fitting is typically used on small tubing under 3/4 inch in diametre.

Fittings and Valves

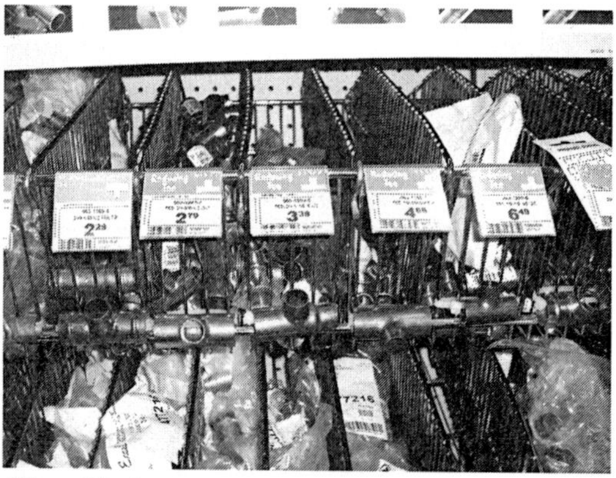

Figure: Copper Pipe Fittings

Fittings are also used to split or join a number of pipes together, and for other purposes. A broad variety of standardised pipe fittings are available; they are generally broken down into either a tee, an elbow, a branch, a reducer/enlarger, or a wye. Valves control fluid flow and regulate pressure. The piping and plumbing fittings and valves articles discuss them further.

Pipe Support

Pipes are either supported from below or hung from above. These devices are called pipe supports. Supports may be as simple as a pipe "shoe" which is akin to a half of an I-beam welded to the bottom of the pipe; they may be "hung" using a clevis or trapeze type of devices called pipe hangers. Pipe Supports of any kind may incorporate springs, snubbers and/or dampers to compensate for thermal expansion, or to provide vibration isolation, shock control, or vibration excitation of the pipe due to earthquake motion. Some dampers are simply fluid dashpots whereas other dampers may be active hydraulic devices that have sophisticated systems that act to dampen peak displacements due to forcing functions. The forcing functions may be process derived (such as in a fluidised bed reactor) or from a natural phenomenon such as an earthquake (design basis event or DBE).

Cleaning

The inside of pipes can be cleaned with the tube cleaning process, if they are contaminated with debris or fouling. This depends on the process that the pipe will be used for and the cleanliness needed for the process. In some cases the pipes are cleaned using a displacement device formally known as a Pipeline Inspection Gauge or "pig"; alternately the pipes or tubes may be chemically flushed using specialised solutions that are pumped through. In some cases, where care has been taken in the manufacture, storage, and installation of pipe and tubing, the lines are blown clean with compressed air or nitrogen.

Other uses

Pipe is widely used in the fabrication of handrails, guardrails and railings.

British Standard Pipe Thread

The British Standard Pipe thread (BSP thread) is a family of standard screw thread types that has been adopted internationally for interconnecting and sealing pipe ends by mating an external (male) with an internal (female) thread.

Types

Two types of threads are distinguished:

- Parallel ('straight') threads, British Standard Pipe Parallel thread (BSPP; originally also known as British Standard Pipe Fitting thread/BSPF and British Standard Pipe Mechanical thread/BSPM), which have a constant diametre; denoted by the letter G.
- Taper threads, British Standard Pipe Taper thread (BSPT), whose diametre increases or decreases along the length of the thread; denoted by the letter R.

These can be combined into two types of joints:

- Jointing threads: These are pipe threads where pressure-tightness is made through the mating of two threads together. They always use a taper male thread, but can have either parallel or taper female threads. (In Europe, taper female pipe threads are not commonly used.)
- Longscrew threads: These are parallel pipe threads used where a pressure-tight joint is achieved by the compression of a soft material (such as an o-ring seal or a washer) between the end

face of the male thread and a socket or nipple face, with the tightening of a backnut.

Thread Form

- the threadform follows the British Standard Whitworth standard
- symmetrical V-thread in which the angle between the flanks is 55° (measured in an axial plane)
- one-sixth of this sharp V is truncated at the top and the bottom
- the threads are rounded equally at crests and roots by circular arcs ending tangentially with the flanks
- the theoretical depth of the thread is therefore 0.64 times the nominal pitch

Pipe Thread Sizes

A list of 16 thread sizes are defined by the standards, ranging from $^{1}D_{16}$ to 6. The size number was originally based on the inner diametre measured in inches of a steel tube for which the thread was intended, but in the modern metric version of the standard is simply a size number.

The major diametre listed is the outer diametre of the external thread. For a taper thread, it is the diametre at the "gauge length" from the small end of the thread. The taper is 1 to 16, meaning that for each 16 units of measurement increase in the distance from the end, the diametre increases by 1 unit of measurement.

Pipe Thread Designations

These standard pipe threads are formally referred to by the following sequence of blocks:

- the words "Pipe thread",
- the document number of the standard (e.g., "ISO 7" or "EN 10226")
- the symbol for the pipe thread type:
- G = external + internal parallel (ISO 228)
- R = external taper (ISO 7)
- Rp = internal parallel (ISO 7/1)
- Rc = internal taper (ISO 7)
- Rs = external parallel
- the thread size

Threads are normally right-hand. For left-hand threads, the letters "LH" are appended.

Example: Pipe thread EN 10226 Rp 2½

The terminology for the use of G and R originated from Germany (G for Gas, as it was originally designed for use on gas pipes ; R for Rohr (meaning pipe).)

Cast Iron Pipe

Cast iron pipe is a historical pipe which found widespread use as a pressure pipe for transmission of water, gas and sewage, and as a water drainage pipe, during the 19th and 20th centuries. It comprises predominantly a gray cast iron tube and was frequently used uncoated, although later developments did result in various coatings and linings to reduce corrosion and improve hydraulics. Cast iron pipe was gradually superseded by ductile iron pipe, which is a direct development, with most existing manufacturing plants transitioning to the new material during the 1970s and 1980s. There is currently almost no new manufacture of cast iron pipe.

History

Early Use

The first use of cast iron pipe is not recorded. Cast iron tubes were first manufactured in the 14th century in Europe for cannon and it is presumed that similar tubes would have found use as water pipes at the same time, however, it was not until 1455 in Siegerland that the first officially recorded cast iron water pipe was produced for use in Dillenburg Castle (since destroyed).

The oldest extant water pipes date from the 17th century and were installed to distribute water throughout the gardens of the Chateau de Versailles. These amount to some 35 km of pipe, typically 1m lengths with flanged joints. The extreme age of these pipes make them of considerable historical value. Following extensive refurbishment in 2008 by Saint-Gobain PAM, 80% remain original.

Manufacture

Vertically Cast

The first cast iron pipe was produced in horizontal molds which resulted in an uneven distribution of metal around the pipe circumference. In 1845, the first pipe was cast vertically and by the turn of the century, all pipe was manufactured to this method.

Centrifugally Cast

Subsequent to its invention by Dimitri Sensaud deLavaud, a French-Brazilian, in 1918, much cast iron pipe manufacture shifted to the dramatically different technique of centrifugal casting. Modern ductile iron pipe production continues to use this method of casting.

Standardisation

The first standardisation of cast iron water pipes in Britain occurred in 1917 with the publishing of BS 78. This standard specified a dimensionless nominal size, which approximately corresponded with the internal diametre in inches of the pipe, and four pressure classes, Class A, Class B, Class C and Class D, each with a specified wall thickness and outer diametre. It is noted that the outer diametre is identical between classes with the exception of sizes 12 to 27, where Classes A and B share one diametre and Classes C and D have another, larger diametre.

BS 78 was finally superseded when the U.K. harmonised with incompatible European standards, however, the specified outer dimensions continue to remain in effect (albeit in metric form) as the standard pipe outer diametre for ductile iron pipe in Australia and New Zealand through the descendant specification, AS/NZS 2280.

Pipe Materials

The earliest known evidence of drain tile being used for plumbing was found in Mesopotamia and is estimated to have been made around 3000 BC. The tiles were made from clay mixed with short lengths of straw. Both brass and copper pipes have been found in Egypt believed to have been made close to 2500 BC. The Romans made extensive use of lead pipe by joining sheets of lead into piping to carry their water supply and waste. During the Dark Ages following the fall of the Roman Empire, plumbing development virtually ceased for centuries except for isolated cases of plumbing installed in palaces and castles. In the 13th century, blacksmiths formed sheets of iron and lap welded the seam to create iron pipe. Though it is unclear as to when galvanized iron pipe was first used, a French chemist named Melouin is credited with developing the process in 1742. The earliest known use for cast iron pipe is for the water supply to a fountain in Langensalza, Germany, built around 1560. In 1819 the first cast iron pipe constructed in the US, was manufactured in Weymouth, New Jersey. Before that time, cast iron pipe and fittings had to be imported from Europe. It was not until the 1960s that the hubless cast iron pipe was brought to the U.S. from Europe by way of Canada. During the early 1900s, heavy-walled copper joined with threaded fittings was in use, but limited to public buildings because of its high cost. However, during the 1930s light-gauge Copper tube and fittings were developed which made copper economically feasible and increased its popularity. Polyvinyl Chloride(PVC) was produced experimentally in the 19th century but did not become practical to manufacture until 1926, when Waldo Semon of BF Goodrich Co. developed a method to plasticize PVC, making it easier to process. PVC pipe began to be manufactured in the 1940s and was in wide use during the DWV reconstruction of Germany and Japan following WWII. In the 1950s, plastics manufacturers in Western Europe and Japan began producing acrylonitrile butadiene styrene (ABS) pipe. The methods for producing cross-linked polyethylene (PEX) was also developed in the 1950s. Plastic supply pipes have become increasingly common, with a variety of materials and fittings employed, however plastic water pipes do not keep water as clean as copper and brass piping does. Copper pipe plumbing is bacteriostatic. This means that bacteria can't grow in the copper pipes. Plumbing codes define which materials may be used, and all materials must be proven by ASTM, UL, and/or NFPA testing.

Steel

Galvanized steel potable water supply and distribution pipes are commonly found with nominal diametres from 3/8" to 2". It is rarely used today for new construction residential plumbing. Steel pipe has National Pipe Thread (NPT) standard tapered male threads, which connect with female tapered threads on elbows, tees, couplers, valves, and other fittings. Galvanized steel (often known simply as "galv" or "iron" in the plumbing trade) is relatively expensive, difficult to work with due to weight and requirement of a pipe threader. It remains in common use for repair of existing "galv" systems and to satisfy building code non-combustibility requirements typically found in hotels, apartment buildings and other commercial applications. It is also extremely durable. Black lacquered steel pipe is the most widely used pipe material for fire sprinklers and natural gas.

Most single family homes' systems typically won't require supply piping larger than 3/4". In addition to expense, another downside is it suffers from a tendency to become obstructed due to internal rusting and mineral deposits forming on the inside of the pipe over time after the internal galvanizing zinc coating has degraded. In potable water distribution service, galvanized steel pipe has a service life of about

30 to 50 years, although it is not uncommon for it to be less in geographic areas with corrosive water contaminants.

Copper

Copper Tubing Sizes (CTS) for Plumbing

Nominal size	*Outside diametre (OD) [in (mm)]*	*Inside diametre (ID) [in (mm)]*		
		Type K	*Type L*	*Type M*
3/8	1?2 (12.7)	0.402 (10.211)	0.430 (10.922)	0.450 (11.430)
1/2	5?8 (15.875)	0.528 (13.411)	0.545 (13.843)	0.569 (14.453)
5/8	3?4 (19.05)	0.652 (16.561)	0.668 (16.967)	0.690 (17.526)
3/4	7?8 (22.225)	0.745 (18.923)	0.785 (19.939)	0.811 (20.599)
1	11?8 (28.575)	0.995 (25.273)	1.025 (26.035)	1.055 (26.797)
11/4	13?8 (34.925)	1.245 (31.623)	1.265 (32.131)	1.291 (32.791)
11/2	15?8 (41.275)	1.481 (37.617)	1.505 (38.227)	1.527 (38.786)
2	21?8 (53.975)	1.959 (49.759)	1.985 (50.419)	2.009 (51.029)
21/2	25?8 (66.675)	2.435 (61.849)	2.465 (62.611)	2.495 (63.373)
3	31?8 (79.375)	2.907 (73.838)	2.945 (74.803)	2.981 (75.717)

Sizes

Common wall-thicknesses of copper tubing in the USA are "Type K", "Type L" and "Type M":

- Type K has the thickest wall section of the three types of pressure rated tubing and is commonly used for deep underground burial such as under sidewalks and streets, with a suitable corrosion protection coating or continuous polyethylene sleeve as required by code.
- Type L has a thinner pipe wall section, and is used in residential and commercial water supply and pressure applications.
- Type M has the thinnest wall section, and is generally suitable for condensate and other drains, but sometimes illegal for pressure applications, depending on local codes.

Types K and L are generally available in both hard drawn "sticks" and in rolls of soft annealed tubing, whereas type M is usually only available in hard drawn "sticks".

In the plumbing trade the size of copper tubing is measured by its nominal diametre (average inside diametre). Some American trades, heating and cooling technicians for instance, use the outside diametre

(OD) to designate copper tube sizes. The HVAC tradesman also use this different measurement to try and not confuse water pipe with copper pipe used for the HVAC trade, as pipe used in the air-conditioning trade uses copper pipe that is made at the factory without processing oils that would be incompatible with the oils used to lubricate the compressors in the AC system. The OD of copper tube is ^{1}D $_{8}$th inch larger than its nominal size. Therefore, 1 inch nominal copper tube and 1 ^{1}D $_{8}$th inch ACR tube are exactly the same tube with different size designations. The wall thickness of the tube, as mentioned above, never affects the sizing of the tube. Type K ^{1}D $_{2}$ inch nominal tube, is the same size as Type L ^{1}D $_{2}$ inch nominal tube (^{5}D $_{8}$ inch ACR).

Common wall-thicknesses in Europe are "Type X", "Type Y" and "Type Z", defined by the EN 1057 standard.

- Type X is the most common, and is used in above ground services including drinking water supply, hot and cold water systems, sanitation, central heating and other general purpose applications.
- Type Y is a thicker walled pipe, used for underground works and heavy duty requirements including hot and cold water supply, gas reticulation, sanitary plumbing, heating and general engineering.
- Type Z is a thinner walled pipe, also used for above groundservices including drinking water supply, hot and cold water systems, sanitation, central heating and other general purpose applications.

In the plumbing trade the size of copper tubing is measured by its outside diametre in millimetres. Common sizes are 15 mm and 22 mm.

Thin-walled types used to be relatively inexpensive, but since 2002 copper prices have risen considerably due to rising global demand and a stagnant supply.

Lead Leaching

Generally, copper tubes are soldered directly into copper or brass fittings, although compression, crimp, or flare fittings are also used. Formerly, concerns with copper supply tubes included the lead used in the solder at joints (50% tin and 50% lead). Some studies have shown significant "leaching" of the lead into the potable water stream, particularly after long periods of low usage, followed by peak demand periods. In hard water applications, shortly after installation, the interior of the pipes will be coated with the deposited minerals that had been dissolved in the water, and therefore the vast majority of exposed lead is prevented from entering the potable water. Building codes now require lead-free solder. Building Codes throughout the U.S. require the use of virtually "lead-free" (<.2% lead) solder or filler metals in plumbing fittings and appliances as well.

Corrosion

Copper water tubes are susceptible to: cold water pitting caused by contamination of the pipe interior typically with soldering flux; erosion corrosion caused by high speed or turbulent flow; and stray current corrosion, caused by poor electrical wiring technique, such as improper grounding and bonding.

Pin holes due to poor plumbing electrical grounding and/or bonding

Pin-hole leaks can occur anytime copper piping is improperly grounded and/or bonded; nonmetal piping, such as Pex or PVC, does not suffer from this problem. The phenomenon is known technically as stray current corrosion or electrolytic pitting. Pin-holing due to poor grounding or poor bonding occurs typically in homes where the original plumbing has been modified; homeowners may find a new plastic water filtration device or plastic repair union has interrupted the water pipe's electrical continuity to ground when they start seeing pinhole water leaks after a recent install. Damage occurs rapidly, usually being seen about six months after the ground interruption. Correctly installed plumbing appliances will have a copper bonding jumper cable connecting the interrupted pipe sections. Pinhole leaks from stray current corrosion can result in thousands of dollars in plumbing bills, and sometimes necessitating the replacement of the entire affected line. The cause is an electrical problem, not a plumbing problem; once the plumbing damage is repaired, an electrician should be consulted to evaluate the grounding and bonding of the entire plumbing system.

The difference between a ground and a bond is subtle. See Ground, for a complete description.

Stray current corrosion occurs because: 1) the piping system is connected accidentally or intentionally to a DC voltage source; 2) the piping does not have metal-to-metal electrical continuity; 3) if the voltage source is AC, one or more naturally occurring minerals coating the pipe interior act as a rectifier, converting AC current to DC. The DC voltage forces the water within the piping to act as an electrical conductor (an electrolyte). Electric current leaves the copper pipe, moves though the water across the nonconductive section (the plastic filter housing in the example above), and reenters the pipe on the opposite side.

Pitting occurs at the electrically negative side (the cathode), which may be upstream or downstream with respect to the water flow direction. Pitting occurs because the electrical voltage ionizes the pipe's interior copper metal, which reacts chemically with dissolved minerals in the water creating copper salts; these copper salts are soluble in water and wash away. Pits eventually grow and consolidate to form pin holes. Where there is one, there are almost certainly more. A complete discussion of stray current corrosion can be found in chapter 11, section 11.4.3, of Handbook of Corrosion Engineering, by Pierre Roberge.

Detecting and eliminating poor bonding is relatively straightforward. Detection is accomplished by use of a simple voltmetre set to DC with the leads placed in various places in the plumbing. Typically, a probe on a hot pipe and a probe on a cold pipe will tell you if there is improper grounding.

Anything beyond a few millivolts is important, potentials of 200 mV are common. A missing bond will show up best in the area of the gap, as potential disperses as the water runs. Since the missing bond is usually seen near the water source, as filtration and treatment equipment are added, pinhole leaks can occur anywhere downstream. It is usually the cold water pipe, as this is the one that gets the treatment devices.

Correcting the problem is a simple matter of either purchasing a copper bonding jumper kit, composed of copper cable at least #6 AWG in diametre and two bronze ground clamps for affixing it the plumbing. See NFPA 70, the U.S. National Electrical Code Handbook (NEC), section on bonding and ground for details on selecting the correct bonding conductor wire size.

A similar bonding jumper wire can also be seen crossing gas metres, but for a different reason.

Note, if homeowners are experiencing shocks or sparks from plumbing fixtures or pipes, it is more than a missing bond, it is likely a live electrical wire is bridging to the plumbing and the plumbing system is not grounded. This is an electrical shock hazard and potential fire danger; consult an electrician immediately!

Plastics

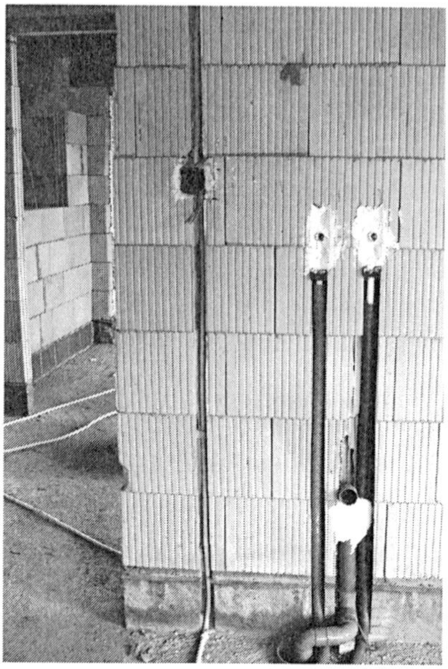

Figure: Plastic piping placed for a sink

Plastic pipe is in wide use for domestic water supply and drainage, waste, and vent (DWV) pipe. For example, polyvinyl chloride (PVC), chlorinated polyvinyl chloride (CPVC), polypropylene (PP), polybutylene(PB), and polyethylene (PE) may be allowed by code for certain uses. Some examples of plastics in water supply systems are:

- PVC/CPVC - rigid plastic pipes similar to PVC drain pipes but with thicker walls to deal with municipal water pressure, introduced around 1970. PVC should be used for cold water only, or venting. CPVC can be used for hot and cold potable water supply. Connections are made with primers and solvent cements as required by code.
- PP - The material is used primarily in housewares, food packaging, and clinical equipment, but since the early 1970s has seen increasing use worldwide for both domestic hot and cold water. PP pipes are heat fused, preventing the use of glues, solvents, or mechanical fittings. PP pipe is often used in green building projects.
- PBT - flexible (usually gray or black) plastic pipe which is attached to barbed fittings and secured in place with a copper crimp ring. The primary manufacturer of PBT tubing and fittings was driven into bankruptcy by a class-action lawsuit

over failures of this system. However, PB and PBT tubing has returned to the market and codes, typically first for 'exposed locations' such as risers.

- PEX - cross linked polyethylene system with mechanically joined fittings employing barbs and crimped steel or copper fittings.
- Polytanks - plastic polyethylene cisterns, underground water tanks, above ground water tanks, are made of linear polyethylene suitable as a potable water storage tank, provided in white, black or green, approved by NSF and made of FDA approved materials.
- Aqua - known as PEX-Al-PEX, for its PEX/aluminium sandwich - aluminium pipe sandwiched between layers of PEX and connected with brass compression fittings. In 2005, a large number of their fittings were recalled.

Fittings and Valves

Potable water supply systems require not only pipe, but also many fittings and valves which add considerably to their functionality as well as cost. The Piping and plumbing fittings and Valves articles discuss them further.

Regulation and Compliance

Before a water supply system is constructed or modified, the designer and contractor need to consult the local plumbing code and obtain a building permits prior to construction. Even replacing an existing water heater may require a permit and inspection of the work. NSF 61 is the U.S. national standard for potable water piping guidelines. National and local fire codes should be integrated in the design phase of the water system too to prevent "failure comply with regulations" notices. Some areas of the United States require on-site water reserves of potable and fire water by law.

Double-walled Pipe

A double-walled pipe is a secondary contained piping system. It is a pipe within a pipe, or encased in an outer covering, with an annulus (interstitial space) between the two diametres. The inner pipe is the primary or carrier pipe and the outer pipe is called the secondary or containment pipe. The great majority of double-walled piping applications involve wastewater, groundwater, and process safety.

U.S. Environmental Protection Agency (EPA) standards, published in 40 CFR Parts 280 and 281, mandate double-walled piping for many below-ground transport systems in wastewater treatment plants and at sanitary or hazardous-waste landfills or remediation sites. These and other regulations affect below-grade transport of hazardous materials in fuel storage systems, tank farms, drainage or runoff from process plants, and some food-processing and related applications. Double-walled containment systems are also used for certain classes of toxic or corrosive chemicals, mostly gases, used in semiconductor fabrication. Double-walled pipes may also be used where a waste waterpipe traverses a drinking water catchment area.

Double-walled pipe system may use plastic pressure pipe systems materials of high-density polyethylene (HDPE), polyvinyl chloride (PVC), chlorinated polyvinyl chloride (CPVC), polypropylene (PP), polyvinylidene fluoride (PVDF), and ethylene chlorotri fluoroethylene copolymer (ECTFE). Dual-wall stainless steel tubing is most commonly used in semiconductor plants for the containment of highly toxic gases.

A leak detection system in the containment pipe indicates if the carrier pipe is leaking. Double wall pipes are usually hydrostatically tested, or tested to hold and maintain a target gas pressure before the first use.

A different special application of a double-walled pipe is a jacketed pipe, which is used to make highviscosity liquids flow at elevated temperature through the carrier pipe. The outer pipe circulates hot fluids that heat up the interior carrier pipe and its contents in turn.

Ductile Iron Pipe

Ductile iron pipe is a pipe commonly used for potable water distribution. The predominant wall material is ductile iron, a spheroidized graphite cast iron, although an internal cement mortar lining usually serves to inhibit corrosion from the fluid being distributed, and various types of external coating are used to inhibit corrosion from the environment. Ductile iron pipe is a direct development of earlier cast iron pipe which it has superseded. Ductile iron has proven to be a better pipe material, being stronger and more fracture resistant; however, like most ferrous materials, it is susceptible to corrosion and retains some brittle characteristics. Relatively recent developments such as polyethylene sleeving and/or zinc and polymer coatings have done much to mitigate corrosion concerns, and a typical life expectancy is in excess of 75 years.

Dimensions

Ductile iron pipe is sized according to a dimensionless term known as the Pipe Size or Nominal Diametre (known by its French abbreviation, DN). This is roughly equivalent to the pipe's internal diametre in inches or millimetres. However, it is the external diametre of the pipe that is kept constant between changes in wall thickness, in order to maintain compatibility in joints and fittings. Consequently the internal diametre varies, sometimes significantly, from its nominal size. Nominal pipe sizes vary from 3 inches up to 64 inches, in increments of at least 1 inch, in the USA.

Pipe dimensions are standardised to the mutually incompatible AWWA C151 (U.S. Customary Units) in the USA, ISO 2531 / EN 545/ 598 (metric) in Europe, and AS/NZS 2280 (metric) in Australia and New Zealand. Although both metric, European and Australian are not compatible and pipes of identical nominal diametres have quite different dimensions.

North America

Pipe dimensions according to the American AWWA C-151

Pipe Size	*Outside Diametre [in (mm)]*
3	3.96 (100.584)
4	4.80 (121.92)
6	6.90 (175.26)
8	9.05 (229.87)
10	11.10 (281.94)
12	13.20 (335.28)

Pipe Size	*Outside Diametre [in (mm)]*
14	15.30 (388.62)
16	17.40 (441.96)
18	19.50 (495.3)
20	21.60 (548.64)
24	25.80 (655.32)
30	32.00 (812.8)

Europe

European pipe is standardised to ISO 2531 and its descendent specifications EN 545 (potable water) and EN 598 (sewage). European pipes are sized to approximately match the internal diametre of the pipe, following internal lining, to the nominal diametre. ISO 2531 maintains dimensional compatibility with older German cast iron pipes. Older British pipes, however, which used the incompatible imperial standard, BS 78, require adapter pieces when connecting to newly installed pipe. Coincidentally, the British harmonization with European pipe standards occurred at approximately the same time as its transition to ductile iron, so almost all cast iron pipe is imperial and all ductile pipe is metric.

DN	*Outside Diametre [mm (in)]*		*Wall thickness [mm (in)]*	
		Class 40	*K9*	*K10*
40	56 (2.205)	4.8 (0.189)	6.0 (0.236)	6.0 (0.236)
50	66 (2.598)	4.8 (0.189)	6.0 (0.236)	6.0 (0.236)
60	77 (3.031)	4.8 (0.189)	6.0 (0.236)	6.0 (0.236)
65	82 (3.228)	4.8 (0.189)	6.0 (0.236)	6.0 (0.236)
80	98 (3.858)	4.8 (0.189)	6.0 (0.236)	6.0 (0.236)
100	118 (4.646)	4.8 (0.189)	6.0 (0.236)	6.0 (0.236)
125	144 (5.669)	4.8 (0.189)	6.0 (0.236)	6.0 (0.236)
150	170 (6.693)	5.0 (0.197)	6.0 (0.236)	6.5 (0.256)
200	222 (8.740)	8.4 (0.331)	6.3 (0.248)	7.0 (0.276)
250	274 (10.787)	5.8 (0.228)	6.8 (0.268)	7.5 (0.295)
300	326 (12.835)	6.2 (0.244)	7.2 (0.283)	8.0 (0.315)
350	378 (14.882)	7.0 (0.276)	7.7 (0.303)	8.5 (0.335)
400	429 (16.890)	7.8 (0.307)	8.1 (0.319)	9.0 (0.354)
450	480 (18.898)	-	8.6 (0.339)	9.5 (0.374)

DN	*Outside Diametre [mm (in)]*		*Wall thickness [mm (in)]*	
		Class 40	*K9*	*K10*
500	532 (20.945)	-	9.0 (0.354)	10.0 (0.394)
600	635 (25.000)	-	9.9 (0.390)	11.1 (0.437)
700	738 (29.055)	-	10.9 (0.429)	12.0 (0.472)
800	842 (33.150)	-	11.7 (0.461)	13.0 (0.512)
900	945 (37.205)	-	12.9 (0.508)	14.1 (0.555)
1000	1,048 (41.260)	-	13.5 (0.531)	15.0 (0.591)
1100	1,152 (45.354)	-	14.4 (0.567)	16.0 (0.630)
1200	1,255 (49.409)	-	15.3 (0.602)	17.0 (0.669)
1400	1,462 (57.559)	-	17.1 (0.673)	19.0 (0.748)

1500	1,565 (61.614)	-	18.0 (0.709)	20.0 (0.787)
1600	1,668 (65.669)	-	18.9 (0.744)	51.0 (2.008)
1800	1,875 (73.819)	-	20.7 (0.815)	23.0 (0.906)
2000	2,082 (81.969)	-	22.5 (0.886)	25.0 (0.984)

Australia

Australian and New Zealand pipes are sized to an independent specification, AS/NZS 2280, that is not compatible with European pipes even though the same nomenclature is used.

Australia adopted at an early point the imperial British cast iron pipe standard BS 78, and when this was retired on British adoption of ISO 2531, rather than similarly harmonizing with Europe, Australia opted for a 'soft' conversion from imperial units to metric, published as AS/NSZ 2280, with the physical outer diametres remaining unchanged, allowing continuity of manufacture and backwards compatibility.

Therefore the inner diametres of lined pipe differ widely from the nominal diametre.

Nominal (DN) Size	*Outside Diametre [mm (in)]*	*Nominal Wall Thickness [mm (in)]*		*Flange Class*
		PN 20	*PN 35*	
375	426 (16.772)	5.1 (0.201)	7.3 (0.287)	10.0
450	507 (19.961)	5.6 (0.220)	8.3 (0.327)	11.0
500	560 (22.047)	6.0 (0.236)	9.0 (0.354)	12.0
600	667 (26.260)	6.8 (0.268)	310.3 (0.406)	13.0
750	826 (32.520)	7.9 (0.311)	12.2 (0.480)	15.0

and hydraulic calculations require some knowledge of the pipe standard.

Nominal Size (DN)	*Outside Diametre [mm (in)]*	*Nominal Wall Thickness [mm (in)]*		*Flange Class*
		PN 20	*PN 35*	
100	122 (4.803)	-	5.0 (0.197)	7.0
150	177 (6.969)	-	5.0 (0.197)	8.0
200	232 (9.134)	-	5.0 (0.197)	8.0
225	259 (10.197)	5.0 (0.197)	5.2 (0.205)	9.0
250	286 (11.260)	5.0 (0.197)	5.6 (0.220)	9.0
300	345 (13.583)	5.0 (0.197)	6.3 (0.248)	10.0

Joints

Individual lengths of ductile iron pipe are joined either by flanges, couplings, or some form of spigot and socket arrangement.

Flanges

Flanges are flat rings around the end of pipes which mate with an equivalent flange from another pipe, the two being held together by bolts usually passed through holes drilled through the flanges. A deformable gasket, usually elastomeric, placed between raised faces on the mating flanges provides the seal. Flanges are designed to a large number of specifications that differ because of dimensional variations in pipes sizes and pressure requirements, and because of independent standards development. In the U.S. flanges are either threaded or welded onto the pipe. In the European market flanges are usually welded on to the pipe. In the U.S. flanges are available in a standard 125 lb. bolt pattern as well as a 250 lb (and heavier) bolt pattern (steel bolt pattern). Both are usually rated at 250 psi (1,700 kPa). A flanged joint is rigid and can bear both tension and compression as well as a limited degree of shear and bending. It also can be dismantled after assembly. Due to the rigid nature of the joint and the risk of excessive bending moment being imposed, it is advised that flanged pipework is not buried.

Current flange standards used in the water industry are ANSI B16.1 in the USA, EN 1092 in Europe, and AS/NZS 4087 in Australia and New Zealand.

Spigot and Socket

Spigot and sockets involve a normal pipe end, the spigot, being inserted into the socket or bell of another pipe or fitting with a seal being made between the two within the socket. Normal spigot and socket joints do not allow direct metal to metal contact with all forces being transmitted through the elastomeric seal. They can consequently flex and allow some degree of rotation, allowing pipes to shift and relieve stresses imposed by soil movement. The corollary is that unrestrained spigot and socket joints transmit essentially no compression or tension along the axis of the pipe and little shear. Any bends, tees or valves therefore require either a restrained joint or, more commonly, thrust blocks, which transmit the forces as compression into the surrounding soil.

A large number of different socket and seals exist. The most modern is the 'push-joint' or 'slip-joint', whereby the socket and rubber seal is designed to allow the pipe spigot to be, after lubrication, simply pushed into the socket. Push joints remain proprietary designs. Also available are locking gasket systems. These locking gasket systems allow the pipe to be pushed together but do not allow the joint to come apart without using a special tool or torch on the gasket.

The earliest spigot and socket cast iron pipes were jointed by filling the socket with a mixture of water, sand, iron filings and sal-ammoniac (ammonium chloride.) A gaskin ring was pushed into te socket round the spigot to contain the mixture which was pounded into the socket with a caulking tool and then pointed off. This took several weeks to set and produced a completely rigid joint. Such pipe systems are often to be seen in nineteenth century churches in the heating system.

Manufacture

Ductile iron pipe is produced by a technique known as centrifugal casting, originally developed for cast iron pipe in 1918. The molten ductile iron is poured into a rapidly spinning water-cooled mold. Centrifugal force results in an even spread of iron around the circumference.

Internal Linings

Ductile iron pipe is somewhat resistant to internal corrosion in potable water and less aggressive forms of sewage. However, even where pipe material loss and consequently pipe wall reduction is slow, the deposition of corrosion products on the internal pipe wall can dramatically reduce the effective internal diametre and effectively choke flow, increasing pumping costs and lowering system pressure, long before the pipe itself is at risk of failure. A variety of linings are available to reduce or eliminate corrosion, including cement mortar, polyurethane and polyethylene. Of these, cement mortar lining is by far the most common.

Polyurethane (PUR)

Polyurethane protects piping made of ductile cast iron against corrosion and ensures meeting hygienic standards for drinking water at the same time. Polyurethane is used for both the inside lining and the outside coating. Because of polyurethane's elasticity, the coating remains intact even if the pipe is deformed. The PUR coating was developed in 1972. In comparison with other coatings, the internal polyurethane lining exhibits a high resistance to various different media such as drinking water, wastewater, de-mineralised water, industrial water and gas, as well as to aggressive solutions such as sulphuric acid. The PUR outside coating is suitable for all kinds of soil.

Specific Properties of Polyurethane

Composition Polyurethane is composed of a two-component resin. Its three-dimensionally linked molecular structure gives it its mechanical stability. Polyurethane is a thermosetting plastic with no solvents. It meets EN 545 and ISO 2531 standards.

Wall thickness Polyurethane PUR inside lining

DN 80-150 = 1.3 mm

DN 200-700 = 1.5 mm

Polyurethane PUR outside coating (ecopur®)

DN 80-700 = 0.9 mm

Colour inside: green

outside: black

Density 1.4 – 1.5 kg/dm3

Continuity continuous coating, no cracks

Bonding strength > 14 Mpa

(EMPA recommends: 2.5 Mpa on a saturated sample)

Bonding strength is tested at regular intervals by our laboratories.

Dielectric resistance > 108 Ùm2

Temperature Water: up to 40°C (constant); up to 80°C (short term)

Air: 120°C

Impact resistance 40 Nm at 20°C

Effect of salt spray no effect after 1000 hours

Expansion > 10%

Friction coefficient k > 0.01 mm

Resistance to chemicals - Acidic or basic with pH values between 1 and 14

- Anorganic solvents
- Sulphuric acid (wasterwater)
- industrial wastewater

Thermal expansion 20 x 10-6 1/K

Deposits None

Use of chlorine The chlorine concentration in drinking water and the selective amount used when disinfecting have no influence on the quality of the polyurethane. Quality Assurance Contract with EMPA in Dübendorf, Switzerland Certification for PUR - Swiss Association for Gas and Water (SVGW)

- Swiss Federal Office of Health (BAG)
- Water Byelaws Advisory Service
- Singapore Institute of Standards and Industrial Research
- for: Bulgaria, Spain, Italy, Lithuania, Poland, Czech Republic, Rumania etc.

Cement Mortar

The predominant form of lining for water applications is cement mortar centrifugally applied during manufacturing. The cement mortar comprises a mixture of cement and sand to a ratio of between 1:2 and 1:3.5. For potable water, portland cement is used; for sewage it is common to use sulfate resisting or high alumina cement.

Cement mortar linings have been found to dramatically reduce internal corrosion. A DIPRA survey has demonstrated that the Hazen-Williams factor of cement lining remains between 130 and 151 with only slight reduction with age.

External Coatings

Unprotected ductile iron, similarly to cast iron, is intrinsically resistant to corrosion in most, although not all, soils. Nonetheless, because of frequent lack of information on soil aggressiveness and to extend the installed life of buried pipe, ductile iron pipe is commonly protected by one or more external coatings. In the U.S. and Australia, loose polyethylene sleeving is preferred. In Europe, standards recommend a more sophisticated system of directly bonded zinc coatings overlaid by a finishing layer be used in conjunction with polyethylene sleeving.

Polyethylene

Polyethylene sleeving was first developed by CIPRA (since 1979, DIPRA) in the U.S. in 1951 for use in highly corrosive soil in Birmingham, Alabama. It was employed more widely in the U.S. in the late 1950s and first employed in the UK in 1965 and Australia in the mid-1960s. Polyethylene sleeving comprises a loose sleeve of polyethylene sheet that completely wraps the pipe, including the bells of any joints. Sleeving inhibits corrosion by a number of mechanisms. It physically separates the pipe from soil particles, preventing direct galvanic corrosion. By providing an impermeable barrier to ground water, the sleeve also inhibits the diffusion of oxygen to the ductile iron surface and limits the availability of electrolytes that would accelerate corrosion. It provides a homogeneous environment along the pipe surface so that corrosion occurs evenly over the pipe. The sleeve also restricts the availability of nutrients which could support sulfate-reducing bacteria, in hibiting microbially induced corrosion. Sleeving is not designed to be completely water-tight but rather to greatly restrict the movement of water to and from the pipe surface. Water present beneath the sleeve and in contact with the pipe surface is rapidly deoxygenated and depleted of nutrients and forms a stable environment in which limited further corrosion occurs. An improperly installed sleeve that continues to allow the free flow of ground water is not effective in inhibiting corrosion.

Polyethylene sleeves are available in a number of materials. The most common contemporary compositions are linear low-density polyethylene film which requires an 8 mil or 200 μm thickness and high-density cross-laminated polyethylene film which requires only a 4 mil or 100 μm thickness. The latter may or may not be reinforced with a scrim layer.

Polyethylene sleeving does have limitations. In European practice, its use in the absence of additional zinc and epoxy protective coatings is discouraged where natural soil resistivity is below 750 ohm/cm, where resistivity is below 1500 ohm/cm and where the pipe is installed at or below the water table, where there are additional artificial soil contaminants or where there are stray currents. Because of the vulnerability of polyethylene to UV degradation, sleeving, or sleeved pipe should not be stored in sunlight, although carbon pigments included in the sleeving can provide some limited protection.

Polyethylene sleeving is standardised according to ISO 8180 internationally, AWWA C105 in the U.S., BS 6076 in the UK and AS 3680 and AS 3681 in Australia.

Zinc

In Europe, ductile iron pipe is typically manufactured with a zinc coating overlaid by either a bituminous, polymer, or epoxy finishing layer. EN 545/598 mandates a minimum zinc content of 200 g/m^2 (with local minima of 110 g/m^2 at 99.99% purity) and a minimum average finishing layer thickness of 70 μm (with local minima of 50 μm).

No current AWWA standards are available for bonded coatings (zinc, coal tar epoxy, tape-wrap systems as seen on steel pipe) for ductile iron pipe, DIPRA does not endorse bonded coatings, and AWWA M41 generally views them unfavourably, recommending they be used only in conjunction with cathodic protection.

Bituminous Coatings

Zinc coatings are generally not employed in the U.S. and Australia. In order to protect ductile iron pipe prior to installation, pipe is instead supplied with a temporary 1 mil or 25 μm thick bituminous coating. This coating is not intended to provide protection once the pipe is installed.

Industry Associations

In the United States ductile iron pipe is often promoted to municipalities and consulting engineers by the Ductile Iron Pipe Research Association. Their focus is to promote the benefits of using ductile iron pipe on utility projects (water & sewer) over alternate products such as PVC, PCP, and HDPE.

Environmental

Ductile iron pipe in the developed world is normally manufactured exclusively from scrap steel. Ductile iron pipe can be recycled. In the U.S. with the growing environmental movement ductile iron pipe is in a natural position to regain market share lost to its largest competitor, the PVC industry, over the past 40 years.

Hollow Structural Section

A hollow structural section (HSS) is a type of metal profile with a hollow tubular cross section. In some countries they are referred to instead as a structural hollow section (SHS). Most HSS are of circular or rectangular section, although other shapes are available, such as elliptical. HSS is only composed of structural steel per code. HSS is sometimes mistakenly referenced as hollow structural steel. Rectangular HSS are also called tube steel or structural tubing. Circular HSS are sometimes mistakenly called steel pipethough true steel pipe is actually dimensioned and classed differently than HSS. The corners of HSS are heavily rounded, or chamfered, at radii approximately twice the wall thickness. The wall thickness is uniform around the section. In the UK, the terms are circular and rectangular hollow section (CHS and RHS). However, the dimensions and tolerances differ slightly from HSS.

Use in Structures

HSS, especially rectangular sections, are commonly used in welded steel frames where members experience loading in multiple directions. Square and circular HSS have very efficient shapes for this multiple-axis loading as they have uniform geometric and thus uniform strength characteristics along two or more cross-sectional axes; this makes them good choices for columns. They also have excellent resistance to torsion. HSS can also be used as beams, although wide flange or I-beam shapes are in many cases a more efficient structural shape for this application. However, the HSS has superior resistance to lateral torsional buckling. The flat square surfaces of rectangular HSS can ease construction, and they are sometimes preferred for architectural aesthetics in exposed structures, although elliptical HSS are becoming more popular in exposed structures for the same aesthetic reasons. HSS is commonly available in mild steel, such as A500 grade B.

Manufacture

Square HSS is made the same way as pipe. During the manufacturing process flat steel plate is gradually changed in shape to become round where the edges are presented ready to weld. The edges are then welded together to form the mother tube. During the manufacturing process the mother tube goes through a series of shaping stands which form the round HSS (mother tube) into the final square or rectangular shape. Most American manufacturers adhere to the ASTM A500 standard, while Canadian manufacturers follow both ASTM A500 and CSA G40.21. European hollow sections are generally in accordance with the EN 10210 standard. It is one or the structural steel element in a construction.

Filled HSS

HSS is often filled with concrete to improve fire rating, as well as robustness. When this is done, the product is referred to as a "Lally column" after its inventor John Lally of Waltham, Massachusetts. Often times, however, the pronunciation is corrupted to "lolly column" by those that don't know any better. For example, barriers around parking areas, bollards, made of HSS are often filled, to at least bumper height, with concrete. This is an inexpensive (when replacement costs are factored in) way of adding compressive strength to the bollard, which can help prevent unsightly local denting, though does not generally significantly increase the overall structural properties of the bollard.

Hydraulic Machinery

Figure: An excavator; main hydraulics: Boom cylinders, swingdrive, cooler fan and trackdrive

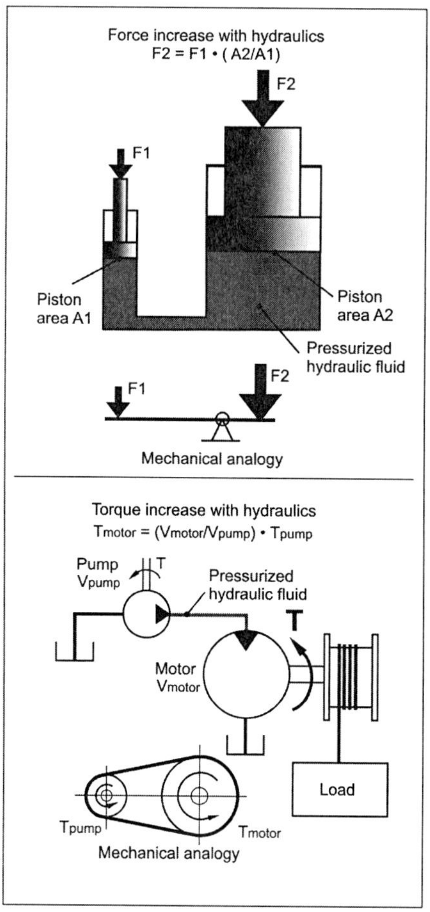

Figure: Fundamental features of using hydraulics compared to mechanics for force and torque increase/decrease in a transmission.

Hydraulic machines are machinery and tools that use liquid fluid power to do simple work. Heavy equipment is a common example.

In this type of machine, hydraulic fluid is transmitted throughout the machine to various hydraulic motors and hydraulic cylinders and which becomes pressurised according to the resistance present. The fluid is controlled directly or automatically by control valves and distributed through hoses and tubes.

The popularity of hydraulic machinery is due to the very large amount of power that can be transferred through small tubes and flexible hoses, and the high power density and wide array of actuators that can make use of this power.

Hydraulic machinery is operated by the use of hydraulics, where a liquid is the powering medium.

Chapter 4

Force and Torque Multiplication

A fundamental feature of hydraulic systems is the ability to apply force or torque multiplication in an easy way, independent of the distance between the input and output, without the need for mechanical gears or levers, either by altering the effective areas in two connected cylinders or the effective displacement (cc/rev) between a pump and motor. In normal cases, hydraulic ratios are combined with a mechanical force or torque ratio for optimum machine designs such as boom movements and track drives for an excavator.

Examples

Two Hydraulic Cylinders Interconnected

Cylinder C1 is one inch in radius, and cylinder C2 is ten inches in radius. If the force exerted on C1 is 10 lbf, the force exerted by C2 is 1000 lbf because C2 is a hundred times larger in area ($S = \pi r^2$) as C1. The downside to this is that you have to move C1 a hundred inches to move C2 one inch. The most common use for this is the classical hydraulic jack where a pumping cylinder with a small diametre is connected to the lifting cylinder with a large diamete.

Pump and Motor

If a hydraulic rotary pump with the displacement 10 cc/rev is connected to a hydraulic rotary motor with 100 cc/rev, the shaft torque required to drive the pump is 10 times less than the torque available at the motor shaft, but the shaft speed (rev/min) for the motor is 10 times less than the pump shaft speed. This combination is actually the same type of force multiplication as the cylinder example (1) just that the linear force in this case is a rotary force, defined as torque.

Both these examples are usually referred to as a hydraulic transmission or hydrostatic transmission involving a certain hydraulic "gear ratio".

Hydraulic Circuits

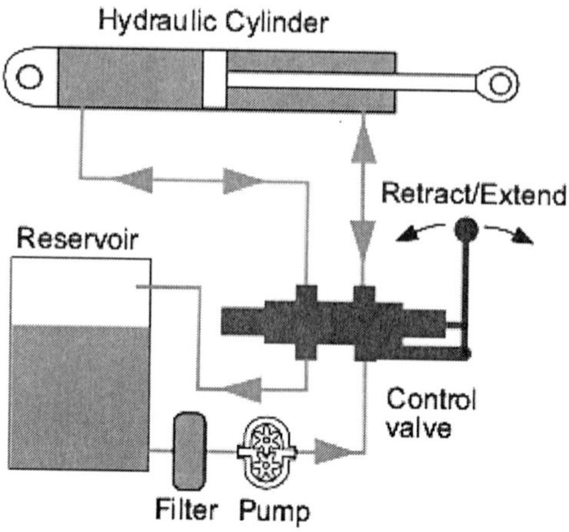

Figure: A simple open centre hydraulic circuit.

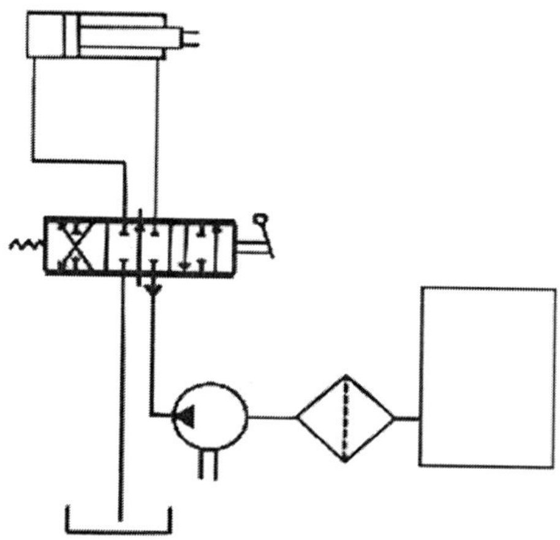

Figure: The equivalent circuit schematic.

For the hydraulic fluid to do work, it must flow to the actuator and or motors, then return to a reservoir. The fluid is then filtered and re-pumped. The path taken by hydraulic fluid is called a hydraulic circuit of which there are several types. Open centre circuits use pumps which supply a continuous flow. The flow is returned to tank through the control valve's open centre; that is, when the control valve is centred, it provides an open return path to tank and the fluid is not pumped to a high pressure. Otherwise, if the control valve is actuated it routes fluid to and from an actuator and tank. The fluid's pressure will rise to meet any resistance, since the pump has a constant output. If the pressure rises too high, fluid returns to tank through a pressure relief valve. Multiple control valves may be stacked in series . This type of circuit can use inexpensive, constant displacement pumps.

Closed centre circuits supply full pressure to the control valves, whether any valves are actuated or not. The pumps vary their flow rate, pumping very little hydraulic fluid until the operator actuates a valve. The valve's spool therefore doesn't need an open centre return path to tank. Multiple valves can be connected in a parallel arrangement and system pressure is equal for all valves.

Constant Pressure and Load-sensing Systems

The closed centre circuits exist in two basic configurations, normally related to the regulator for the variable pump that supplies the oil:

Constant pressure systems (CP-system), standard. Pump pressure always equals the pressure setting for the pump regulator. This setting must cover the maximum required load pressure. Pump delivers flow according to required sum of flow to the consumers. The CP-system generates large power losses if the machine works with large variations in load pressure and the average system pressure is much lower than the pressure setting for the pump regulator. CP is simple in design. Works like a pneumatic system. New hydraulic functions can easily be added and the system is quick in response.

Constant pressure systems (CP-system), unloaded. Same basic configuration as 'standard' CP-system but the pump is unloaded to a low stand-by pressure when all valves are in neutral position. Not so fast response as standard CP but pump lifetime is prolonged.

Load-sensing systems (LS-system) generates less power losses as the pump can reduce both flow and pressure to match the load requirements, but requires more tuning than the CP-system with respect to system stability. The LS-system also requires additional logical valves and compensator valves in the directional valves, thus it is technically more complex and more expensive than the CP-system. The LS-system system generates a constant power loss related to the regulating pressure drop for the pump regulator:

$$Power\ loss = \Delta p_{LS} \cdot Q_{tot}$$

The average Δp_{LS} is around 2 MPa (290 psi). If the pump flow is high the extra loss can be considerable. The power loss also increases if the load pressures vary a lot. The cylinder areas, motor displacements and mechanical torque arms must be designed to match load pressure in order to bring down the power losses. Pump pressure always equals the maximum load pressure when several functions are run simultaneously and the power input to the pump equals the (max. load pressure + Δp_{LS}) x sum of flow.

Five basic types of load-sensing systems

(1) Load sensing without compensators in the directional valves. Hydraulically controlled LS-pump.

(2) Load sensing with up-stream compensator for each connected directional valve. Hydraulically controlled LS-pump.

(3) Load sensing with down-stream compensator for each connected directional valve. Hydraulically controlled LS-pump.

(4) Load sensing with a combination of up-stream and down-stream compensators. Hydraulically controlled LS-pump.

(5) Load sensing with synchronised, both electric controlled pump displacement and electric controlled valve flow area for faster response, increased stability and less system losses. This is a new type of LS-system, not yet fully developed.

Technically the down-stream mounted compensator in a valveblock can physically be mounted "up-stream", but work as a down-stream compensator.

System type (3) gives the advantage that activated functions are synchronised independent of pump flow capacity. The flow relation between 2 or more activated functions remains independent, of load pressures even if the pump reach the maximum swivel angle. This feature is important for machines that often run with the pump at maximum swivel angel and with several activated functions that must be synchronised in speed, such as with excavators. With type (4) system, the functions with up-stream compensators have priority. Example: Steering-function for a wheel loader. The system type with down-stream compensators usually have a unique trademark depending on the manufacturer of the valves, for example "LSC" (Linde Hydraulics), "LUDV" (Bosch Rexroth Hydraulics) and "Flowsharing" (Parker Hydraulics) etc. No official standardised name for this type of system has been established but Flowsharing is a common name for it

Open and Closed Circuits

Open-loop: Pump-inlet and motor-return (via the directional valve) are connected to the hydraulic tank. The term loop applies to feedback; the more correct term is open versus closed "circuit".

Closed-loop: Motor-return is connected directly to the pump-inlet.

To keep up pressure on the low pressure side, the circuits have a charge pump (a small gearpump) that supplies cooled and filtered oil to the low pressure side. Closed-loop circuits are generally used for hydrostatic transmissions in mobile applications. Advantages: No directional valve and better response, the circuit can work with higher pressure. The pump swivel angle covers both positive and negative flow direction.

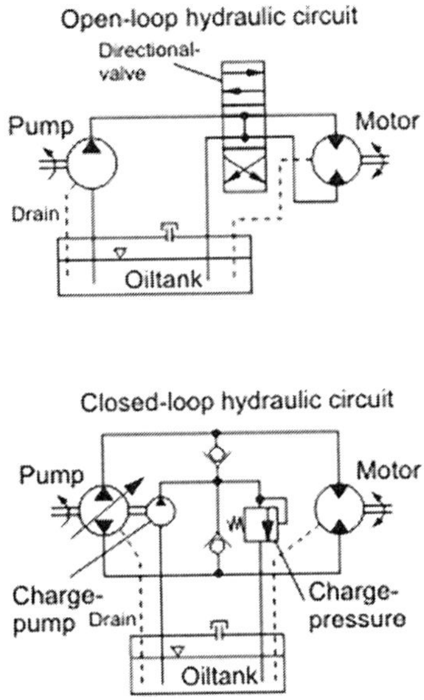

Figure: Open Loop and Closed Loop Circuits

Disadvantages: The pump cannot be utilised for any other hydraulic function in an easy way and cooling can be a problem due to limited exchange of oil flow. High power closed loop systems generally must have a 'flush-valve' assembled in the circuit in order to exchange much more flow than the basic leakage flow from the pump and the motor, for increased cooling and filtering. The flush valve is normally integrated in the motor housing to get a cooling effect for the oil that is rotating in the motor housing itself. The losses in the motor housing from rotating effects and losses in the ball bearings can be considerable as motor speeds will reach 4000-5000 rev/min or even more at maximum vehicle speed. The leakage flow as well as the extra flush flow must be supplied by the charge pump. Large charge pumps is thus very important if the transmission is designed for high pressures and high motor speeds. High oil temperatures, is usually a major problem when using hydrostatic transmissions at high vehicle speeds for longer periods, for instance when transporting the machine from one work place to the other. High oil temperatures for long periods will drastically reduce the lifetime for the transmission. To keep down the oil temperature, the system pressure during transport must be lowered, meaning that the minimum displacement for the motor must be limited to a reasonable value. Circuit pressures during transport around 200-250 bar is recommended.

Closed loop systems in mobile equipment are generally used for the transmission as an alternative to mechanical and hydrodynamic (converter) transmissions. The advantage is a stepless gear ratio (continuously variable speed/torque) and a more flexible control of the gear ratio depending on the load and operating conditions. The hydrostatic transmission is generally limited to around 200 kW maximum power, as the total cost gets too high at higher power compared to a hydrodynamic transmission. Large wheel loaders for instance and heavy machines are therefore usually equipped with converter transmissions. Recent technical achievements for the converter transmissions have improved the efficiency and developments in the software have also improved the characteristics, for example selectable gear shifting programs during operation and more gear steps, giving them characteristics close to the hydrostatic transmission.

Hydrostatic transmissions for earth moving machines, such as for track loaders, are often equipped with a separate 'inch pedal' that is used to temporarily increase the diesel engine rpm while reducing the vehicle speed in order to increase the available hydraulic power output for the working hydraulics at low speeds and increase the tractive effort. The function is similar to stalling a converter gearbox at high engine rpm. The inch function affects the preset characteristics for the 'hydrostatic' gear ratio versus diesel engine rpm.

Components

Hydraulic Pump

Hydraulic pumps supply fluid to the components in the system. Pressure in the system develops in reaction to the load. Hence, a pump rated for 5,000 psi is capable of maintaining flow against a load of 5,000 psi.

Pumps have a power density about ten times greater than an electric motor (by volume). They are powered by an electric motor or an engine, connected through gears, belts, or a flexible elastomeric coupling to reduce vibration.

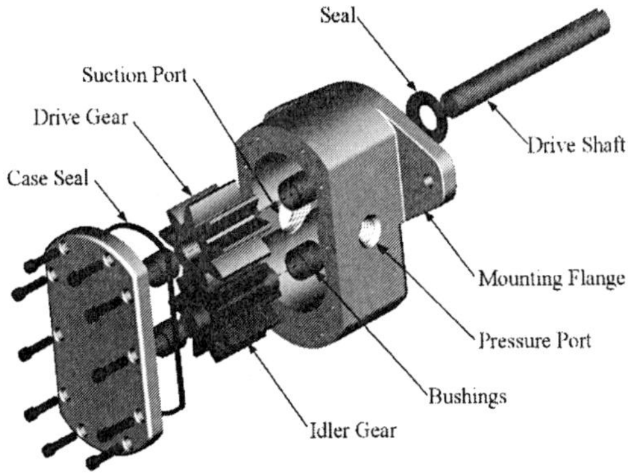

Figure: An exploded view of an external gear pump.

Common types of hydraulic pumps to hydraulic machinery applications are;

- Gear pump: cheap, durable, simple. Less efficient, because they are constant (fixed) displacement, and mainly suitable for pressures below 20 MPa (3000 psi).
- Vane pump: cheap and simple, reliable (especially in g-rotor form). Good for higher-flow low-pressure output.
- Axial piston pump: many designed with a variable displacement mechanism, to vary output flow for automatic control of pressure. There are various axial piston pump designs, including swashplate (sometimes referred to as a valveplate pump) and checkball (sometimes referred to as a wobble plate pump). The most common is the swashplate pump. A variable-angle swashplate causes the pistons to reciprocate a greater or lesser distance per rotation, allowing output flow rate and pressure to be varied (greater displacement angle causes higher flow rate, lower pressure, and vice versa).
- Radial piston pump A pump that is normally used for very high pressure at small flows.

Piston pumps are more expensive than gear or vane pumps, but provide longer life operating at higher pressure, with difficult fluids and longer continuous duty cycles. Piston pumps make up one half of a hydrostatic transmission.

Control Valves

Directional control valves route the fluid to the desired actuator. They usually consist of a spool inside a cast iron or steel housing. The spool slides to different positions in the housing, intersecting grooves and channels route the fluid based on the spool's position.

The spool has a central (neutral) position maintained with springs; in this position the supply fluid is blocked, or returned to tank. Sliding the spool to one side routes the hydraulic fluid to an actuator and provides a return path from the actuator to tank. When the spool is moved to the opposite direction the supply and return paths are switched. When the spool is allowed to return to neutral (centre) position the actuator fluid paths are blocked, locking it in position.

Directional control valves are usually designed to be stackable, with one valve for each hydraulic cylinder, and one fluid input supplying all the valves in the stack. Tolerances are very tight in order to handle the high pressure and avoid leaking, spools typically have aclearance with the housing of less than a thousandth of an inch (25 μm). The valve block will be mounted to the machine's frame with a three point pattern to avoid distorting the valve block and jamming the valve's sensitive components.

The spool position may be actuated by mechanical levers, hydraulic pilot pressure, or solenoids which push the spool left or right. A seal allows part of the spool to protrude outside the housing, where it is accessible to the actuator.

The main valve block is usually a stack of off the shelf directional control valves chosen by flow capacity and performance. Some valves are designed to be proportional (flow rate proportional to valve position), while others may be simply on-off. The control valve is one of the most expensive and sensitive parts of a hydraulic circuit.

- Pressure relief valves are used in several places in hydraulic machinery; on the return circuit to maintain a small amount of pressure for brakes, pilot lines, etc... On hydraulic cylinders, to prevent overloading and hydraulic line/seal rupture. On the hydraulic reservoir, to maintain a small positive pressure which excludes moisture and contamination.
- Pressure regulators reduce the supply pressure of hydraulic fluids as needed for various circuits.
- Sequence valves control the sequence of hydraulic circuits; to ensure that one hydraulic cylinder is fully extended before another starts its stroke, for example.
- Shuttle valves provide a logical or function.
- Check valves are one-way valves, allowing an accumulator to

charge and maintain its pressure after the machine is turned off, for example.

- Pilot controlled Check valves are one-way valve that can be opened (for both directions) by a foreign pressure signal. For instance if the load should not be held by the check valve anymore. Often the foreign pressure comes from the other pipe that is connected to the motor or cylinder.
- Counterbalance valves are in fact a special type of pilot controlled check valve. Whereas the check valve is open or closed, the counterbalance valve acts a bit like a pilot controlled flow control.
- Cartridge valves are in fact the inner part of a check valve; they are off the shelf components with a standardised envelope, making them easy to populate a proprietary valve block. They are available in many configurations; on/off, proportional, pressure relief, etc. They generally screw into a valve block and are electrically controlled to provide logic and automated functions.
- Hydraulic fuses are in-line safety devices designed to automatically seal off a hydraulic line if pressure becomes too low, or safely vent fluid if pressure becomes too high.
- Auxiliary valves in complex hydraulic systems may have auxiliary valve blocks to handle various duties unseen to the operator, such as accumulator charging, cooling fan operation, air conditioning power, etc. They are usually custom valves designed for the particular machine, and may consist of a metal block with ports and channels drilled. Cartridge valves are threaded into the ports and may be electrically controlled by switches or a microprocessor to route fluid power as needed.

Actuators

- Hydraulic cylinder
- Swashplates are used in 'hydraulic motors' requiring highly accurate control and also in 'no stop' continuous (360°) precision positioning mechanisms. These are frequently driven by several hydraulic pistons acting in sequence.
- Hydraulic motor (a pump plumbed in reverse)
- hydrostatic transmission
- Brakes.

Reservoir

The hydraulic fluid reservoir holds excess hydraulic fluid to accommodate volume changes from: cylinder extension and contraction, temperature driven expansion and contraction, and leaks. The reservoir is also designed to aid in separation of air from the fluid and also work as a heat accumulator to cover losses in the system when peak power is used. Design engineers are always pressured to reduce the size of hydraulic reservoirs, while equipment operators always appreciate larger reservoirs. Reservoirs can also help separate dirt and other particulate from the oil, as the particulate will generally settle to the bottom of the tank. Some designs include dynamic flow channels on the fluid's return path that allow for a smaller reservoir.

Accumulators

Accumulators are a common part of hydraulic machinery. Their function is to store energy by using pressurised gas. One type is a tube with a floating piston. On one side of the piston is a charge of pressurised gas, and on the other side is the fluid. Bladders are used in other designs. Reservoirs store a system's fluid.

Examples of accumulator uses are backup power for steering or brakes, or to act as a shock absorber for the hydraulic circuit.

Hydraulic Fluid

Also known as tractor fluid, hydraulic fluid is the life of the hydraulic circuit. It is usually petroleum oil with various additives. Some hydraulic machines require fire resistant fluids, depending on their applications. In some factories where food is prepared, either an edible oil or water is used as a working fluid for health and safety reasons. In addition to transferring energy, hydraulic fluid needs to lubricate components, suspend contaminants and metal filings for transport to the filter, and to function well to several hundred degrees Fahrenheit or Celsius.

Filters

Filters are an important part of hydraulic systems. Metal particles are continually produced by mechanical components and need to be removed along with other contaminants.

Filters may be positioned in many locations. The filter may be located between the reservoir and the pump intake. Blockage of the filter will cause cavitation and possibly failure of the pump. Sometimes the filter is located between the pump and the control valves. This arrangement is more expensive, since the filter housing is pressurised, but eliminates cavitation problems and protects the control valve from pump failures. The third common filter location is just before the return line enters the reservoir. This location is relatively insensitive to blockage and does not require a pressurised housing, but contaminants that enter the reservoir from external sources are not filtered until passing through the system at least once.

Tubes, Pipes and Hoses

Hydraulic tubes are seamless steel precision pipes, specially manufactured for hydraulics. The tubes have standard sizes for different pressure ranges, with standard diametres up to 100 mm. The tubes are supplied by manufacturers in lengths of 6 m, cleaned, oiled and plugged. The tubes are interconnected by different types of flanges (especially for the larger sizes and pressures), welding cones/nipples (with o-ring seal), several types of flare connection and by cut-rings. In larger sizes, hydraulic pipes are used. Direct joining of tubes by welding is not acceptable since the interior cannot be inspected.

Hydraulic pipe is used in case standard hydraulic tubes are not available. Generally these are used for low pressure. They can be connected by threaded connections, but usually by welds. Because of the larger diametres the pipe can usually be inspected internally after welding. Black pipe is non-galvanized and suitable for welding.

Hydraulic hose is graded by pressure, temperature, and fluid compatibility. Hoses are used when pipes or tubes can not be used, usually to provide flexibility for machine operation or maintenance. The hose is built up with rubber and steel layers. A rubber interior is surrounded by multiple layers of woven wire and rubber. The exterior is designed for abrasion resistance. The bend radius of hydraulic hose is carefully designed into the machine, since hose failures can be deadly, and violating the hose's minimum bend radius will cause failure. Hydraulic hoses generally have steel fittings swaged on the ends. The weakest part of the high pressure hose is the connection of the hose to the fitting. Another disadvantage of hoses is the shorter life of rubber which requires periodic replacement, usually at five to seven year intervals.

Tubes and pipes for hydraulic applications are internally oiled before the system is commissioned. Usually steel piping is painted outside. Where flare and other couplings are used, the paint is removed under the nut, and is a location where corrosion can begin. For this reason, in marine applications most piping is stainless steel.

Seals, Fittings and Connections

In general, valves, cylinders and pumps have female threaded bosses for the fluid connection, and hoses have female ends with captive nuts. A male-male fitting is chosen to connect the two. Many standardised systems are in use.

Fittings serve several purposes;

1. To bridge different standards; O-ring boss to JIC, or pipe threads to face seal, for example.
2. To allow proper orientation of components, a 90°, 45°, straight, or swivel fitting is chosen as needed. They are designed to be positioned in the correct orientation and then tightened.
3. To incorporate bulkhead hardware.
4. A quick disconnect fitting may be added to a machine without modification of hoses or valves

A typical piece of heavy equipment may have thousands of sealed connection points and several different types:

- Pipe fittings, the fitting is screwed in until tight, difficult to orient an angled fitting correctly without over or under tightening.
- O-ring boss, the fitting is screwed into a boss and orientated as needed, an additional nut tightens the fitting, washer and o-ring in place.
- Flare fittings, are metal to metal compression seals deformed with a cone nut and pressed into a flare mating.
- Face seal, metal flanges with a groove and o-ring are fastened together.
- Beam seals are costly metal to metal seals used primarily in aircraft.
- Swaged seals, tubes are connected with fittings that are swaged permanently in place. Primarily used in aircraft.

Elastomeric seals (O-ring boss and face seal) are the most common types of seals in heavy equipment and are capable of reliably sealing 6000+ psi (40+ MPa) of fluid pressure.

Basic Calculations

Hydraulic power is defined as flow times pressure. The hydraulic power supplied by a pump:

$$\text{Power} = (P \times Q) \div 600$$

where power is in kilowatts [kW], P pressure in bars, and Q is the flow in liters per minute. For example, a pump delivers 180 lit/min and the pressure equals 250 bar, therefore the power of the pump is 75 kW.

When calculating the power input to the pump, the total pump efficiency η_{total} must be included. This efficiency is the product of volumetric efficiency, η_{vol} and the hydromechanical efficiency, η_{hm}. Power input = Power output ÷ η_{total}. The average for axial piston pumps, $\eta_{total} = 0.87$. In the example the power source, for example a diesel engine or an electric motor, must be capable of delivering at least $75 \div 0.87 = 86$ [kW]. The hydraulic motors and cylinders that the pump supplies with hydraulic power also have efficiencies and the total system efficiency (without including the pressure drop in the hydraulic pipes and valves) will end up at approx. 0.75. Cylinders normally have a total efficiency around 0.95 while hydraulic axial piston motors 0.87, the same as the pump. In general the power loss in a hydraulic energy transmission is thus around 25% or more at ideal viscosity range 25-35 [cSt].

Calculation of the required max. power output for the diesel engine, rough estimation:

(1) Check the max. powerpoint, i.e. the point where pressure times flow reach the max. value.

(2) $E_{diesel} = (P_{max} \cdot Q_{tot}) \div \eta$.

Q_{tot} = calculate with the theoretical pump flow for the consumers not including leakages at max. power point.

P_{max} = actual pump pressure at max. power point.

Note: η is the total efficiency = (output mechanical power ÷ input mechanical power). For rough estimations, $\eta = 0.75$. Add 10-20% (depends on the application) to this power value.

(3) Calculate the required pump displacement from required max. sum of flow for the consumers in worst case and the diesel engine rpm in this point. The max. flow can differ from the flow used for calculation of the diesel engine power. Pump volumetric efficiency average, piston pumps: $\varsigma_{vol} = 0.93$.

Pumpdisplacement $V_{pump} = Q_{tot} \div n_{diesel} \div 0.93$.

(4) Calculation of prel. cooler capacity: Heat dissipation from hydraulic oil tanks, valves, pipes and hydraulic components is less than a few percent in standard mobile equipment and the cooler capacity must include some margins. Minimum cooler capacity, $E_{cooler} = 0.25E_{diesel}$

At least 25% of the input power must be dissipated by the cooler when peak power is utilised for long periods. In normal case however, the peak power is used for only short periods, thus the actual cooler capacity required might be considerably less. The oil volume in the hydraulic tank is also acting as a heat accumulator when peak power is used. The system efficiency is very much dependent on the type of hydraulic work tool equipment, the hydraulic pumps and motors used and power input to the hydraulics may vary a lot. Each circuit must be evaluated and the load cycle estimated. New or modified systems must always be tested in practical work, covering all possible load cycles. An easy way of measuring the actual average power loss in the system is to equip the machine with a test cooler and measure the oil temperature at cooler inlet, oil temperature at cooler outlet and the oil flow through the cooler, when the machine is in normal operating mode. From these figures the test cooler power dissipation can be calculated and this is equal to the power loss when temperatures are stabilised. From this test the actual required cooler can be calculated to reach specified oil temperature in the oil tank. One problem can be to assemble the measuring equipment inline, especially the oil flow metre. Hydraulic transmission has existed since antiquity. Practical applications originate with hydraulic power networks during the 19th century, transmitting water from a steam engine. Hydraulic transmissions with hydraulic fluids were later adopted to internal-combustion engines. Hydraulic brakes came around in 1918.

National Pipe Thread

Figure: Threaded pipe and elbow. The pipe has amale NPT thread, while the elbow's is female.

National Pipe Thread Tapered Thread (NPT) is a U.S.standard for tapered threads used on threaded pipes andf ittings. The taper rate for all NPT threads is ^{1}D $_{16}$ (^{3}D $_{4}$ inch per foot) measured by the change of diametre (of the pipe thread) over distance. The angle between the taper and the centre axis of the pipe is 1° 472 243 (1.7899°). Commonly-used sizes are ^{1}D $_{8}$, ^{1}D $_{4}$, ^{3}D $_{8}$, ^{1}D $_{2}$, ^{3}D $_{4}$, 1, 1^{1}D $_{4}$, 1 ^{1}D $_{2}$, and 2 inch, appearing on pipes and fittings by most U.S. suppliers. Sizes smaller than ^{1}D $_{8}$ are occasionally used for compressed air, while sizes larger than 2 in are uncommon, due to the use of alternative methods of joining that are used with these larger sizes. NPT is defined by ANSI/ ASME standard B1.20.1.

Nominal Pipe Size is loosely related to the inside diametre of Schedule 40 pipe. Because of the pipe wall thickness, the actual diametre of the threads is larger than the NPS, considerably so for small NPS. Other schedules of pipe have different wall thickness but the OD (outer diametre) and thread profile remain the same, so the inside diametre of the pipe is therefore different from the nominal diametre.

Threaded pipes can provide an effective seal for pipes transporting liquids, gases, steam, and hydraulic fluid. These threads are now used in materials other than steel and brass, including PTFE, PVC, nylon, bronze and cast iron.

The taper on NPT threads allows them to form a seal when torqued as the flanks of the threads compress against each other, as opposed to parallel/straight thread fittings or compression fittings in which the threads merely hold the pieces together and do not provide the seal. However, a clearance remains between the crests and roots of the threads, resulting in a leakage around this spiral. This means that NPT fittings must be made leak free with the aid of thread seal tape or a thread sealant compound. (The use of tape or sealant will also help to limit corrosion on the threads, which otherwise can make future disassembly nearly impossible.) There is also a semi-compatible variant called National Pipe Taper Fuel (NPTF) also called Dryseal American National Standard Taper Pipe Thread, defined by ANSI B1.20.3, designed to provide a more leak-free seal without the use of teflon tape or other sealant compound. NPTF threads are the same basic shape but with crest and root heights adjusted for an interference fit, eliminating the spiral leakage path.

Sometimes NPT threads are referred to as MPT ('Male Pipe Thread'), MNPT, or NPT(M) for male(external) threads; and FPT ('Female Pipe Thread'), FNPT, or NPT(F) for female (internal) threads.

An equivalent designation is MIP (Male iron pipe) and FIP (Female iron pipe). Also the terms NPS and NPSM are sometimes used to designate a straight, not tapered, thread. (this should not be confused with NPS meaning Nominal Pipe Size)

Thread Form

NPT and NPS threads have a 60° included angle and have a Sellers thread form (flattened peaks and valleys).

Terminology

Sometimes these terms may be used:

- MIP: stands for Male Iron Pipe, or Male International Pipe, or MPT Male Pipe Thread. It is a term for pipe fittings.
- FIP: stands for Female Iron Pipe, or Female International Pipe, or FPT. It is a term for pipe fittings that MIP fittings fit into.

The difference between FIP, MIP and Compression fittings is in the tapering of the thread. FIP has taper of 1:16 (6.25% slope), compression has taper of 13:96(13.5% slope).

A female iron connection has a tapered thread, which thins out to the end of the pipe. As the fitting is tightened, the ever-decreasing thread depth means that the connection becomes watertight. To properly seal the fitting, paste thread sealant or PTFE (polytetrafluoroethylene) tape wrapped around the thread is required.

MIP fittings are usually found in thicker copper pipe, iron pipe, steel pipe, gas pipe, gas stove, gas oven, gas heater, and gas water heater connections.

Panzergewinde

An assortment of PG cable glands. Similar glands with DIN/ISO metric threads are also available.

The Stahlpanzerrohrgewinde standard for screw threads, more often called by the shorter name Panzergewinde (pronunciation), is a technical standard created in Germany and subsequently used in various countries, such as the German-speaking zone (Germany, Switzerland, Austria) and neighbouring European countries. The

German name “Stahlpanzerrohrgewinde” translates to English as ‘steel conduit thread’. The thread is used to join pieces of electrical conduit and cable glands. Alternate stylings of the German name are Stahl-Panzer-Rohr-Gewinde and an abbreviated form, StaPa-Rohr-Gewinde. The standard, codified by the Deutsches Institut für Normung (DIN, German Institute for Standardisation), is DIN 40430. Panzergewinde sizes are named with the prefix PG plus a nominal number which approximately corresponds to the maximum cable diametre (in millimetres) that can be passed through the conduit.

Because the walls of the conduit are usually relatively thin, the thread depth should not be very large. Thus a thread angle of 80° is used. The Verband der Elektrotechnik, Elektronik und Informations technik (VDE) (which began as a trade association for standardisation in electrical engineering) originally standardised (and named) the thread for use with conduit and cable glands that were made of steel, although today the thread is used with both steel (typically plated with combinations of nickel, zinc, or tin to resist rusting) and polyvinyl chloride (PVC). Beginning in 2000, the VDE standard for cable glands (VDE 0619) was formally replaced by EN 50262. After a transitional period of several years during which it could still be used, it was replaced in 2003 by a final metric fine thread with 1.5 mm pitch. Similarly, conduit threads were replaced by EN 60423. Even today, Panzergewinde cable glands are still often found on chemical reactors and bioreactors (for example, PG13.5 thread for screwing in sensors) and various other equipment, enclosures, junction boxes, and connectors.

Tube Bending

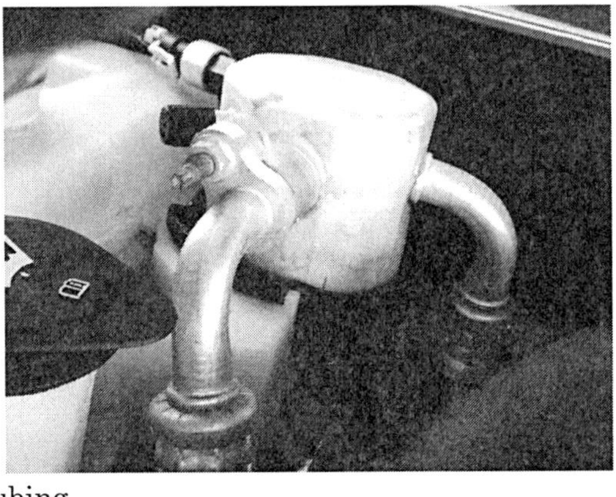

Figure: Bent tubing

Tube bending is the umbrella term for metal forming processes used to permanently form pipes or tubing. One has to differentiate between form-bound and freeform-bending procedures, as well as between heat supported and cold forming procedures.

Form bound bending procedures like "press bending" or "rotary draw bending" are used to form the work piece into the shape of a die. Straight tube stock can be formed using a bending machine to create a variety of single or multiple bends and to shape the piece into the desired form. This processes can be used to form complex shapes out of different types of ductile metal tubing. Freeform-bending processes, like three-roll-pushbending, shape the workpiece kinematically, thus the bending contour is not dependent on the tool geometry.

Generally, round stock is what is used in tube bending. However, square and rectangular tubes and pipes may also be bent to meet job specifications. Other factors involved in the tube bending process is the wall thickness, tooling and lubricants needed by the pipe and tube bender to best shape the material.

Geometry

A tube can be bent in multiple directions and angles. Common simple bends consist of forming elbows, which are bends that range from 2 to 90°, and U-bends, which are 180° bends. More complex geometries include multiple two-dimensional (2D) bends and three-dimensional (3D) bends. A 2D tube has the openings on the same plane; a 3D has openings on different planes.

One side effect of bending the workpiece is the wall thickness changes; the wall along the inner radius of the tube becomes thicker and the outer wall becomes thinner. To overcome this the tube may be supported internally or externally to preserve the cross section. Depending on the bend angle, wall thickness, and bending process the inside of the wall may wrinkle.

Processes

Tube bending as a process starts with loading a tube into a pipe bender and clamping it into place between two dies, the clamping block and the forming die. The tube is also loosely held by two other dies, the wiper die and the pressure die.

The process of tube bending involves using mechanical force to push stock material pipe or tubing against a die, forcing the pipe or tube to conform to the shape of the die. Often, stock tubing is held firmly in place while the end is rotated and rolled around the die. Other forms of processing including pushing stock through rollers that bend it into a simple curve. For some tube bending processing, a mandrel is placed inside the tube to prevent collapsing. The tube is also held in tension by a wiper die to prevent any creasing during stress. A wiper die is usually made of a softer alloy i.e. aluminium, brass to avoid scratching or damaging the material being bent.

Much of the tooling is made of hardened steel or tooled steel to maintain and prolong the tools life. However wherever there is a concern of scratching or gouging the work piece, a softer material such as aluminium or bronze is utilised. For example, the clamping block, rotating form block and pressure die are often formed from the hardened steel because the tubing is not moving past these parts of the machine. On the other hand, the pressure die and the wiping die are formed from aluminium or bronze to maintain the shape and surface of the workpiece as it slides by.

Pipe bending machines are typically human powered, pneumatic powered, hydraulic assisted, hydraulic driven, or electric servomotor.

Press Bending

Probably will be the first bending process used on cold pipes and tubing. In this process a die in the shape of the bend is pressed against the pipe forcing the pipe to fit the shape of the bend. Because the pipe is not supported internally there is some deformation of the shape of the pipe giving an ovular cross section. This process is used where a consistent cross section of the pipe is not required. Although a single die can produce various shapes, it only works for one size tube and radius.

Rotary Draw Bending

Rotary draw bending (RDB) are precise in that they bend using tooling or "die sets" which have a constant centre line radius (CLR). The die set consists of two parts: The former die creates the shape to which the material will be bent. The counter die does the work of pushing the material into the former die while travelling the length of the bend. Rotary draw benders can be programmable to store multiple bend jobs with varying degrees of bending. Often a positioning index table (IDX) is attached to the bender allowing the operator to reproduce complex bends which can have multiple bends and differing planes. Rotary draw benders are the most popular machines for use in bending tube, pipe and solids for applications like: handrails, frames, motor vehicle roll cages, handles, lines and much more. Rotary draw benders create aesthetically pleasing bends when the right tooling is matched to the application.

Three-Roll Pushbending

The Three-Roll Pushbending (TRPB) is the most commonly used freeform-bending process to manufacture bending geometries consisting of several plain bending curves. Nevertheless, a 3D-shaping is possible. The profile is guided between bending-roll and supporting-roll(s), while being pushed through the tools. The position of the forming-roll defines the bending radius. The bending point is the tangent-point between tube and bending-roll. To change the bending plane, the pusher rotates the tube around its longitudinal axis. Generally, a TRPB tool kit can be applied on a conventional rotary draw bending machine.

Bending contours defined as spline- or polynomial-functions can be manufactured.

Heat-induction

An induction coil is placed around a small section of the pipe at the bend point. It is then heated to between 800 and 2,200 degrees Fahrenheit (430 and 1,200 °C). While the pipe is hot, pressure is placed on the pipe to bend it. The pipe is then quenched with either air or water spray. Heat-Induction bending is used on large pipes such as freeway signs, power plants, and petroleum pipe lines.

Roll Benders

During the roll bending process the pipe, extrusion, or solid is passed through a series of rollers (typically 3) that apply pressure to the pipe gradually changing the bend radius in the pipe. The pyramid style roll benders have one moving roll, usually the top roll. Double pinch type roll benders have two adjustable rolls, usually the bottom rolls, and a fixed top roll. This method of bending causes very little deformation in the cross section of the pipe. This process is suited to producing coils of pipe as well as long gentle bends like those used in truss systems.

Sand-packing/hot-slab Forming

In the sand packing process the pipe is filled with fine sand and the ends are capped. The pipe is then heated in a furnace to 1,600 °F (870 °C) or higher. The pipe is then placed on a slab with pins set in it. The pipe is then bent around the pins using a winch, crane, or some other mechanical force. The sand in the pipe minimises distortion in the pipe cross section.

Mandrels

A mandrel is a steel rod or linked ball inserted into the tube while it is being bent to give the tube extra support to reduce wrinkling and breaking the tube during this process. The different types of mandrels are as follows.

1. Plug mandrel, a solid rod used on normal bends.
2. Form mandrel, a solid rod with curved end used on bend when more support is need.
3. Ball mandrel without cable, unlinked steel ball bearings inserted into tube, used on critical and precise bends.
4. Ball mandrel with cable, linked ball bearings inserted into tube, used on critical bend and precise bends.
5. Sand, sand packed into tube.

In production of a product where the bend is not critical a plug mandrel can be used. A form type tapers the end of the mandrel to provide more support in the bend of the tube. When precise bending is needed a ball mandrel (or ball mandrel with steel cable) should be used. The conjoined ball-like disks are inserted into the tubing to allow for bending while maintaining the same diametre throughout. Other styles include using sand, cerrobend, or frozen water. These allow for a somewhat constant diametre while providing an inexpensive alternative to the aforementioned styles.

Performance automotive or motorcycle exhaust pipe is a common application for a mandrel.

Bending Springs

These are strong but flexible springs inserted into a pipe to support the pipe walls during manual bending. They have diametres only slightly less than the internal diametre of the pipe to be bent. They are only suitable for bending 15-and-22 mm (0.6-and-0.9 in) soft copper pipe (typically used in household plumbing).

The spring is pushed into the pipe until its centre is roughly where the bend is to be. A length of flexible wire can be attached to the end of the spring to facilitate its removal. The pipe is generally held against the flexed knee, and the ends of the pipe are pulled up to create the bend. To make it easier to retrieve the spring from the pipe, it is a good idea to bend the pipe slightly more than required, and then slacken it off a little. They are less cumbersome than rotary benders, but are not suitable for bending short lengths of piping when it is difficult to get the required leverage on the pipe ends.

Bending springs for smaller diametre pipes (10 mm copper pipe) slide over the pipe instead of inside.

Piping and Plumbing Fittings

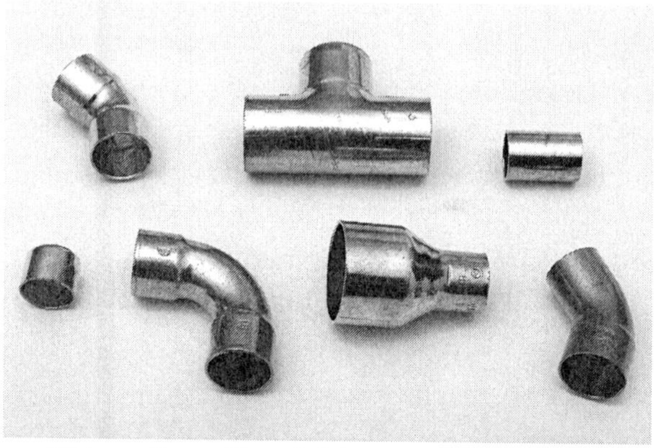

Figure: Copper fittings for soldered joints

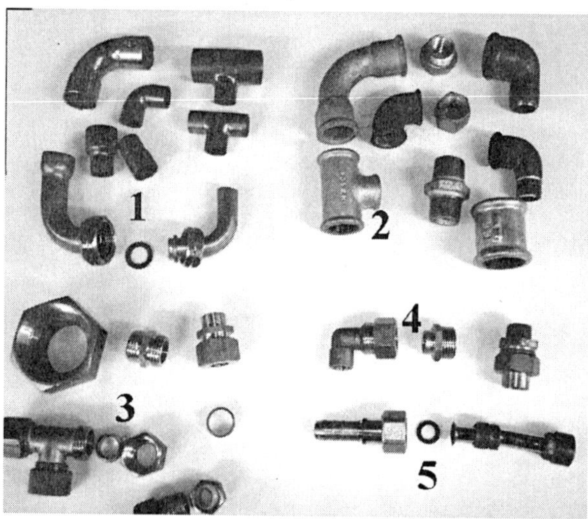

Figure: Assorted pipe fittings: 1.Copper, solder; 2.Iron or brass, threaded; 3.Brass, compression; 4.Brass, transition from compression to solder; 5.Brass, special adaptors

Fittings are used in pipe and plumbing systems to connect straight pipe or tubing sections, to adapt to different sizes or shapes, and for other purposes, such as regulating or measuring fluid flow. The termplumbing is generally used to describe conveyance of water, gas, or liquid waste in ordinary domestic or commercial environments, whereas piping is often used to describe high-performance (e.g. high pressure, high flow, high temperature, hazardous materials) conveyance of fluids in specialised applications. The term tubing is sometimes used for lighter-weight piping, especially types that are flexible enough to be supplied in coiled form.

Fittings (especially uncommon types) require money, time, materials, and tools to install, so they are a non-trivial part of piping and plumbing systems. Valves are technically fittings, but are usually discussed separately.

Standards

There are certain standard codes that need to be followed while designing or manufacturing any piping system. Organisations that promulgate piping standards include:

- ASME - The American Society of Mechanical Engineers
- ASTM - American Society for Testing and Materials
- API - American Petroleum Institute
- AWS - American Welding Society
- AWWA - American Water Works Association
- MSS – Manufacturers' Standardisation Society
- ANSI - American National Standards Institute
- NFPA - National Fire Protection Association
- EJMA - Expansion Joint Manufacturers Association

For example pipes need to conform to the dimensional requirements of :

- ASME B36.10M - Welded and Seamless Wrought Steel Pipe
- ASME B36.19M - Stainless Steel Pipe
- ASME B31.3 2008 - Process Piping

The B31.3 code has requirements for piping found in petroleum refineries; chemical, pharmaceutical, textile, paper, semiconductor, and cryogenic plants; and related processing plants and terminals. This code specifies requirements for materials and components, design, fabrication, assembly, erection, examination, inspection, and testing of piping. This Code is applicable to piping for all fluids including: (1) raw, intermediate, and finished chemicals; (2) petroleum products; (3) gas, steam, air and water; (4) fluidised solids; (5) refrigerants; and (6) cryogenic fluids.

Materials

The material with which a pipe is manufactured often forms as the basis for choosing any pipe. Materials that are used for manufacturing pipes include:

- Carbon Steel (CS)
- Low Temperature Service Carbon Steel (LTCS)
- Stainless Steel (SS)
- Non-Ferrous Metals (Inconel, Incoloy, Cupro-nickel etc.)
- Non-Metallic (ABS, GRE, PVC, HDPE, tempered glass, etc.)
- Chrome-molybdenum steel (Alloy steel) — Generally used for high temperature service

The bodies of fittings for pipe and tubing are most often of the same base material as the pipe or tubing being connected, for example, copper, steel, polyvinyl chloride (PVC), chlorinated polyvinyl chloride (CPVC), or acrylonitrile butadiene styrene (ABS). However, any material that is allowed by the plumbing, health, or building code (as applicable) may be used, but must be compatible with the other materials in the system, the fluids being transported, and the temperatures and pressures inside and outside of the system. For example, brass- or bronze-bodied fittings are common in otherwise copper piping and plumbing systems. Fire hazards, earthquake resistance, and other factors also influence choice of fitting materials.

Gaskets

Gaskets are used for sealing of flange joints. In general, gaskets should not be reused. Various types of gaskets are available depending upon their construction, materials, and features. The following are the type of gaskets commonly used:

- Non-Metallic Gaskets (ASME B 16.21)
- Spiral-Wound Gaskets (ASME B 16.20)
- Ring Joint Gaskets (ASME B 16.20)

Non-Metallic Gaskets are used with flat face or raised face flanges. Spiral-Wound Gaskets are used with raised face flanges. They are available with an inner ring and outer ring, which is also known as the cantering ring. Ring Joint Gaskets are used with Ring Type Joint (RTJ) flanges. They are available in octagonal or oval cross sections.

A very high surface stress is developed between an RTJ gasket and the flange groove when RTJ is bolted up in a flange. This leads to plastic deformation of the gasket. Thus, the hardness of the gasket is kept less than the hardness of the groove to achieve coining i.e bringing two metal surfaces of different hardness so tightly together that the softer surface deforms to match harder surface exactly in shape and finish. Hence RTJ gaskets are recommended for reuse.

Gender of Fittings

Piping or tubing are usually (but not always) inserted into fittings to make connections. To avoid confusion, connections are conventionally assigned a gender of male or female, respectively abbreviated as "M" or "F".

Common Fittings for Both Piping and Plumbing

While there are hundreds of specialised fittings manufactured, some common types of fittings are used widely in piping and plumbing systems.

Elbow

Figure: Short radius or regular 45° elbow (copper sweat)

Figure: Long radius or sweep 90° elbow (copper sweat)

An elbow is a pipe fitting installed between two lengths of pipe or tubing to allow a change of direction, usually a 90° or 45° angle, though 22.5° elbows are also made. The ends may be machined for butt welding, threaded (usually female), or socketed, etc. When the two ends differ in size, the fitting is called a reducing elbow or reducer elbow.

Elbows are categorised based on various design features as below:

- Long Radius (LR) Elbows – radius is 1.5 times the pipe diametre
- Short Radius (SR) Elbows – radius is 1.0 times the pipe diametre
- 90 Degree Elbow – where change in direction required is 90°
- 45 Degree Elbow – where change in direction required is 45°

A 90 degree elbow is also called a "90 bend" or "90 ell". It is a fitting which is bent in such a way to produce 90 degree change in the direction of flow in the pipe. It used to change the direction in piping and is also sometimes called a "quarter bend".

A 90 degree elbow attaches readily to plastic, copper, cast iron, steel and lead. It can also attach to rubber with stainless steel clamps. It is available in many materials like silicone, rubber compounds, galvanized steel, etc. The main application of an elbow (90 degree) is to connect hoses to valves, water pressure pumps, and deck drains. These elbows can be made from tough nylon material or NPT thread.

A 45 degree elbow is also called a "45 bend" or "45 ell". It is commonly used in water supply facilities, food industrial pipeline networks, chemical industrial pipeline networks, electronic industrial pipeline networks, air conditioning facility pipeline, agriculture and garden production transporting system, pipeline network for solar energy facility, etc.

Most elbows are available in short radius or long radius variants. The short radius elbows have a centre-to-end distance equal to the Nominal Pipe Size (NPS) in inches, while the long radius is 1.5 times the NPS in inches. Short elbows are widely available, and are typically used in pressurised systems.

Long elbows are typically used in low-pressure gravity-fed systems and other applications where low turbulence and minimum deposition of entrained solids are of concern. They are readily available in acrylonitrile butadiene styrene (ABS plastic), polyvinyl chloride (PVC) for DWV, sewage and central vacuums, chlorinated polyvinyl chloride (CPVC) and copper for 1950s to 1960s houses with copper drains.

Coupling

A coupling connects two pipes to each other. If the size of the pipe is not the same , the fitting may be called a reducing coupling or reducer, or an adapter. By convention, the term "expander" is not generally used for a coupler that increases pipe size; instead the term "reducer" is used.

Union

Figure: A combination pipe union and reducer fitting (brass threaded)

A union is similar to a coupling, except it is designed to allow quick and convenient disconnection of pipes for maintenance or fixture replacement. While a coupling would require either solvent welding, soldering or being able to rotate with all the pipes adjacent as with a threaded coupling, a union provides a simple transition, allowing easy connection or disconnection at any future time. A standard union pipe is made in three parts consisting of a nut, a female end, and a male end. When the female and male ends are joined, the nuts then provide the necessary pressure to seal the joint. Since the mating ends of the union are interchangeable, changing of a valve or other device can be achieved with a minimum loss of time. Pipe unions are essentially a type of flange connector, as discussed further below.

In addition to a standard union, there exist dielectric unions which are used to separate dissimilar metals (such as copper and galvanized steel) to avoid the damaging effects of galvanic corrosion. When two dissimilar metals are in contact with an electrically conductive solution (even tap water is conductive), they will form a battery and generate a voltage by electrolysis. When the two metals are in direct contact with each other, the electric current from one metal to the other will cause a movement of ions from one to the other, dissolving one metal and depositing it on the other. A dielectric union breaks the electric current path with a plastic liner between two halves of the union, thus limiting galvanic corrosion.

Reducer

A reducer allows for a change in pipe size to meet hydraulic flow requirements of the system, or to adapt to existing piping of a different size. Reducers are usually concentric but eccentric reducers are used when required to maintain the same top- or bottom-of-pipe level.

Olets

Whenever branch connections are required in size where reducing tees are not available and/or when the branch connections are of smaller size as compared to header size, olets are generally used. The following are few configurations of olet connections :

- Flanged Olet
- Socket-Weld & Threaded Olet
- Lateral & Elbow Olets
- Nipple Olet
- Butt-Weld Olet
- Swage Nipples

Tee

Figure: Pipe tee (copper sweat)

A tee is the most common pipe fitting. It is available with all female thread sockets, all solvent weld sockets, or with opposed solvent weld sockets and a side outlet with female threads. It is used to either combine or split a fluid flow. It is a type of pipe fitting which is T-shaped having two outlets, at 90° to the connection to the main line. It is a short piece of pipe with a lateral outlet. A tee is used for connecting pipes of different diametres or for changing the direction of pipe runs. They are made of various materials and available in various sizes and finishes. They are extensively used in pipeline networks to transport two-phase fluid mixtures. They are categorised as:

- Equal
- Unequal.

When the size of the branch is same as header pipes, equal tee is used and when the branch size is less than that of header size, reduced tee will be used. Most common are tees with the same inlet and outlet sizes. Some of the industrial tees are Straight Tee, Reducing Tee, Double Branch Tee, Double Branch Reducing Tee, Conical Tee, Double Branch Conical Tee, Bullhead Tee, Conical Reducing Tee, Double Branch Conical Reducing Tee, Tangential Tee, and Double Branch Tangential Tee. The above tees are categorised on the basis of their shapes and structure. They can also be classified on the basis of the application they are required to perform. The three outlet sizes should be named in order (e.g. left, middle, right; measuring 15-22-15).

Cross

Cross fittings are also called 4-way fittings. If a branch line passes completely through a tee, the fitting becomes a cross. A cross has one inlet and three outlets, or vice versa. They often have solvent welded socket ends or female threaded ends. Cross fittings can generate a huge amount of stress on pipe as temperature changes, because they are at the centre of four connection points. A tee is more steady than a cross, as a tee behaves like a three-legged stool, while a cross behaves like a four-legged stool. Crosses are common in fire sprinkler systems, but not in plumbing, due to their extra cost as compared to using two tees.

Cap

Figure: Pipe cap (copper sweat)

A type of pipe fitting, usually liquid or gas tight, which covers the end of a pipe. A cap is used like plug, except that the pipe cap screws or attaches on the male thread of a pipe or a nipple. A cap may have a solvent weld socket end or a female threaded end and the other end closed off. If a solvent weld cap is used to provide for a future connection point, several inches of pipe must be left before the cap. This is because when the cap is cut off for the future connection, enough pipe must remain to allow a new fitting to be glued onto it. In plumbing systems that use threads, the cap has female threads. Industrial caps can be round, square, rectangular, U-shaped, I-shaped and have a round hand grip or a flat hand grip.

Plug

A plug closes off the end of a pipe. It is similar to a cap but it fits inside the fitting it is mated to. In a threaded iron pipe plumbing system, plugs have male threads. Some of the popular types of plugs are:

- Mechanical pipe plug
- Pneumatic disk pipe plug
- Single size pneumatic all rubber pipe plug
- Multi-size pneumatic pipe plug
- Multi-size flow-through pipe plug
- High pressure pipe plug

Nipple

A short stub of pipe, usually threaded steel, brass, chlorinated polyvinyl chloride (CPVC) or copper; occasionally just bare copper. A nipple is defined as being a short stub of pipe which has a male pipe thread at each end, for connecting two other fittings. Nipples are commonly used for plumbing and hoses, and second as valves for funnels and pipes.

Barb

A barb is used to connect flexible hoses to pipes. A barb clamp fitting has a male-threaded end used to mate with the female-threads. The other end of the fitting has either a single or multiple barbed tube having a tapered stub with ridges which is inserted into the flexible hose to secure it. An adjustable worm driver screw clamp helps to keep the hose from slipping off the barbed tube. Barb fittings can be made of plastic or brass. Brass is used for hot water applications while plastic is used for cold. The barb fitting can be either elbow-shaped or straight.

Valves

Valve is equipment designed to stop or regulate flow of any fluid (liquid, gas, condensate, stem, slurry etc.) in its path. Valves are categorised depending on their applications like isolation, throttling and non-return. It is installed in the piping system based on its requirement. Various type of valves are available depending upon the type of construction as follows:

- Gate valve - used for isolation only
- Plug valve - used for isolation only
- Globe valve - used for throttling
- Butterfly valve - used for isolation as well as throttling
- Check valve - used for preventing reverse flow (non-return)
- Diaphragm valve - used for isolation as well as throttling
- Ball valve - used for isolation only.

Drain-waste-vent (DWV) and Related Fittings

Because they operate at low pressure and rely on gravity to move fluids (and often entrained solids), drain-waste-vent systems use fittings designed to be as smooth as possible on their interior surfaces.

The fittings may be "belled" or expanded slightly in diametre, or otherwise shaped to accommodate insertion of pipe or tubing, without forming a sharp interior ridge that might catch debris or accumulate buildup of material and cause clogging. The absence of interior snags also makes it much easier to "snake out" or "rod out" a clogged pipe using long flexible tools made for this purpose.

Underground piping systems for landscaping drainage, or disposal of stormwater or groundwater, similarly use gravity flow at low pressure, often with entrained solids. Piping fittings used for these systems bear a strong resemblance to DWV fittings, though often at a larger scale. When high peak flow volumes are involved, the design and construction of these systems are closely inter-related tosewer design.

Fittings for central vacuum systems are very similar to DWV fittings, though usually of thinner and lighter construction, since the weight of the materials conveyed through the system is much less. Vacuum system designs share with DWV designs a concern about eliminating internal ridges, burrs, sharp turns, or other obstructions to smooth flow that might cause build-up of material into pipe blockages.

Sweep Elbow

DWV elbows are usually long radius or sweep types, to reduce flow resistance and solids deposition when the direction of flow is changed. A well-designed system will often employ multiple 45° elbows in preference over 90° elbows (even sweep elbows), to reduce flow disruption as much as possible.

Central vacuum system inlet fittings are intentionally designed with a tighter radius of curvature than any other bends in the system. This is done to insure that if any vacuumed debris becomes stuck, it will jam right at the inlet, where it is easiest to discover and to remove.

Closet Flange

The closet flange is the drain pipe flange to which a water closet (toilet) is attached. It is a specialised type of flange connection designed to sit flush with the floor, allowing a standard toilet to be installed above it.

Clean-Outs

Clean-outs are fittings with removable elements that allow access to drains without requiring removal of plumbing fixtures. They are used for allowing an auger or plumber's snake to clean out a plugged drain. Clean-outs should be placed in accessible locations at regular intervals throughout a drainage system, often including outside the building, because clean-out augers have limited length. The minimum requirement is typically at the end of each branch in piping, just ahead of each water closet, at the base of each vertical stack, and both inside and outside the building in the building main drain/sewer. Clean-outs normally have screw-on caps or screw-in plugs. Clean-outs are also known as rodding eyes from the eye-shaped cover plates often used on external versions.

Trap Primers

Trap primers regularly inject water into traps so that "water seals" are maintained, as necessary to keep sewer gases out of buildings. The trap primer must be installed in a readily available place for easy access for adjustments, replacement, and repair. Strictly speaking, a trap primer is a specialised valve, and it is usually connected to a clean water supply, in addition to a DWV system. Because of this dual connection, the design usually must be certified to resist accidental backflow of contaminated water.

Combo-tee

A combination tee (combo tee) is a tee with a gradually curving centre connecting joint. It is used in drain systems to provide a smooth, gradually curving path to reduce the likelihood of clogs, and to ease pushing a plumber's snake through a drain system. The "combo" is a combination of a wye and a 1/8 bend or 45° elbow.

Sanitary Tee

A sanitary tee is a tee with a curved centre section designed to minimise the possibility of siphon action that could draw water out of a trap. The centre connection is generally connected to the pipe which leads to a trap (the trap arm).

Double Sanitary Tee (Sanitary Cross)

Similar to a cross. This fitting differs from a standard cross in that two of the ports have curved inlets. The fitting has been used in the past for connecting the drains of back-to-back fixtures (such as back-to-back bathroom sinks). Some current codes (including the 2006 UPC) prohibit the use of this fitting for that purpose, instead requiring a special "double fixture fitting".

Wye Fitting

A fitting with three openings, a wye is used to create branch lines. It is a type of waste fitting tee which has the side inlet pipe entering at a 45° angle, or an angle other than 90 degrees. A standard wye is a "Y" shaped fitting which allows one pipe to be joined to another at a 45 degree angle.

Wyes are similar to tees except that the branch line is angled to reduce friction and turbulence that could hamper the flow. The connection is typically at a 45-degree angle rather than a 90-degree angle. If a branch turns out further at the end to be perpendicular to the through line, the fitting becomes a "tee wye" (TY).

PVDF Corrosive Waste Piping Systems utilise wye fittings that feature a smooth inside diametre without any irregularities, for uninterrupted flow path. These fittings should be able to withstand acids, bases, and solvents. Flame-retardant systems withstand intermittent corrosive fluids up to high temperatures. Polypropylene and PVDF Double Wyes and Double Reducing Wyes come in various sizes and are largely used for this purpose.

A wye branch allows to split a branch line equally in two directions. The opening sizes can vary for different situations, for instance in situation where a large main line needs to be split into two smaller branches.

Low-priced wyes are often spot-welded together, whereas industrial wyes have a continuous weld at each seam. In long-distance pipeline applications, a true Wye fitting is also engineered for closed system instrumentation pigging configurations or wherever a smooth pipe branch is required.

Double-tapped Bushing

A double-tapped bushing is a fitting that has opposing threads on the inside diametre of the bushing.ggf

Hydraulic Fittings

Hydraulic systems use extremely high fluid pressures to create useful work, such as in the hydraulic actuators for powered machinery such as bulldozers and backhoes. Therefore, hydraulic fittings are designed and rated for much greater pressures than those experienced in general piping systems, and they are generally not compatible for use in general plumbing. Fittings are designed and constructed to resist leakage and sudden explosive failure. More information on hydraulics and their specialised fittings can be found in the hydraulic machinery article.

Connection Types

Much of the work of installing a piping or plumbing system involves making leakproof, reliable connections. Depending on the technology used, basic skills may be required, or specialised skills and professional licensure may be required.

Fastener

A fastener is a hardware device that mechanically joins or affixes two or more objects together. The following are the type of fasteners commonly used:

- Stud bolt with nut
- Machine bolt with nut

Usually the stud bolts are used with full threading and with two heavy hexagonal nuts.

Threaded Pipe

A threaded pipe is a pipe with a screw thread at one or both ends for assembly. Steel pipe is often joined using threaded connections, where tapered threads (in the US/Canada see National Pipe Thread, in Commonwealth Nations such as the UK, Australia, New Zealand see British standard pipe thread) are cut into the end of the tubing segment, sealant is applied in the form of thread sealing compound or thread seal tape (also known as PTFE or Teflon tape), and it is then threaded into a corresponding threaded fitting using a pipe wrench.

Figure: MIP pipe and FIP elbow fitting (steel threaded).

Threaded steel pipe is still widely used in many homes and businesses to convey natural gas or propane fuel, and is a popular choice in fire sprinkler systems due to its high heat resistance. Threaded brass pipe was once used in a similar fashion, and was considered superior to steel for carrying drinking water, but is now effectively obsolete.

Assembling threaded steel pipe takes some skill, plus careful planning to allow lengths of pipe to be screwed together in proper sequence. Most threaded pipe systems require occasional use of pipe union fittings to allow final assembly.

Solvent Welding

A solvent is applied to PVC, CPVC, ABS, or other plastic piping, to partially dissolve and fuse the adjacent surfaces of piping and fitting. Solvent welding is usually used with a sleeve-type joint, to connect pipe and fittings made of the same (or closely compatible) material.

Unlike regular welding of metals, solvent welding is relatively easy to perform, although care is still needed to produce reliable joints. Solvents typically used for plastics are usually toxic, may be carcinogenic, and may also be flammable, requiring adequate ventilation.

Soldering

To make a solder connection, a chemical flux is applied to the inner sleeve of a sleeve type joint, and the pipe is inserted. The joint is then heated using a propane torch or MAPP gas torch, solder is applied to the heated joint, and the melted solder is drawn into the joint by capillary action as the flux vapourises. Sweating is an alternate term sometimes used to describe soldering of pipe joints.

In situations where many connections must be made in a short period of time (such as plumbing of a new building), solder offers much quicker and much less expensive joinery than compression or flare fittings. A degree of skill is needed to rapidly make large numbers of reliable soldered joints. If flux residues are thoroughly cleaned, soldering can produce a long-lasting connection at low cost. However, the use of open flames for heating joints can present fire and health hazards to occupants of a building being worked on, and requires adequate ventilation.

Brazing

Brazing is a thermal joining process in which the two pieces of the base metal are joined when a molten brazing filler metal is allowed to be drawn into a capillary gap between them. It is used to join most metals and alloys commonly used in engineering. Brazing filler metals have very high melting points, but always below the melting point of the metals being joined. It is closely related to soldering, but uses harder materials and higher temperatures. Brazing can be used to join pipes, rods, metal pieces or any other shape as long as the pieces fit neatly against each other without leaving large gaps. It is also capable of joining tungsten carbide, ceramics and similar non-metallic materials.

Successfully brazed joints are as strong as the parent metal pieces being joined and can withstand demanding service conditions i.e they are strong and ductile. Well brazed joints have smooth, neat fillets and they offer good electrical conductivity.

Welding

Welding of metals differs from soldering and brazing, in that the connection is made without adding a special low-melting-point material (e.g. solder) to complete a joint. Instead, the material of the pipe or tubing is itself partially melted in a carefully controlled manner, and the fitting and piping are directly fused together. Generally this requires that the piping and the fitting be made of the same (or closely compatible) material. Pipe welding is a specialised skill, and is often performed by specially licensed workers who are tested periodically for the quality of their work. For ultra-critical applications, every joint is tested using non-destructive testing methods.

Properly welded and inspected joints are considered to be very reliable, robust, and long-lasting. Proper ventilation is essential to remove dangerous metallic fumes from welding operations. Because of the expensive skilled labour required, welded pipe joints are usually restricted to high-performance applications, such as in shipbuilding, chemical reactors, and nuclear reactors.

Compression Fittings

Compression connectors on an isolating valve. Red metal is a copper compression ring.

Compression fittings consist of a tapered concave conical seat, a hollow barrel-shaped compression ring, and a compression nut which is threaded onto the body of the fitting and tightened to make a leakproof connection. Fittings are typically made of brass or plastic, but stainless steel or other materials may be used.

Compression connections do not typically have the long life that sweat connections offer, but are advantageous in many cases because they are easy to make using basic tools. A disadvantage of compression connections is that they take longer to make than sweated joints, and sometimes require retightening over time to stop leaks.

Flare Fittings

Flared connectors should not be confused with compression connectors, with which they are generally not interchangeable. Flared connectors lack a compression ring, but do use a threaded nut. A special flaring tool is used to enlarge tubing into a tapered "bellmouth" shape that matches the tapered projecting conical shape of the flare fitting. The flare nut, which has previously been installed over the tubing, is then tightened onto the fitting. Fittings are typically made of brass or plastic, but stainless steel or other materials may be used.

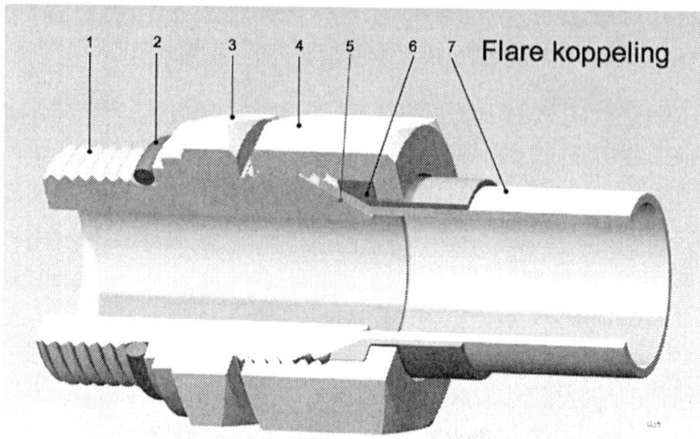

Flare connection: 1 Screw thread, 2 O-ring, 3 Body, 4 Nut, 5 Seal interface, 6 Support ring (sleeve), 7 Flared tubing.

Flare connections are a labour intensive method of making connections, but are quite reliable over the course of many years. Flared fittings are sometimes thought to be more secure against leaks and sudden failures, and are often preferred for safety-critical connections, such as in hydraulic brake systems.

Flange Fittings

Flanges are generally used when there is a connection to valves, in-line instruments and/or connection to equipment nozzles is required. Flange fittings generally involve pressing two surfaces to be joined tightly together, by means of threaded bolts, wedges, clamps, or other means of applying high compressive forces. Often, a gasket, packing, or an O-ring is installed between the flanges to prevent leakage, but it is sometimes possible to use only a special grease, or nothing at all, if the mating surfaces are precisely formed. Flanges are designed to the following pressure ratings: 150lb, 300lb, 400lb, 600lb, 900lb, 1500lb and 2500lb or 10 Bar, 15Bar, 25Bar, 40Bar, 64Bar, 100Bar and 150Bar. Various types of flanges are available depending upon the type of their constructional features. The following are types of flanges generally used in piping. These flanges are available with different facing like raised face, flat face, ring joint face etc.

- Orifice
- Threaded
- Slip-On
- Blind

- Weld Neck
- Socket
- Lap Joint
- Reducing

Slip On flanges are slipped over the pipe and then welded from both inside and outside to provide sufficient strength and prevent leakage. This flange is used instead of weld necks by many users because of its lower cost and also the fact that it requires less accuracy when cutting pipe to length.

Blind flanges do not have a bore and are used to shut off a piping system or vassal opening. Its design permits easy access to vassal or piping system for inspection purpose. It can be supplied with or without hubs at the manufacturer's option.

Weld Neck Flanges are designed to be joined to a piping system by butt welding. They are expensive because of its long neck, but is preferred for high stresses to the pipe, reducing stress applications. The neck, or hub, transmits stress concentration at the base of the flange. The gradual transition of thickness from the base of the hub to the wall thickness at the butt weld provides important reinforcement of the flange. Turbulence and erosion are reduced due to the matching bore size of the pipe and flange.

Socket Flanges are similar to a slip on flanges in outline, but the bore is counter-bored to accept pipe. The diametre of the remaining bore is same as the inside diametre of the pipe. A fillet weld around the hub of the flange attaches the flange to the pipe. An optional interval weld may be applied in high stress applications. Its biggest use is in high pressure system such as hydraulic and steam lines.

Lap Joint flange is again similar to a slip flange, but it has radius at the intersection of the bore and the flange face to accommodate a lap stub end. The face on the stub end forms the gasket face of the flange. Its applications are where sections of piping systems need to be dismantled quickly and easily for inspection or replacement.

Flanged connections tend to be more bulky than other connections, but can perform well in demanding applications, such as large water mains and hydroelectric power systems.

Mechanical Fittings

Manufacturers such as Victaulic or Grinnell produce special sleeve clamp fittings that are increasingly replacing classic flanged connections. They typically attach to the end of a pipe segment by using circumferential grooves pressed (or cut, in older designs) around the end of the pipe to be joined. Mechanical connectors are widely used on larger steel pipes, but can also be used with other materials.

Figure: Multiple mechanical sleeve clamps connecting piping

The chief advantage of these newer connectors is that they can be installed in the field after cutting the pipe to length, which is much faster than traditional flanged connections, which must be factory-welded or field-welded to pipe segments.

Crimped or Pressed Fittings

Crimped or pressed connections use special fittings which are permanently attached to tubing with a powered crimper. The special fittings, manufactured with sealant already inside, slide over the tubing to be connected. High pressure is used to deform the fitting and compress the sealant against the inner tubing, creating a leakproof seal. The advantages of this method are that it should last as long as the tubing, it takes less time to complete than other methods, it is cleaner in both appearance and the materials used to make the connection, and no open flame is used during the connection process. The disadvantages are that the fittings used are harder to find and cost significantly more than sweat type fittings.

Plastic Pipework

Plastic pipework is used for the conveyance of drinking water, waste water, chemicals, heating and cooling fluids, foodstuffs, ultra-pure liquids, slurries, gases, compressed air and vacuum system applications.

Materials Used

- ABS (Acrylonitrile Butadiene Styrene)
- UPVC (Unplasticized Polyvinyl Chloride)
- CPVC (Post Chlorinated Polyvinyl Chloride)
- PP (Polypropylene)
- PE (Polyethylene) also known as LDPE, MDPE and HDPE (low, medium, and high density)
- PVDF (Polyvinylidene fluoride).

Material Characteristics

ABS (Acrylonitrile Butadiene Styrene)

ABS is used for the conveyance of potable water, slurries and chemicals.

UPVC (Unplasticized Polyvinyl Chloride) and CPVC (Post Chlorinated Polyvinyl Chloride)

UPVC has excellent chemical resistance across its operating temperature range, with a broad band of operating pressures. Due to its long-term strength characteristics, high stiffness and cost effectiveness, UPVC systems account for a large proportion of plastic piping installations.

CPVC is resistant to many acids, bases, salts, paraffinic hydrocarbons, halogens and alcohols. It is not resistant to solvents, aromatics and some chlorinated hydrocarbons.

PP (Polypropylene)

Polypropylene is suitable for use with foodstuffs, potable and ultra pure waters, as well as within the pharmaceutical and chemical industries.

PE (Polyethylene)

Polyethylene has been successfully used for the safe conveyance of gases for many years.

PVDF (Polyvinylidene Fluoride)

PVDF has excellent chemical resistance which means that it is widely used in the chemical industry as a piping system for aggressive liquids.

Components of Plastic Pressure Pipe Systems

Pipes, fittings, valves, and accessories make up a plastic pressure pipe system. The range of pipe diametres for each pipe system does vary. However, the size ranges from 12 to 400 mm (0.472 to 15.748 in) and ^{3}D $_{8}$ to 16 in (9.53 to 406.40 mm).

Pipes are extruded and are generally available in: 3 m (9.84 ft),4 m (13.12 ft), 5 m (16.40 ft), and 6 m (19.69 ft) straight lengths and 25 m (82.02 ft), 50 m (164.04 ft), 100 m (328.08 ft), and 200 m (656.17 ft) coils for LDPE and MDPE

Pipe fittings are moulded and come in many sizes: Tee 90° equal (straight and reducing), Tee 45°, Cross equal, Elbow 90° (straight and reducing), Elbow 45°, Short Radius Bend 90° Socket/coupler (straight and reducing), Union, End Caps, Reducing Bush, and Stub, Full face, and blanking flanges.

Valves are moulded and also come in many types: Ball Valve (also multiport), Butterfly valve, spring-, ball-, and swing-check non-return valves, Diaphragm valve, Knife Gate Valve, Globe Valve and Pressure Relief/Reduction Valves

Accessories are solvents, cleaners, glues, clips, backing rings, and gaskets.

Plumbing

Figure: Water and sewage pipes of a Jerusalem building built around 1930

Plumbing is the system of pipes and drains installed in a building for the distribution of potable drinking water and the removal of waterborne wastes, and the skilled trade of working with pipes, tubing and plumbing fixtures in such systems. A plumber is someone who installs or repairspiping systems, plumbing fixtures and equipment such as water heaters. The plumbing industry is a basic and substantial part of every developed economy due to the need for clean water, and proper collection and transport of wastes. The word "plumbing" comes from the Latinplumbum for lead, as pipes were once made from lead.

Plumbing is usually distinguished from water and sewage systems, in that a plumbing system serves one building, while water and sewage systems serve a group of buildings or a city. Plumbing fixtures are exchangeable devices that can be connected to a building's plumbing system.

Plumbing was extremely rare until the growth of modern cities in the 19th centuries. At about the same time public health authorities began pressing for better waste disposal systems to be installed. Earlier, the waste disposal system merely consisted of collecting waste and dumping it on ground or into a river.

History

Figure: Roman lead pipe with a folded seam, at the Roman Baths in Bath, England

Plumbing originated during the ancient civilizations such as the Greek, Roman, Persian, Indian, and Chinese civilizations as they developed public baths and needed to provide potable water, and drainage of wastes. Standardised earthen plumbing pipes with broad flanges making use of asphalt for preventing leakages appeared in the urban settlements of the Indus Valley Civilization by 2700 B.C.. The Romans used lead pipe inscriptions to prevent water theft.

Improvement in plumbing systems was very slow, with virtually no progress made from the time of the Roman system of aqueducts and lead pipes until the growth of modern cities. Until then waste disposal systems merely consisted of collecting waste and dumping it on ground or into a river. In the 19th century, public health authorities began pressing for better waste disposal systems to be installed. Eventually the development of separate, underground water and sewage systems eliminated open sewage ditches and cesspools.

Most large cities today pipe solid wastes to treatment plants in order to separate and partly purify the water before emptying into streams or other bodies of water. For potable water use, galvanized iron piping was commonplace in the United States from the late 1800s until around 1960. After that period, copper took over, first with soft copper with flared fittings, then with rigid copper tubing utilising soldered fittings. The use of lead for potable water declined sharply after World War II because of the dangers of lead poisoning. At this time, copper piping was introduced as a better and safer alternative to lead pipes.

Materials

Water systems of ancient times relied on gravity for the supply of water, using pipes or channels usually made of clay, lead, bamboo wood or stone. Another material used for plumbing pipes, particularly water main, was hollowed wooden logs wrapped in steel banding. Logs used for water distribution were used in England close to 500 years ago. The US cities began using hollowed logs in the late 18th through the 19th centuries.

Present-day water-supply systems use a network of high-pressure pumps, and pipes are now made of copper, brass, plastic, or other nontoxic material. Lead is not used in modern water-supply piping due to its toxicity. Drain and vent lines are made of plastic, steel, cast-iron, and lead.

The "straight" sections of plumbing systems are pipes or tubes. A pipe is typically formed via casting or welding, where a tube is made through extrusion. Pipe normally has thicker walls and may be threaded or welded, where tubing is thinner-walled and requires special joining techniques such as brazing, compression fitting, crimping, or for plastics, solvent welding.

Components

In addition to the straight pipe or tubing, many fittings are required in plumbing systems, such as valves, elbows, tees, and unions. The piping and plumbing fittings and valves articles discuss these features further. Plumbing fixtures are designed for the end-users. Some examples of fixtures include water closets (also known as toilets), urinals, bidets, showers, bathtubs, utility and kitchen sinks, drinking fountains, ice makers, humidifiers, air washers, fountains, and eye wash stations.

Equipment

Plumbing equipment, not present in all systems, include, for example, water metres, pumps, expansion tanks, backflow preventers, filters, UV lights, water softeners, water heaters, wrenches, heat exchangers, flaring pliers, gauges, and control systems.

Now there is more equipment that is technologically advanced and helps plumbers fix problems without the usual hassles. For example, plumbers use video cameras for inspections of hidden leaks or problems, they use hydro jets, and high pressure hydraulic pumps connected to steel cables for trench-less sewer line replacement.

Systems

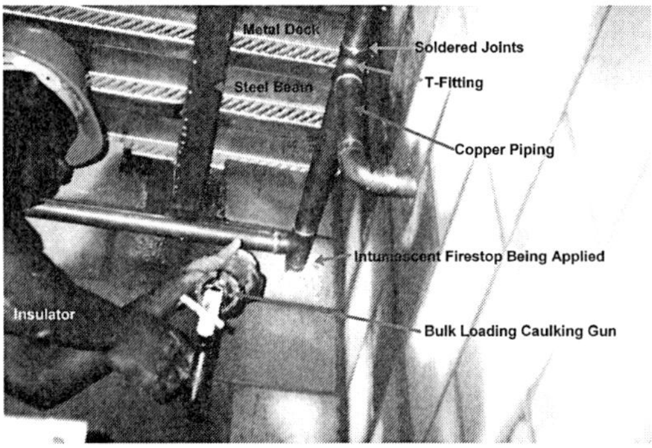

Figure: Copper piping system in a building with intumescent firestop being installed by an insulator, Vancouver, Canada

The major categories of plumbing systems or subsystems are:

- Potable cold and hot water supply
- Plumbing drainage venting
- Septic systems
- [Rainwater, surface, and subsurface water drainage
- Fuel gas piping

For their environmental benefit and sizable energy savingshot water heat recycling units are growing in use throughout the residential building sectors. Further ecological concern has seen increasing interest in grey-water recovery and treatment systems.

The New York City steam system is an example of a large district heating system.

Firestopping

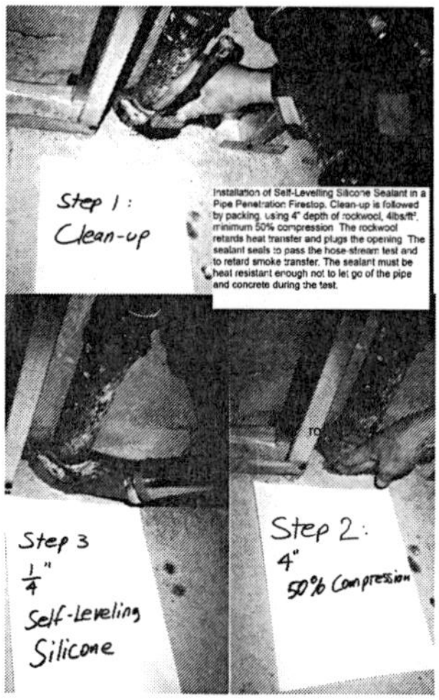

Figure: Self-levelling silicone firestop installation in mechanical service penetration in 2 hour rated concrete floor

Firestopping is required where mechanical penetrantstraverse fire-resistance rated wall and floor assemblies, or membranes thereof. This work is usually done worldwide by the insulation trade and/or speciality firestop subcontractors.

Regulation

Much of the plumbing work in populated areas is regulated by government or quasi-government agencies due to the direct impact on the public's health, safety, and welfare. Plumbing installation and repair work on residences and other buildings generally must be done according to plumbing and building codes to protect the inhabitants of the buildings and to ensure safe, quality construction to future buyers. If permits are required for work, plumbing contractors typically secure them from the authorities on behalf of home or building owners. In the United Kingdom the professional body is the Chartered Institute of Plumbing and Heating Engineering (educational charity status) and it is true that the trade still remains virtually ungoverned; there are no systems in place to monitor or control the activities of unqualified plumbers or those home owners who choose to undertake installation and maintenance works themselves, despite the health and safety issues which arise from such works when they are undertaken incorrectly; see Health Aspects of Plumbing (HAP) published jointly by the World Health Organisation (WHO) and the World Plumbing Council (WPC). WPC has subsequently appointed a representative to the World Health Organisation to take forward various projects related to Health Aspects of Plumbing.

Reinforced Thermoplastic Pipe

RTP is a generic term referring to a reelable high strength synthetic fibre (such as Glass, Aramid or Carbon) reinforced thermoplastic pipe, initially developed in the early 1990s by Wavin Repox, Akzo Nobel and by Tubes d'Aquitaine from France, who developed the first pipes reinforced with synthetic fibre to replace medium pressure steel pipes in response to growing demand for non-corrisive conduits for application in the onshore oil and gas industry, particularly in the Middle East.

Typically, the materials used in the construction of the pipe might be Polyethylene (PE), Polyamide-11 or PVDF and may be reinforced with Aramid or Polyester fibre although other combinations are used.

More recently the technology of producing such pipe, including the marketing, rests with a few key companies, one of which is Pipelife with Soluforce where it is available in coils up to 400 m (1,312 ft) length. These pipes are available in pressure ratings from 30 to 90 bar (3 to 9 MPa; 435 to 1,305 psi). Over the last few years this type of pipe has been acknowledged as a standard alternative solution to steel for oilfield flowline applications by certain oil companies and operators. The great advantage of this pipe is also its very fast installation time compared to steel pipe when considering the welding time as average speeds up to 1,000 m (3,281 ft)/day have been reached installing RTP in ground surface.

Primarily, the pipe provides benefit to applications where steel may rupture due to corrosion and installation time is an issue.

Technology and History

The idea of synthetic fibre reinforced pipe has origins in the flexible hose and offshore industry where it has been frequently used for applications such as control lines in umbilicals and production flowlines for over 30 years. However, the commercialisation and realisation of a competitive product for the onshore oil industry came from a partnership between Akzo Nobel (supplier of Aramid fibre) and Wavin Repox (manufacturer of reinforced thermoset pipes), where Bert Dalmolen initiated a project to develop such a pipe. He was later employed by Pipelife where a state of the art production line was developed to produce RTP. Pipelife also developed a pipe reinforced with steel wire to achieve even higher pressure ratings of over 150 bar (15 MPa; 2,176 psi) using steel reinforcement. Mr Chevrier (Tubes d'Aquitaine) also developed machinery that could produce such pipes, but was not successful in commercialising RTP.

Trap (plumbing)

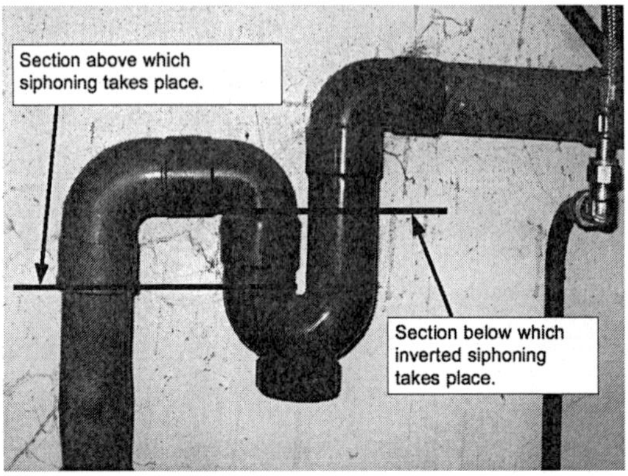

Figure: Water seal under a sink. Siphoning takes place

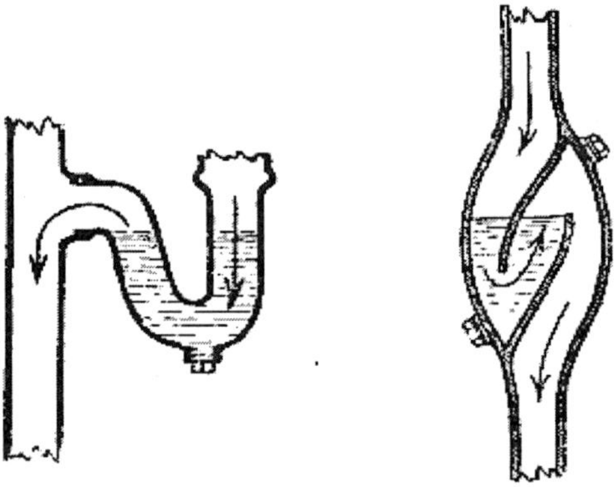

Figure: Examples of traps

In plumbing, a trap is a U-, S-, or J-shaped pipe located below or within a plumbing fixture. An S-shaped trap is also known as the S-bend invented by Alexander Cummings in 1775 but became known as the U-bend following the introduction of the U-shaped trap by Thomas Crapper in 1880. The new U-bend could not jam, so, unlike the S-bend, it did not need an overflow. The bend is used to prevent sewer gases from entering buildings. In refinery applications, it also prevents hydrocarbons and other dangerous gases from escaping outside through drains.

The most common of these traps in houses is referred to as a P-trap. It is the addition of a 90 degree fitting on the outlet side of a U-bend, thereby creating a P-like shape. It can also be referred to as a sink trap due to the fact it is installed under most house sinks.

Because of its shape, the trap retains a small amount of water after the fixture's use. This water in the trap creates a seal that prevents sewer gas from passing from the drain pipes back into the occupied space of the building. Essentially all plumbing fixtures including sinks, bathtubs, and toilets must be equipped with either an internal or external trap.

Because it is a localised low-point in the plumbing, sink traps also tend to capture heavy objects (such as jewellery) that are inadvertently dropped into the sink. Traps also tend to collect hair, sand, and other debris and limit the ultimate size of objects that will pass on into the rest of the plumbing, thereby catching over-sized objects. For all of these reasons, most traps can either be disassembled for cleaning or they provide some sort of cleanout feature.

Venting and Auxiliary Devices

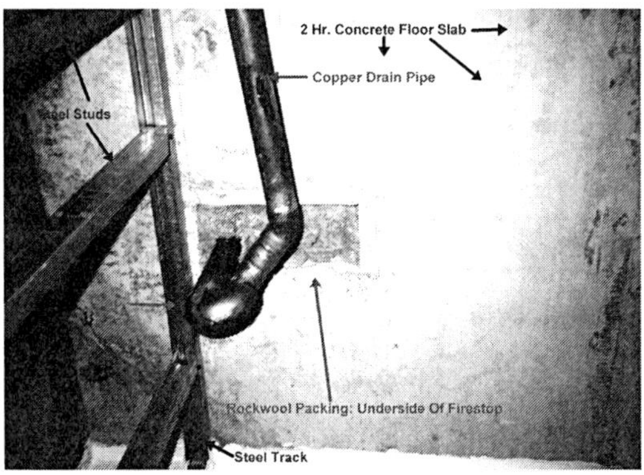

Figure: Trap with copper drain pipe at underside of firestop packing in 2 hour fire-resistance rated concrete floor slab

Maintaining the water seal is critical to trap operation; traps can and do dry out, and poorventing cansiphon or blow water out of the traps. This is usually avoided by venting the drain pipes downstream of the trap; by being vented to the atmosphere outside the building, the drain lines never operate at a pressure much higher or lower than atmospheric pressure. Plumbing codes usually provide strict limitations on how far a trap may be located from the nearest vent stack.

When a vent cannot be provided, codes may allow the use of an air admittance valve instead. These devices avoid negative pressure in the drain pipe by venting room air into the drain pipe (behind the trap). A "Chicago Loop" is another alternative.

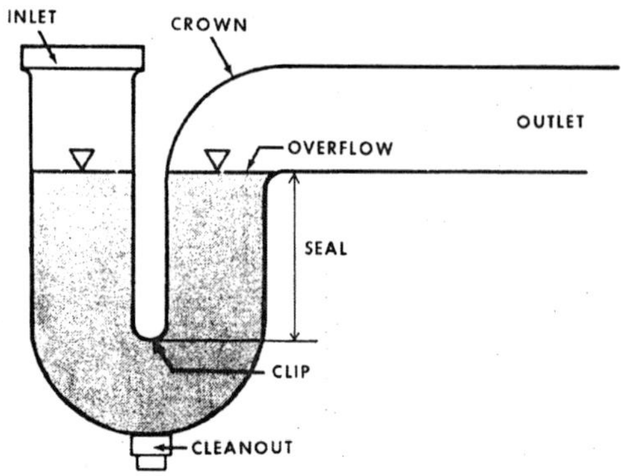

Figure: Typical P-trap

When a trap is installed on a fixture that is not routinely used, the eventual evaporation of the water in the trap must be considered. In these cases, a trap primer may be installed; these are devices that automatically recharge traps with water to maintain their water seals.

Accepted Traps

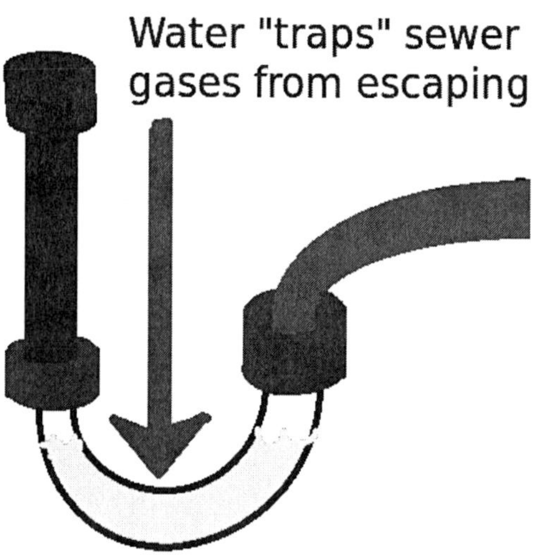

Figure: Gravity holds water to act as a barrier, preventing the escape of sewer gases. If a plumbing fixture is not used over a month or so, the water in the trap will evaporate, allowing noxious odours to escape.

In many locations, "S" traps are no longer accepted by the plumbing codes and are even illegal, as these traps tend to easily siphon dry even when well-vented. It is often possible to tell whether a household uses an S or U-bend by checking for the presence of an overflow pipe outlet. Certain drum-styled traps are also discouraged or banned.

Tube Beading: Metal Forming Process

Tube beading is a metal forming process that forms a bead on the end of a tube. Tube beads can be used to help hold a hose on the end of a tube or to strengthen the end of the tube. There are two forming processes: internal roll forming and ram forming.

Internal Roll Forming

Internal roll forming is generally slower than ram forming but it holds tight tolerances. Serrated clamp jaws are used to hold the tube while an axial pressure is applied to the end of the tube to form the bead.

Ram Forming

Ram forming is quicker and usually preferred when speed of production is a concern. The automotive industry usually uses this process over internal roll-forming. A clamp is used to hold the tube, the tube is expanded to desired diametre by an expansion punch, and then another punch is used to reduce the tube back to pilot diametre. Multiple beads are possible using this process.

Water Pipe

Figure: A system of copper water tubes used in a radiator heating system.

Water pipes are pipes or tubes, frequently made of polyvinyl chloride (PVC/uPVC), ductile iron, steel, cast iron, polypropylene, polyethylene, or copper, that carry pressurised and treated fresh water to buildings (as part of a municipal water system), as well as inside the building.

History

For many centuries, lead was the favoured material for water pipes, due to its malleability (this use was so common that the word "plumbing" derives from the Latin word for lead). This was a source of lead-related health problems in the years before the health hazards of ingesting lead were fully understood; among these were still births and high rates of infant mortality. Lead water pipes were still in common use in the early 20th century and remain in many households. Lead-tin alloy solder was commonly used to join copper pipes, but modern practice uses tin-antimony alloy solder to join copper in order to eliminate lead hazards.

Despite the Romans common use of lead pipes, their aqueducts rarely poisoned people. Unlike other parts of the world where lead pipes cause poisoning, the Roman water had so much calcium in it, that a layer of plaque prevented the water contacting the lead itself. What often causes confusion is the large amount of evidence of widespread lead poisoning, particularly amongst those who would have had easy access to piped water.

This was an unfortunate result of lead being used in cookware and as an additive to processed food and drink, such as a preservative in wine. Roman lead pipe inscriptions provided information on the owner to prevent water theft.

Wooden pipes were used in London and elsewhere during the 16th and 17th centuries. The pipes were hollowed-out logs, which were tapered at the end with a small hole in which the water would pass through. The multiple pipes were then sealed together with hot animal fat. They were often used in Montreal and Boston in the 1800s, and built-up wooden tubes were widely used in the USA during the 20th century. These pipes, used in place of corrugated iron or reinforced concrete pipes, were made of sections cut from short lengths of wood. Locking of adjacent rings with hardwood dowel pins produced a flexible structure. About 100,000 feet of these wooden pipes were installed during WW2 in drainage culverts, storm sewers and conduits, under highways and at army camps, naval stations, airfields and ordnance plants.

Cast iron and ductile iron pipe was long a lower-cost alternative to copper, before the advent of durable plastic materials but special non-conductive fittings must be used where transitions are to be made to other metallic pipes, except for terminal fittings, in order to avoid corrosion owing to electrochemical reactions between dissimilar metals (see galvanic cell).

Bronze fittings and short pipe segments are commonly used in combination with various materials.

Pipe vs. Tube

The difference between pipe and tube is simply in the way it is sized. PVC pipe for plumbing applications and galvanized steel pipe for instance, are measured in IPS (iron pipe size). Copper tube, CPVC, PeX and other tubing is measured nominally, which is basically an average diametre. These sizing schemes allow for universal adaptation of transitional fittings. For instance, 1/2" PeX tubing is the same size as 1/2" copper tubing. 1/2" PVC on the other hand is not the same size as 1/2" tubing, and therefore requires either a threaded male or female adapter to connect them. When used in agricultural irrigation, the singular form "pipe" is often used as a plural.

Piping is available in rigid "joints", which come in various lengths depending on the material. Tubing, in particular copper, comes in rigid hard tempered "joints" or soft tempered (annealed) rolls. PeX and CPVC tubing also comes in rigid "joints" or flexible rolls. The temper of the copper, that is whether it is a rigid "joint" or flexible roll, does not affect the sizing.

The thicknesses of the water pipe and tube walls can vary. Pipe wall thickness is denoted by various schedules or for large bore polyethylene pipe in the UK by the Standard Dimension Ratio (SDR) being the ratio of the pipe diametre to its wall thickness. Pipe wall thickness increases with schedule, and is available in schedules 20, 40, 80, and higher in special cases. The schedule is largely determined by the operating pressure of the system, with higher pressures commanding greater thickness. Copper tubing is available in four wall thicknesses: type DWV (thinnest wall; only allowed as drain pipe per UPC), type 'M' (thin; typically only allowed as drain pipe by IPC code), type 'L' (thicker, standard duty for water lines and water service), and type 'K' (thickest, typically used underground between the main and the metre). Because piping and tubing are commodities, having a greater wall thickness implies higher initial cost. Thicker walled pipe generally implies greater durability and higher pressure tolerances.

Wall thickness does not affect pipe or tubing size. 1/2" L copper has the same outer diametre as 1/2" K or M copper. The same applies to pipe schedules. As a result, a slight increase in pressure losses is realised due to a decrease in flowpath as wall thickness is increased. In other words, 1 foot of 1/2" L copper has slightly less volume than 1 foot of 1/2 M copper.

Demand for copper products have fallen due to the dramatic increase in the price of copper, resulting in increased demand for alternative products including PEX and stainless steel.

Chapter 5

Types and Applications of Pressure Vessels

Gas Compressor

A gas compressor is a mechanical device that increases the pressure of a gas by reducing its volume.

Compressors are similar to pumps: both increase the pressure on a fluid and both can transport the fluid through a pipe.

As gases are compressible, the compressor also reduces the volume of a gas. Liquids are relatively incompressible; while some can be compressed, the main action of a pump is to pressurise and transport liquids.

Types of Compressors

The main types of gas compressors are illustrated and discussed below:

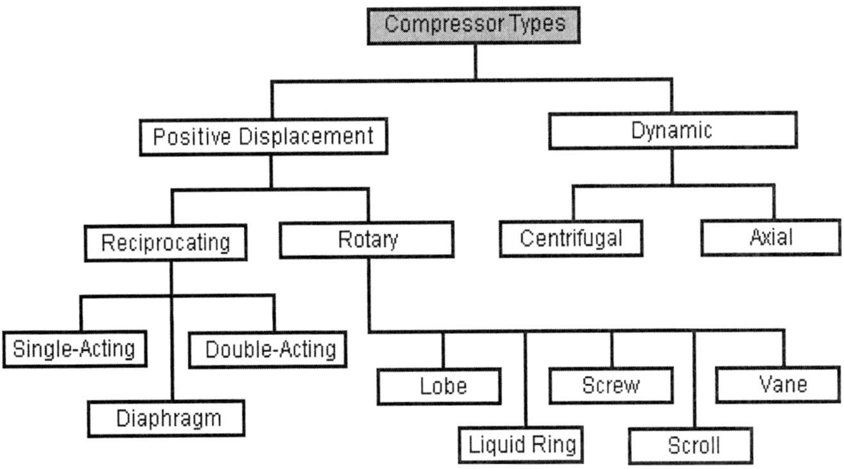

Figure: Hermetically Sealed, Open, or Semi-hermetic

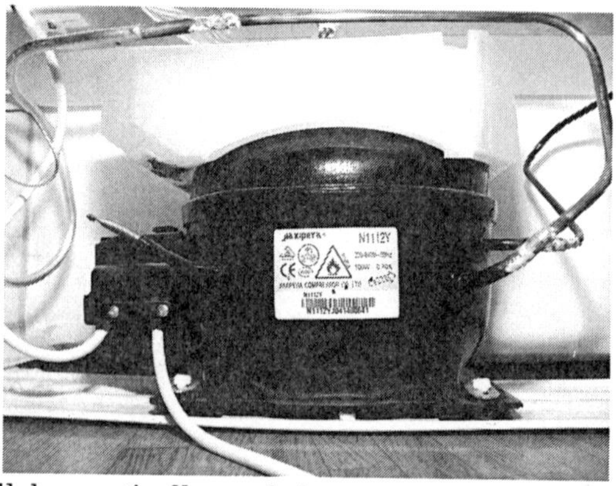

Figure: A small hermetically sealed compressor in a common consumer refrigerator or freezer; it typically has a rounded steel outer shell that is permanently welded shut, and which seals operating gases inside the system. There is no route for gases to leak, such as around motor shaft seals. On this model, the plastic top section is part of an auto-defrost system which uses motor heat to evaporate the water.

Compressors are often described as being either open, hermetic, or semi-hermetic, to describe how the compressor and motor drive is situated in relation to the gas or vapour being compressed. The industry name for a hermetic is hermetically sealed compressor, while a semi- is commonly called a semi-hermetic compressor.

In hermetic and most semi-hermetic compressors, the compressor and motor driving the compressor are integrated, and operate within the pressurised gas envelope of the system. The motor is designed to operate and be cooled by the gas or vapour being compressed.

The difference between the hermetic and semi-hermetic, is that the hermetic uses a one-piece welded steel casing that cannot be opened for repair; if the hermetic fails it is simply replaced with an entire new unit. A semi-hermetic uses a large cast metal shell with gasketed covers that can be opened to replace motor and pump components.

The primary advantage of a hermetic and semi-hermetic is that there is no route for the gas to leak out of the system. Open compressors rely on either natural leather or synthetic rubber seals to retain the internal pressure, and these seals require a lubricant such as oil to retain their sealing properties.

An open pressurised system such as an automobile air conditioner can leak its operating gases, if it is not operated frequently enough. Open systems rely on lubricant in the system to splash on pump components and seals. If it is not operated frequently enough, the lubricant on the seals slowly evaporates, and then the seals begin to leak until the system is no longer functional and must be recharged. By comparison, a hermetic system can sit unused for years, and can usually be started up again at any time without requiring maintenance or experiencing any loss of system pressure.

The disadvantage of hermetic compressors is that the motor drive cannot be repaired or maintained, and the entire compressor must be removed if a motor fails. A further disadvantage is that burnt out windings can contaminate whole systems requiring the system to be entirely pumped down and the gas replaced. Typically hermetic compressors are used in low-cost factory-assembled consumer goods where the cost of repair is high compared to the value of the device, and it would be more economical to just purchase a new device.

An advantage of open compressors is that they can be driven by non-electric power sources, such as an internal combustion engine or turbine. However, open compressors that drive refrigeration systems are generally not totally maintenance free throughout the life of the system, since some gas leakage will occur over time.

Centrifugal Compressors

Figure : A single stage centrifugal compressor

Centrifugal compressors use a rotating disk or impeller in a shaped housing to force the gas to the rim of the impeller, increasing the velocity of the gas. A diffuser (divergent duct) section converts the velocity energy to pressure energy. They are primarily used for continuous, stationary service in industries such as oil refineries, chemical and petrochemical plants and natural gas processing plants. Their application can be from 100 horsepower (75 kW) to thousands of horsepower. With multiple staging, they can achieve extremely high output pressures greater than 10,000 psi (69 MPa). Many large snowmaking operations (like ski resorts) use this type of compressor. They are also used in internal combustion engines as superchargers and turbochargers. Centrifugal compressors are used in small gas turbine engines or as the final compression stage of medium sized gas turbines. Sometimes the capacity of the compressors is written in NM3/hr. Here 'N' stands for normal temperature pressure (20°C and 1 atm) for example 5500 NM3/hr.

Diagonal or Mixed-flow Compressors

Diagonal or mixed-flow compressors are similar to centrifugal compressors, but have a radial and axial velocity component at the exit from the rotor. The diffuser is often used to turn diagonal flow to an axial rather than radial direction.

Axial-flow Compressors

Figure: An animation of an axial compressor.

Axial-flow compressors are dynamic rotating compressors that use arrays of fan-like airfoils to progressively compress the working fluid. They are used where there is a requirement for a high flow rate or a compact design.

The arrays of airfoils are set in rows, usually as pairs: one rotating and one stationary. The rotating airfoils, also known as blades or rotors, accelerate the fluid. The stationary airfoils, also known as stators or vanes, decelerate and redirect the flow direction of the fluid, preparing it for the rotor blades of the next stage. Axial compressors are almost always multi-staged, with the cross-sectional area of the gas passage diminishing along the compressor to maintain an optimum axial Mach number. Beyond about 5 stages or a 4:1 design pressure ratio, variable geometry is normally used to improve operation. Axial compressors can have high efficiencies; around 90% polytropic at their design conditions. However, they are relatively expensive, requiring a large number of components, tight tolerances and high quality materials. Axial-flow compressors can be found in medium to large gas turbine engines, in natural gas pumping stations, and within certain chemical plants.

Reciprocating Compressors

Figure: A motor-driven six-cylinder reciprocating compressor that can operate with two, four or six cylinders.

Reciprocating compressors use pistons driven by a crankshaft. They can be either stationary or portable, can be single or multi-staged, and can be driven by electric motors or internal combustion engines. Small reciprocating compressors from 5 to 30 horsepower (hp) are commonly seen in automotive applications and are typically for intermittent duty. Larger reciprocating compressors well over 1,000 hp (750 kW) are commonly found in large industrial and petroleum applications. Discharge pressures can range from low pressure to very high pressure (>18000 psi or 180 MPa). In certain applications, such as air compression, multi-stage double-acting compressors are said to be the most efficient compressors available, and are typically larger, and more costly than comparable rotary units. Another type of reciprocating compressor is the swash plate compressor, which uses pistons which are moved by a swash plate mounted on a shaft - see Axial Piston Pump.

Household, home workshop, and smaller job site compressors are typically reciprocating compressors 1½ hp or less with an attached receiver tank.

Rotary Screw Compressors

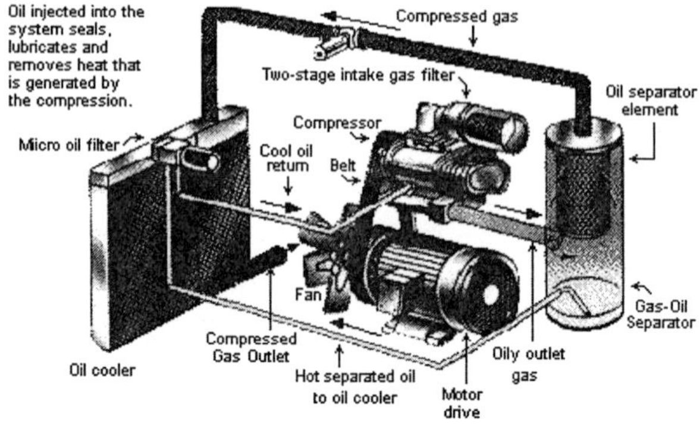

Figure: Diagram of a rotary screw compressor

Rotary screw compressors use two meshed rotating positive-displacement helical screws to force the gas into a smaller space. These are usually used for continuous operation in commercial and industrial applications and may be either stationary or portable. Their application can be from 3 horsepower (2.2 kW) to over 1,200 horsepower (890 kW) and from low pressure to moderately high pressure (>1,200 psi or 8.3 MPa).

Rotary screw compressors are commercially produced in Oil Flooded, Water Flooded and Dry type.

Rotary Vane Compressors

Rotary vane compressors consist of a rotor with a number of blades inserted in radial slots in the rotor. The rotor is mounted offset in a larger housing which can be circular or a more complex shape. As the rotor turns, blades slide in and out of the slots keeping contact with the outer wall of the housing. Thus, a series of decreasing volumes is created by the rotating blades. Rotary Vane compressors are, with piston compressors one of the oldest of compressor technologies.

With suitable port connections, the devices may be either a compressor or a vacuum pump. They can be either stationary or portable, can be single or multi-staged, and can be driven by electric motors or internal combustion engines. Dry vane machines are used at relatively low pressures (e.g., 2 bar or 200 kPa; 29 psi) for bulk material movement while oil-injected machines have the necessary volumetric efficiency to achieve pressures up to about 13 bar (1,300 kPa; 190 psi) in a single stage. A rotary vane compressor is well suited to electric motor drive and is significantly quieter in operation than the equivalent piston compressor. Rotary vane compressors can have mechanical efficiencies of about 90%.

Scroll Compressors

A scroll compressor, also known as scroll pump and scroll vacuum pump, uses two interleaved spiral-like vanes to pump or compress fluids such as liquids and gases. The vane geometry may be involute, archimedean spiral, or hybrid curves. They operate more smoothly, quietly, and reliably than other types of compressors in the lower volume range.

Often, one of the scrolls is fixed, while the other orbits eccentrically without rotating, thereby trapping and pumping or compressing pockets of fluid or gas between the scrolls.

This type of compressor was used as the supercharger on Volkswagen G60 and G40 engines in the early 1990s.

Diaphragm Compressors

A diaphragm compressor (also known as a membrane compressor) is a variant of the conventional reciprocating compressor. The compression of gas occurs by the movement of a flexible membrane, instead of an intake element. The back and forth movement of the membrane is driven by a rod and a crankshaft mechanism. Only the membrane and the compressor box come in contact with the gas being compressed.

The degree of flexing and the material constituting the diaphragm affects the maintenance life of the equipment. Generally stiff metal diaphragms may only displace a few cubic centimetres of volume because the metal can not endure large degrees of flexing without cracking, but the stiffness of a metal diaphragm allows it to pump at high pressures. Rubber or silicone diaphragms are capable of enduring deep pumping strokes of very high flexion, but their low strength limits their use to low-pressure applications, and they need to be replaced as plastic embrittlement occurs.

Diaphragm compressors are used for hydrogen and compressed natural gas (CNG) as well as in a number of other applications.

The photograph included in this section depicts a three-stage diaphragm compressor used to compress hydrogen gas to 6,000 psi (41 MPa) for use in a prototype compressed hydrogen and compressed natural gas (CNG) fuelling station built in downtown Phoenix, Arizona by the Arizona Public Service company (an electric utilities company). Reciprocating compressors were used to compress the natural gas.

The prototype alternative fuelling station was built in compliance with all of the prevailing safety, environmental and building codes in Phoenix to demonstrate that such fuelling stations could be built in urban areas.

Figure: A three-stage diaphragm compressor

Air Bubble Compressor

Also known as a trompe. A mixture of air and water generated through turbulence is allowed to fall into a subterranean chamber where the air separates from the water. The weight of falling water compresses the air in the top of the chamber. A submerged outlet from the chamber allows water to flow to the surface at a lower height than the intake. An outlet in the roof of the chamber supplies the compressed air to the surface. A facility on this principal was built on the Montreal River at Ragged Shutes near Cobalt, Ontario in 1910 and supplied 5,000 horsepower to nearby mines.

Temperature

Compression of a gas naturally increases its temperature, often referred to as the heat of compression.

$$W = \int_{V_1}^{V_2} P dV = P_1 V_1^n \int_{V_1}^{V_2} V^{-n} dV \quad \text{where}$$

$$\frac{P_2}{P_1} = \left(\frac{\upsilon_1}{\upsilon_2}\right)^n$$

so

$$W = \frac{P_1 V_1^n}{1-n}\left(V_2^{1-n} - V_1^{1-n}\right)$$

with n taking different values for different compression processes (see below).

- Adiabatic - This model assumes that no energy (heat) is transferred to or from the gas during the compression, and all supplied work is added to the internal energy of the gas, resulting in increases of temperature and pressure. Theoretical temperature rise is: $T_2 = T_1.R_c^{(k-1)/k}$ with T_1 and T_2 in degrees Rankine or kelvins, and k = ratio of specific heats (approximately 1.4 for air). R_c is the compression ratio; being the absolute outlet pressure divided by the absolute inlet pressure. The rise in air and temperature ratio means compression does not follow a simple pressure to volume ratio. This is less efficient, but quick. Adiabatic compression or expansion more closely model real life when a compressor has good insulation, a large gas volume, or a short time scale (i.e., a high power level). In practice there will always be a certain amount of heat flow out of the compressed gas. Thus, making a perfect adiabatic compressor would require perfect heat insulation of all parts of the machine. For example, even a bicycle tire pump's metal tube becomes hot as you compress the air to fill a tire. The relation between temperature and compression ratio described above means that the value of n for an adiabatic process is k (the ratio of specific heats).
- Isothermal - This model assumes that the compressed gas remains at a constant temperature throughout the compression or expansion process. In this cycle, internal energy is removed from the system as heat at the same rate that it is added by the mechanical work of compression. Isothermal compression or expansion more closely models real life when the compressor has a large heat exchanging surface, a small gas volume, or a long time scale (i.e., a small power level). Compressors that utilise inter-stage cooling between compression stages come closest to achieving perfect isothermal compression. However, with practical devices perfect isothermal compression is not attainable. For example, unless you have an infinite number of compression stages with corresponding intercoolers, you will never achieve perfect isothermal compression.

For an isothermal process, n is 1, so the value of the work integral for an isothermal process is:

$$W = -P_1 \upsilon_1 In\left(\frac{P_2}{P_1}\right)$$

When evaluated, the isothermal work is found to be lower than the adiabatic work.

- Polytropic - This model takes into account both a rise in temperature in the gas as well as some loss of energy (heat) to the compressor's components. This assumes that heat may enter or leave the system, and that input shaft work can appear as both increased pressure (usually useful work) and increased temperature above adiabatic (usually losses due to cycle efficiency). Compression efficiency is then the ratio of temperature rise at theoretical 100 percent (adiabatic) vs. actual (polytropic). Polytropic compression will use a value of n between 0 (a constant-pressure process) and infinity (a constant volume process). For the typical case where an effort is made to cool the gas compressed by an approximately adiabatic process, the value of n will be between 1 and k.

Staged Compression

In the case of centrifugal compressors, commercial designs currently do not exceed a compression ratio of more than a 3.5 to 1 in any one stage (for a typical gas). Since compression generates heat, the compressed gas is to be cooled between stages making the compression less adiabatic and more isothermal. The inter-stage coolers typically result in some partial condensation that is removed in vapor-liquid separators. In the case of small reciprocating compressors, the compressor flywheel may drive a cooling fan that directs ambient air across the intercooler of a two or more stage compressor.

Because rotary screw compressors can make use of cooling lubricant to remove the heat of compression, they very often exceed a 9 to 1 compression ratio. For instance, in a typical diving compressor the air is compressed in three stages. If each stage has a compression ratio of 7 to 1, the compressor can output 343 times atmospheric pressure (7 × 7 × 7 = 343 atmospheres). (343 atm/34.8 MPa; 5.04 ksi)

Prime Movers

There are many options for the "prime mover" or motor which powers the compressor:

- gas turbines power the axial and centrifugal flow compressors that are part of jet engines

- steam turbines or water turbines are possible for large compressors
- electric motors are cheap and quiet for static compressors. Small motors suitable for domestic electrical supplies use single phase alternating current. Larger motors can only be used where an industrial electrical three phase alternating current supply is available.
- diesel engines or petrol engines are suitable for portable compressors and support compressors. Common in automobiles and other types of vehicles (including piston-powered aeroplanes, boats, trucks, etc.), diesel or gasoline engines can power compressors using their own crankshaft power (this setup known as a supercharger), or, using their waste exhaust gas to spin a turbine connected to the compressor (this setup known as a turbocharger).

Applications

Gas compressors are used in various applications where either higher pressures or lower volumes of gas are needed:

- in pipeline transport of purified natural gas to move the gas from the production site to the consumer. Often, the compressor in this application is driven by a gas turbine which is fuelled by gas bled from the pipeline. Thus, no external power source is necessary.
- in petroleum refineries, natural gas processing plants, petrochemical and chemical plants, and similar large industrial plants for compressing intermediate and end product gases.
- in refrigeration and air conditioner equipment to move heat from one place to another in refrigerant cycles: see Vapor-compression refrigeration.
- in gas turbine systems to compress the intake combustion air
- in storing purified or manufactured gases in a small volume, high pressure cylinders for medical, welding and other uses.
- in many various industrial, manufacturing and building processes to power all types of pneumatic tools.
- as a medium for transferring energy, such as to power pneumatic equipment.
- in pressurised aircraft to provide a breathable atmosphere of higher than ambient pressure.

- in some types of jet engines (such as turbojets and turbofans) to provide the air required for combustion of the engine fuel. The power to drive the combustion air compressor comes from the jet's own turbines.
- in SCUBA diving, hyperbaric oxygen therapy and other life support devices to store breathing gas in a small volume such as in diving cylinders.
- in surface supplied diving an air compressor is frequently used to supply low pressure air (10 to 20 bar) to the diver for breathing
- in submarines, to store air for later use in displacing water from buoyancy chambers, for adjustment of depth.
- in turbochargers and superchargers to increase the performance of internal combustion engines by increasing mass flow.
- in rail and heavy road transport to provide compressed air for operation of rail vehicle brakes or road vehicle brakes and various other systems (doors, windscreen wipers, engine/ gearbox control, etc.).
- in miscellaneous uses such as providing compressed air for filling pneumatic tires.
- in the case of the fire piston and the heat pump, the desired outcome is the temperature rise of the gas, and compressing the gas is only a means to that end.

Hot Water Storage Tank

A hot water storage tank (also hot water tank, thermal storage tank, hot water thermal storage unit, heat storage tank, hot water cylinder) is a water tank, that is used for storing hot water for space heating or domestic use. A heavily insulated tank can retain heat for days. . Hot water tanks may have a built-in gas or oil burner system, or electric immersion heaters, or may use an external heat exchanger to heat water from another energy source such as district heating, Wood-burning stove, or a district heating system.

Water is a convenient heat storage medium, because it has a high specific heat capacity. Compared with other substances it can store more heat per unit of weight (and volume). Water is non-toxic and low in cost.

Water heaters for washing, bathing, or laundry have thermostat controls to regulate the temperature, in the range of 40 to 60°C, and are connected to the domestic cold water supply. Where the water supply has a high content of dissolved minerals such as limestone, heating the water causes the minerals to precipitate in the tank; a water tank may develop leaks due to corrosion after only a few years. Dissolved oxygen in the water can also accelerate corrosion of the tank and its fittings. A hot water storage tank is wrapped in heat insulation to reduce heat loss and energy consumption.

Solar Hot Water Storage Tank

In a solar heating system, a solar hot water storage tanks stores heat from solar thermal collectors. The tank has a built-in heat-exchanger to heat domestic cold water. These tanks have the same problems with limestone deposit and corrosion, which reduce the efficiency of the heat exchanger.

Hot Water Storage Tank with Closed Water Circuit

Another method to store heat in a hot water storage tank is to heat the water directly in one (or more) heat sources via external heat-exchangers. The heat can later be tapped again to hot water or heating via external heat-exchangers.

The good thing about avoiding sending domestic cold water directly into the hot water storage tank is that only a little limestone is deposited when water is replenished in the closed water circuit.

Another good thing is that the hot water storage tank water only seldom is added domestic cold water with oxygen, which relaxes the materials choices for the hot water storage tank and closed water circuits. The external heat-exchanger used for domestic hot water will have limestone deposits and for that reason need to be exchanged or use some benign descaling agent (temporary pressurised CO_2?) from time to time, but this is cheaper than exchanging the hot water storage tank.

Stratified hot Water Storage Tank with Closed Water Circuit

Another method to store heat in a hot water storage tank is as above, but with some important changes. It have many names: Stratified hot water storage tank with closed water circuit, stratified thermal storage, thermocline tank and water stratified tank storage.

The important changes are as follow. Water mixing within the hot water storage tanks vertical water column must be minimised. These water layers typically has different temperatures. The purpose is to maintain the vertical stratification of the water column. This is achieved by:

- The different heating and cooling loops must send the heated or cooled water in with as low velocity as possible.
- This means that the heating and cooling loops each must have a velocity controlled pump - and the tube ports must have as big inner diametre as is feasible.
- The heat sources hot return water are send into the top - and the heat sources gets cooler water from the bottom.
- Heat usage gets hot water out at the top and cool return water is send in at the bottom.
- There can be "stratification-enhancing"-devices within the hot water storage tank, but if it the water inlet velocity is as low as possible this might not be needed.
- There must be used a more advanced heat control system

When a stratified hot water storage tank has closed water circuits, the water temperatures can be up to 90...95°C at the top - and e.g. 20...40°C at the bottom.

Calm and undisturbed water is a relatively bad heat conductor comparable with glass, bricks, soil.

This is illustrated by a lake, where the surface water is warm but deeper layers are cooler. An arbitrary volume of hot water can be stored, as long as the stratification is kept intact. No vertical metal plates or tubes may be used inside the hot water storage tank, because this would conduct heat between the stratified layers. The hot water storage tank is insulated.

The advantage is that water can be stored at 90...95°C in this hot water storage tank type, giving a higher energy density, and this energy can be stored a long time if the hot water is not diluted.

Optimising water exchanges can allow different levels to be tapped within the tank, matching temperature to the required use.

A practical solar heating system with many parameters can be read as a function of time - incl. the hot water storage tank temperature near the top.

Diving Cylinder

A diving cylinder, scuba tank or diving tank is a gas cylinder used to store and transport high pressure breathing gas as a component of a scuba set. It provides gas to the scuba diver through the demand valve of a diving regulator.

Diving cylinders to be filled at a diving air compressor station

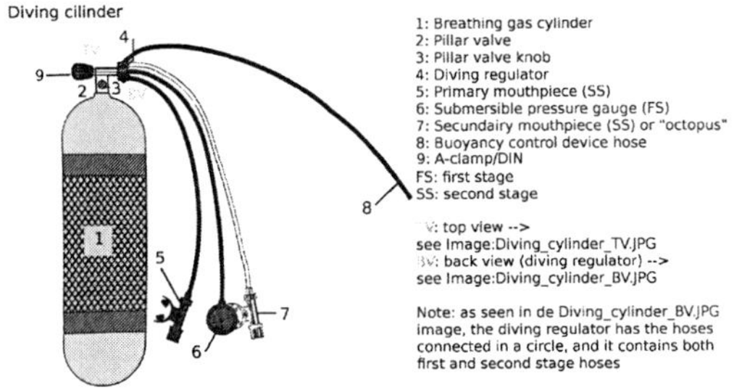

A diving cylinder with its various components, with a two stage open circuit regulator attached

Diving cylinders typically have an internal volume of between 3 and 18 litres (0.11 and 0.64 cu ft) and a maximum pressure rating from 200 to 300 bars (2,900 to 4,400 psi). The internal cylinder volume is also expressed as "water capacity" - the volume of water which could be contained by the cylinder. When pressurised, a cylinder carries a volume of gas greater than its water capacity because gas is compressible. 600 litres (21 cu ft) of gas at atmospheric pressure is compressed into a 3-litre cylinder when it is filled to 200 bar. Cylinders also come in smaller sizes, such as 0.2, 1.5 and 2 litres, however these are not generally used for breathing, instead being used for purposes such as Surface Marker Buoy, drysuit andbuoyancy compensator inflation.

Divers use gas cylinders above water for many purposes including storage of gases for oxygen first aid treatment of diving disorders and as part of storage "banks" for diving air compressor stations. They are also used for many purposes not connected to diving. For these applications they are not diving cylinders.

The term "diving cylinder" tends to be used by gas equipment engineers, manufacturers, support professionals, and divers speaking British English. "Scuba tank" or "diving tank" is more often used colloquially by non-professionals and native speakers of American English. The term "oxygen tank" is commonly used by non-divers when referring to diving cylinders; however, this is a misnomer. These cylinders typically contain (atmospheric) breathing air, or an oxygen-enriched air mix. They rarely contain pure oxygen, except when used for rebreather diving, shallow decompression stops in technical diving or for oxygen therapy. Breathing oxygen at depths greater than 40 feet (12 m) can result in oxygen toxicity, a highly dangerous condition that can trigger seizures and thus lead to drowning.

Parts of a Cylinder

The diving cylinder consists of several parts:

The Pressure Vessel

The pressure vessel is normally made of cold-extruded aluminium or forged steel. An especially common cylinder available at tropical dive resorts is an "aluminium-80" which is an aluminium cylinder with an internal volume of 0.39 cubic feet (11 L) rated to hold about 80 cubic feet (2,300 L) of atmospheric pressure gas at its rated pressure of 3,000 psi (210 bar). Aluminium cylinders are also often used where divers carry many cylinders, such as intechnical diving in warm water where the dive suit does not provide much buoyancy, because the greater buoyancy of aluminium cylinders reduces the extra buoyancy the diver would need to achieve neutral buoyancy. They are also preferred when carried as "sidemount" or "sling" cylinders as the near neutral buoyancy allows them to hang comfortably along the sides of the diver's body, without disturbing trim, and can be handed off to another diver with a minimal effect on buoyancy. In cold water diving, where a diver wearing a highly buoyant thermally insulating dive suit has a large excess of buoyancy, steel cylinders are often used because they are denser than aluminium cylinders. Kevlar wrapped composite cylinders are used in fire fighting breathing apparatus and oxygen first aid equipment, but are rarely used for diving, due to their high positive buoyancy.

The aluminium alloys used for diving cylinders are 6061 and 6351. 6351 alloy is subject to sustained stress cracking and cylinders manufactured of this alloy should be periodically eddy current tested according to national legislation and manufacturer's recommendations.

The neck of the cylinder is internally threaded to fit a cylinder valve. There are several standards for neck threads , these include:

- Taper thread (17E) , which is sealed with thread tape and torqued to between 120 and 150 N.m on steel cylinders, and 75 to 140 N.m on aluminium cylinders
- M25x2 parallel thread, which is sealed by an O-ring and torqued to 100 to 130 N.m on steel, and 95 to 130 N.m on aluminium cylinders
- M18x1.5 parallel thread, which is sealed by an O-ring, and torqued to 100 to 130 N.m on steel cylinders, and 85 to 100 N.m on aluminium cylinders
- 3/4"x14 BSP
- 3/4"x14 NGS (NPSM) sealed by an O-ring, torqued to 40 ft.lbf on aluminium cylinders
- 3/4"x16 UNF, sealed by an O-ring, torqued to 40 ft.lbf on aluminium cylinders

The shoulder of the cylinder carries stamp markings providing required information about the cylinder

The Cylinder Valve

- the pillar valve or cylinder valve is the point at which the pressure vessel connects to the diving regulator. The purpose of the pillar valve is to control gas flow to and from the pressure vessel and to form a seal with the regulator. Some countries require that the pillar valve includes a burst disk, a type of pressure 'fuse', that will fail before the pressure vessel fails in the event of over pressurisation.
- a rubber o-ring forms a seal between the metal of the pillar valve and the metal of the diving regulator. Fluoroelastomer (i.e. "viton") o-rings are used with cylinders storing oxygen-rich gas mixtures to reduce the risk of fire.

Optional Accessories

Types of cylinder valve

- The most commonly used cylinder valve type is the single outlet plain valve, sometimes known as a "K" valve, which allows connection of a single regulator, and has no reserve function. It simply opens to allow gas flow, or closes to shut it off. Several configurations are used, with options of DIN or A-clamp connection, and vertical ir transverse spindle arrangements. The valve is operated by turning a knob, usually rubber or plastic, which affords a comfortable grip. Several turns are required to fully open the valves. Some DIN valves are convertable to A-clamp by use of an insert which is screwed into the opening.
- Y and H cylinder valves have two outlets, each with its own valve, allowing two regulators to be connected to the cylinder. If one regulator "freeflows", which is a common failure mode, or ices up, which can happen in water below about 5°C, its valve can be closed and the cylinder breathed from the regulator connected to the other valve. The difference between an H valve and a Y valve is that the Y valve body splits into two posts roughly 90° to each other and 45° from the vertical axis, looking like a Y, while an H valve is usually assembled from a valve designed as part of a manifold system with an additional valve post connected to the manifold socket, with the valve posts parallel and vertical, which looks a bit like an H.
- Reserve lever or "J-valve" (obsolescent). Until the 1970s, when submersible pressure gauges on regulators came into common use, diving cylinders often used a mechanical reserve mechanism to indicate to the diver that the cylinder was nearly empty. The gas supply was automatically cut-off when the gas pressure reached the reserve pressure. To release the reserve, the diver pulled down on a rod that ran along the side of the tank and which activated a lever on the valve. The diver would then finish the dive before the reserve (typically 300 pounds per square inch (21 bar)) was consumed. On occasion, divers would inadvertently trigger the mechanism while donning gear or performing a movement underwater and, not realising that the reserve had already been accessed, could find themselves out of air at depth with no warning whatsoever. The J-valve got its name from being item number J in one of the first scuba equipment manufacturer catalogues. The

standard non-reserve yoke valve at the time was item K, and is often still referred to as a K-valve. J-valves are still occasionally used by professional divers in zero visibility, where the SPG can not be read.

- Less common in the 1950s thru 1970s was an R-valve which was equipped with a restriction that caused breathing to become difficult as the tank neared exhaustion, but that would allow less restricted breathing if the diver began to ascend and the ambient water pressure lessened, providing a larger pressure differential over the orifice. It was never particularly popular because, were it necessary for the diver to descend (as is often necessary in cave and wreck diving, breathing would become progressively more difficult as the diver went deeper, eventually becoming impossible until the diver could begin his or her ascent.

Manifolds

A cylinder manifold is a tube which connects two cylinders together so that the contents of both can be supplied to one or more regulators. There are three commonly used configurations of manifold:

- The earliest type is a tube with a connector on each end which is attached to the cylinder valve outlet, and an outlet connection in the middle, to which the regulator is attached. A variation on this pattern includes a reserve valve at the outlet connector. The cylinders are isolated from the manifold when closed, and the manifold can be attached or disconnected while the cylinders are pressurised.
- More recently, manifolds have become available which connect the cylinders on the cylinder side of the valve, leaving the outlet connection of the cylinder valve available for connection of a regulator. This means that the connection cannot be made or broken while the cylinders are pressurised, as there is no valve to isolate the manifold from the interior of the cylinder. This apparent inconvenience allows a regulator to be connected to each cylinder, and isolated from the internal pressure independently, which allows a malfunctioning regulator on one cylinder to be isolated while still allowing the regulator on the other cylinder access to all the gas in both cylinders.
- These manifolds may be plain or may include an isolation valve in the manifold, which allows the contents of the cylinders to be isolated from each other. This allows the contents of one

cylinder to be isolated and secured for the diver if a leak at the cylinder neck thread, manifold connection, or burst disk on the other cylinder causes its contents to be lost.

A relatively uncommon manifold system is a connection which screws directly into the neck threads of both cylinders, and has a single valve to release gas to a connector for a regulator. These manifolds can include a reserve valve, either in the main valve or at one cylinder. This system is mainly of historical interest.

Cylinder Bands

Cylinder bands are straps, usually of stainless steel, which are used to clamp two cylinders together as a twin set. The cylinders may be manifolded or independent. It is usual to use a cylinder band near the top of the cylinder, just below the shoulders, and one lower down. The standard distance between centrelines for bolting to a backplate is 11 inches (280 mm).

Cylinder Boot

A cylinder boot is a hard rubber or plastic cover which fits over the base of a diving cylinder to protect the paint from abrasion and impact, to protect the surface the cylinder stands on from impact with the cylinder, and in the case of round bottomed cylinders, to allow the cylinder to stand upright on its base.

Cylinder Net

A cylinder net is a tubular net which is stretched over a cylinder and tied on at top and bottom. The function is to protect the paintwork from scratching, and on booted cylinders it also helps drain the surface between the boot and cylinder, which reduces corrosion problems under the boot. Mesh size is usually about 6 millimetres (0.24 in). Some divers will not use boots or nets as they can snag more easily than a bare cylinder and constitute an entrapment hazard in some environments such as caves and the interior of wrecks.

Cylinder Handle

A cylinder handle may be fitted, usually clamped to the neck, to conveniently carry the cylinder. This can also increase the risk of snagging in an enclosed environment.

Cylinder Capacity

There are two commonly used conventions for describing the capacity of a diving cylinder. One is based on the internal volume of the cylinder. The other is based on nominal volume of gas stored.

Internal Volume

The internal volume is commonly quoted in most countries. It can be measured easily by filling the cylinder with fresh water. This has resulted in the term 'water capacity' (WC) which is often marked on the cylinder shoulder. It's almost always expressed as a volume but sometimes as weight of the water. Fresh water has a density close to one kilogram per litre so the numerical values will be similar.

The usual units are:

- Volume in litres
- Weight in kilograms
- Pressure in bar.

Nominal Volume of Gas Stored

The nominal volume of gas stored is commonly quoted in the USA. It's a measure of the volume of gas that can be released from the cylinder at atmospheric pressure. Terms used for the volume include 'free gas' or 'free gas equivalent'. It depends on the internal volume and the working pressure of a cylinder. If the working pressure is higher, the cylinder will store more gas for the same volume. The working pressure is not neccessarily the same as the actual pressure used. Some cylinders are permitted to exceed the nominal working pressure by 10% and this is indicated by a '+' symbol. This extra pressure allowance is dependant on the cylinder passing the appropriate periodical hydrostatic test and is not generally valid for US cylinders exported to countries with differing standards. For example, common Al80 cylinder is an aluminium cylinder which has a nominal 'free gas' volume of 80 cubic feet (2,300 L) when pressurised to 3,000 pounds per square inch (210 bar). It has an internal volume of 10.94 litres (0.386 cu ft).

Types of Pillar Valve

Figure: A 232 DIN type pillar valve

There are three types of pillar valve in general use for Scuba cylinders containing air:

- A-clamp or yoke - the connection on the regulator surrounds the valve pillar and presses the output O-ring of the pillar valve against the input seat of the regulator. The yoke is screwed down snug by hand (over-tightening can make the yoke impossible to remove later without tools) and the seal is created by pressure when the valve is opened. This type is simple, cheap and very widely used worldwide. It has a maximum pressure rating of 232 bar and the weakest part of the seal, the O-ring, is not well protected from over-pressurisation.
- 232 bar DIN (5-thread, G5/8) - the regulator screws into the pillar valve trapping the O-ring securely. These are more reliable than A-clamps because the O-ring is well protected, but many countries do not use DIN fittings widely on compressors, or cylinders which have DIN fittings, so a European diver with a DIN system abroad in many places will need to take an adaptor.
- 300 bar DIN (7-thread, G5/8) - these are similar to 5-thread DIN fitting but are rated to 300 bar working pressures. The 300 bar pressures are common in European diving and in US cave diving, but their acceptance in U.S. sport diving has been hampered by the fact that United States Department of Transportation rules presently prohibit the transport of metal scuba cylinders on public roads with pressures above about 230 bar, even if the cylinders and air delivery systems have been rated for these pressures by the American agencies which oversee cylinder testing and equipment compatibility for SCUBA (Occupational Safety and Health Administration and Compressed Gas Association).

Adaptors are available to allow connection of DIN regulators to yoke cylinder valves (A-clamp or yoke adaptor), and to connect yoke regulators to DIN cylinder valves. (plug adaptors and block adaptors)

There are also cylinder valves for Scuba cylinders containing gases other than air:

- The new European Norm EN 144-3:2003 introduced a new type of valve, similar to existing 232 bar or 300 bar DIN valves, however, with a metric M 26×2 fitting on both the cylinder and the regulator. These are to be used for breathing gas with oxygen content above that normally found in natural air in the Earth's atmosphere (i.e. 22–100%). From August 2008, these shall be required for all diving equipment used with nitrox or pure oxygen. The idea behind this new standard is to prevent a rich mixture being filled to a cylinder that is not oxygen clean. However even with use of the new system there still remains nothing except human procedural care to ensure that a cylinder with a new valve remains oxygen-clean - which is exactly how the current system works.

- A male thread cylinder valve was supplied with some Dräger semi-closed circuit recreational rebreathers (Dräger Ray) for use with nitrox mixtures.

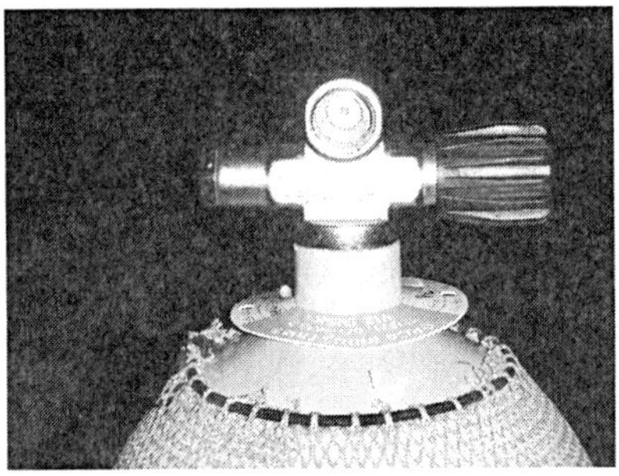

Figure: Left side cylinder valve for barrel seal manifold with blanking plug and DIN connection

Applications and Configurations of Diving Cylinders

Divers may carry one cylinder or multiples, depending on the requirements of the dive. Where diving takes place in low risk areas, where the diver may safely make a free ascent, or where a buddy is available to provide an alternative air supply in an emergency, recreational divers usually carry only one cylinder. An example of this type is coral reef diving where it is possible to do an interesting dive without going deep or needing decompression. Where diving risks are higher, for example where the visibility is low or when recreational divers do deeper or decompression diving, divers routinely carry more than one gas source. An example of this type is north European diving where the temperature is often less than 15 °C(60 °F) and visibility less than 10 m (33 ft) and many interesting dive sites are shipwrecks in deeper water on the sea bed.

Diving cylinders may serve different purposes. One or two cylinders may be used as a primary breathing source which is intended to be breathed from for most of the dive. A smaller cylinder carried in addition to a larger cylinder is called a "pony bottle". A cylinder to be used purely as an independent safety reserve is called a "bailout bottle". A pony bottle is commonly used as a bailout bottle, but this would depend on the time required to surface.

Divers doing technical diving often carry different gases, each in a separate cylinder, for each phase of the dive:

- "travel gas" is used during the descent and ascent. It is typically air or nitrox with an oxygencontent between 21% and 40%. Travel gas is needed when the bottom gas is hypoxic and therefore is unsafe to breathe in shallow water.
- "bottom gas" is only breathed at depth. It is typically a helium-based gas which is low in oxygen (below 21%) or hypoxic (below 17%).
- "deco gas" is used at the decompression stops and is generally a nitrox with a high oxygen content, or pure oxygen, to accelerate decompression.
- a "stage" is a cylinder holding reserve, travel or deco gas. They are usually carried "side slung", clipped on either side of the diver to the harness of the backplate and wing or buoyancy compensator, rather than on the back. Commonly divers use aluminium stage cylinders because they are nearly neutrally buoyant in water and can be removed underwater with less effect on the diver's overall buoyancy.

Rebreathers may use internal cylinders:

- oxygen rebreathers have an oxygen cylinder
- semi-closed circuit rebreathers have a cylinder which usually contains nitrox or a helium based gas.
- closed circuit rebreathers have an oxygen cylinder and a "diluent" cylinder, which contains air, nitrox or a helium based gas.

Rebreathers may also be supplied from "off-board" cylinders, which are not permanently plumbed into the rebreather, but connected to it by a flexible hose and coupling and usually carried side slung. Rebreather divers also often carry a bailout cylinder if the internal diluent cylinder is too small for safe use for bailout.

For safety, divers sometimes carry an additional redundant aqualung (a second scuba tank and scuba valve) to mitigate out-of-air emergencies should the primary breathing source fail. For most common recreational diving (for example dives of 20 m (66 ft) to examine typical coral reefs), such extra equipment is usually not needed or used. This extra cylinder is known as a bailout cylinder, and may be carried in several ways, and can be any size that can hold enough gas to get the diver safely back to the surface.

Open-circuit

For open-circuit divers, there are several options for the combined cylinder and regulator system:

- Single cylinder or single aqualung: consists of a single large cylinder with one first-stage regulator, and usually two secondary regulator/mouthpieces. This configuration is simple and cheap but it is only a single system: it has no redundancy in case of failure. If the cylinder or first-stage regulator fails, the diver is totally out of air and faces an emergency. All training agencies train divers to rely on a buddy to assist them in this situation. The skill of gas sharing is required at the most basic scuba course. This equipment configuration, although common with entry-level divers and for most sport diving, is not recommended for any dive that is deeper than 30 m (100 ft) or where decompression stops are needed, or where there is an overhead environment (wreck diving, cave diving, or ice diving). Generally, these conditions, because they prevent immediate emergency ascent, may define technical diving.
- Single cylinder with dual regulators: consists of a single large cylinder with two first-stage regulators each with a second stage regulator/mouthpiece. This system is used for recreational diving where cold water makes redundancy required. It is common in continental Europe, especially Germany. The advantage is that a regulator failure can be solved under water to bring the dive to a controlled conclusion without buddy breathing or gas sharing. However, it is hard to reach the valves, so there is some reliance on the dive buddy to help close the valves of the free-flowing regulator quickly.
- Main cylinder plus a small independent cylinder: this configuration uses a larger, main cylinder along with an independent smaller cylinder, often called a "pony". The diver has two independent systems, but the total 'breathing system' is now heavier, more expensive to buy and maintain.
- The pony is typically a 2 to 5 litre cylinder. Its capacity determines the depth of dive and decompression duration for which it provides protection. Ponies are generally fixed to the

diver's buoyancy compensator (BC) or main cylinder behind the diver's back. They can also be clipped to the BC at the diver's side or chest. Ponies provided an acceptable emergency supply but are only useful if the diver trains to bail out, i.e. to use one.

- Another type of separate small air source is a hand-held cylinder filled with about 85 litres (0.279 ft) of free air with a diving regulator directly attached, such as the Spare Air. This source provides only a few breaths of gas at depth and is mainly suitable as a shallow water bailout.
- Independent twin set /doubles: this consists of two independent cylinders and two regulators. This system is heavier, more expensive to buy and maintain and more expensive to fill. Also the diver must swap demand valves during dive to preserve a safety reserve of air in each cylinder. If this is not done, then should a cylinder fail the diver may end up having no reserve. Independent twin sets do not work well with air-integrated computers - as they usually only monitor one tank. Many divers feel the complexity of switching regulators periodically to ensure both cylinders are evenly used is offset by the redundancy of two entirely separate breathing supplies. These will normally be mounted as a twin set on the diver's back, but alternatively can be carried in a sidemount configuration where penetration of wrecks or caves requires it.
- Manifolded twin set /doubles with a single regulator: two cylinders are joined at their pillar valves with a manifold but only one regulator is attached to the system. This makes it simple and cheap but means there is no redundant breathing system, only a double gas supply.
- Manifolded twin set /doubles with two regulators: consist of two cylinders with their pillar valves joined with a manifold, with a valve that can isolate the two pillar valves. In the event of a problem with one cylinder the diver may close the isolator valve to preserve gas in the cylinder which has not failed. The pros of this configuration include a large gas supply, no requirement to change regulators underwater, automatic gas supply management, and in most failure situations the diver may close a failed valve or isolate a cylinder in order to leave himself with an emergency supply. On the down side the manifold is another potential point of failure, and there is a danger of losing all air if the manifold valve cannot be closed when a problem occurs. This configuration of cylinders is often

used in Technical diving.

- Sling bottles/cylinders: are a configuration of independent cylinder used for technical diving. They are independent cylinders with their own regulators and as carried clipped to the harness at the side of the diver. Their purpose may be to carry either stage, travel, decompression, or bailout gas while the back mounted cylinder(s) carry bottom gas. Stage cylinders carry gas to extend bottom time, travel gas is used to reach a depth where bottom gas may be safely used if it is hypoxic at the surface, and decompression gas is gas intended to be used during decompression to accelerate the elimination of inert gases. Bailout gas is an emergency supply intended to be used to surface if the main gas supply is lost.
- Side mount cylinders: are sling cylinders mounted at the diver's side which carry bottom gas when the diver does not carry back mount cylinders. They may be used in conjunction with other sling cylinders where necessary.

Closed-circuit

Diving cylinders are used in closed-circuit diving in two roles:

- As part of the rebreather itself. The rebreather must have at least one source of fresh gas stored in a cylinder; many have two and some have more cylinders. Due to the lower gas consumption of rebreathers, these cylinders typically are smaller than those used for equivalent open-circuit dives. See the main article: rebreather.
- In a bail out system: rebreather divers often carry one or more redundant gas sources should the rebreather fail:
- Open-circuit: a simple diving cylinder and regulator. The number of open-circuit bail outs, their capacity and the breathing gases they contain depend on the depth and decompression needs of the dive. So on a deep, technical rebreather dive, the diver will need a bail out "bottom" gas and a bail out "decompression" gas for use. On such a dive, it is the capacity and duration of the bail out that limits the depth and duration of the dive - not the capacity of the rebreather.
- Closed-circuit: a rebreather containing a diving cylinder and regulator. Using another rebreather as a bail out is possible but uncommon. Although the long duration of rebreathers seems compelling for a bail out, rebreathers are relatively bulky,

complex, vulnerable to damage and require more time to start breathing from, than easy-to-use, instantly available, robust and reliable open-circuit equipment.

Gas Calculations

Breathing Gas Endurance

A commonly asked question is 'what is the underwater duration of a particular cylinder?'

There are two parts to this problem:

The cylinder's capacity to store gas

Two features of the cylinder determine its gas carrying capacity:

- working gas pressure : this normally ranges between 200 and 300 bars (2,900 and 4,400 psi)
- internal volume : this normally ranges between 3 litres and 18 litres

To calculate the quantity of gas:

Volume of gas at atmospheric pressure = (cylinder volume) x (cylinder pressure) / (atmospheric pressure)

So a 12 litre cylinder at 232 bar would hold almost 2,784 litres (98.3 cu ft) of air at atmospheric pressure.

In the US and in many diving resorts you might find aluminium cylinders with an internal capacity of 0.39 cubic feet (11 L) filled to 3,000 psi (210 bar); Taking air pressure as 14.7 psi, this gives 0.39 x 3000 / 14.7 = 80 ft^3 These cylinders would be described by US convention as "80 cubic foot cylinders", (the common "aluminium-80") as the US normally refers to cylinder capacity as free-air equivalent at its working pressure, rather than the internal volume of the cylinder, which is the measure commonly used in metric countries.

Up to 200 bar the ideal gas law remains valid and the relationship between the pressure, size of the cylinder and gas contained in the cylinder is linear; at higher pressures there is proportionally less gas in the cylinder. A 3 litre, 300 bar cylinder can only carry up to 810 litres (29 cu ft) of atmospheric pressure gas and not the 900 litres expected from the ideal gas law.

Diver Gas Consumption

There are three factors at work here:

- breathing rate or respiratory minute volume (RMV) of the diver.

In normal conditions this will be between 10 and 25 litres per minute (L/min) for recreational divers who are not working hard. At times of extreme high work rate, breathing rates can rise to 95 L/min.

In the UK, a working breathing rate of 40 litres per minute is used for commercial diving, whilst a figure of 50 litres per minute is used for emergencies. (The Association of Diving Contractors)

- time
- ambient pressure: the depth of the dive determines this. The ambient pressure at the surface is 1 bar (15 psi). For every 10 metres (33 ft) in salt water the diver descends, the pressure increases by 1 bar (15 psi). As a diver goes deeper, the breathing gas is delivered at a pressure equal to ambient water pressure. Thus, it requires twice as much mass of gas to fill the same volume (the diver's lungs) at 10 metres (33 ft) as it does at the surface, and three times as much at 20 metres (66 ft). If a given cylinder consumed at a constant rate would last a diver one hour at the surface, it would last 30 minutes at 10 metres (33 ft), 20 minutes at 20 metres (66 ft) and just 15 minutes at 30 metres (98 ft).

To calculate the quantity of gas consumed:

gas consumed = breathing rate × time × ambient pressure

Thus, a diver with a breathing rate of 20 L/min will consume at 30 metres (4 bar) the equivalent of 80 L/min at 1 bar (e.g. at the surface). If this diver only had a 10 litre 200 bar cylinder to breathe from, the gas in the cylinder would be exhausted after 2000/80 = 25 minutes.

Keeping this in mind, it is not hard to see why technical divers who do long deep dives require multiple cylinders or rebreathers.

Breathing Time

For metric users:

Absolute maximum breathing time (BT) can be calculated as

BT = available air / rate of consumption

which, using the ideal gas law, is

BT = (available cylinder pressure × cylinder volume) / (rate of air consumption at surface) × (ambient pressure)

This may be written as

(1) $BT = \frac{(CP - AP) * CS}{BR * AP}$

with

BT = Breathing Time (in minutes)

CP = Cylinder Pressure (in bars)

CS = Cylinder Size (in liters)

AP = Ambient Pressure (in bars)

BR = Breathing Rate (in liters per minute)

AP is deducted from CP, as the quantity of air represented by AP can in practice not be used for breathing by the diver as she needs it to overcome the pressure of the water (AP) when inhaling.

However, in normal diving usage, a reserve is always factored in. The reserve is a proportion of the cylinder pressure which a diver will not expect to use other than in case of emergency. The reserve may be a quarter or a third of the cylinder pressure or it may be a fixed pressure, common examples are 50 bar and 500 psi. The formula above is then modified to give the usable breathing time as

(2) $BT = \frac{(CP - RP) * CS}{BR * AP}$

where RP is the reserve pressure.

Ambient pressure (AP) is the surrounding water pressure at a given depth and is made up of the sum of the water pressure and the air pressure at the surface. It is calculated as

(3) $AP = \frac{D * g * \rho}{100000} +$ atmospheric pressure

with

D = Depth (in metres)

g = Standard gravity (in metres per second squared)

ñ = Water Density (in kg per cube metre)

In practical terms, this formula can be approximated by

(4) $AP = \frac{D}{10} + 1$

For example (using the first formula (1) for absolute maximum breathing time), a diver at a depth of 15 metres in water with an average density of 1020 kg / m^3 (typical salt water), who breathes at a rate of 20 liters per minute, using a dive cylinder of 18 liters pressurised at 200 bars, can breathe for a period of 72 minutes before the cylinder and supply line pressure has fallen so low as to prevent her from inhaling. In most open circuit scuba systems this happens quite suddenly, from a normal breath to the next abnormal breath, a breath which typically cannot be fully drawn. (There is never any difficulty exhaling). In such circumstances there remains air under pressure in the cylinder, but the diver is unable to breathe it. Some of it can be breathed if the diver ascends, and even without ascent, in some systems a bit of air from the cylinder is available to inflate BCD devices even after it no longer has pressure enough to actuate the mouthpiece valve.

Using the same conditions and a reserve of 50 bar, the formula (2) for usable breathing time is worked thus:

Ambient pressure = water pressure + atmospheric pressure = 15/10 + 1 = 2.5 bar

Usable air = usable pressure * cylinder capacity = (200-50) * 18 = 2700 liters

Rate of consumption = surface air consumption * ambient pressure = 20 * 2.5 = 50 liters/min

Usable breathing time = 2700 liters / 50 liters/min = 54 min

This would give a dive time of 54 min at 15 m before reaching the reserve of 50 bar.

Reserves

It is strongly recommended that a portion of the usable gas of the cylinder be held aside as a safety reserve. The reserve is designed to provide gas for longer than planned decompression stops or to provide time to resolve underwater emergencies.

The size of the reserve depends upon the risks involved during the dive. A deep or decompression dive warrants a greater reserve than a shallow or a no stop dive. In recreational diving for example, it is recommended that the diver plans to surface with a reserve remaining in the cylinder of 500 psi, 50 bar or 25% of the initial capacity, depending of the teaching of the diver training organisation. This is because recreational divers practicing within "no-decompression" limits can normally make a direct ascent in an emergency. On technical dives where a direct ascent is either impossible (due to overhead obstructions) or dangerous (due to the requirement to make decompression stops), divers plan larger margins of safety using the rule of thirds: one third of the gas supply is planned for the outward journey, one third is for the return journey and one third is a safety reserve.

Some training agencies teach the concept of minimum gas and provide a simple calculation that allows a diver to work out an acceptable reserve to get two divers in an emergency to the surface. SeeDIR diving for more information.

Weight of Gas Consumed

The loss of the weight of the gas taken from the cylinder makes the cylinder and diver more buoyant. This can be a problem if the diver is unable to remain neutrally buoyant towards the end of the dive because most of the gas has been breathed from the cylinder.

Table showing the buoyancy of diving cylinders in water when empty and full of air.

Assumes 1 litre of air at atmospheric pressure and 10°C weighs 1.25g.

Cylinder			*Air*		*Weight on land*		*Buoyancy*	
Material	*Volume*	*Pressure*	*Volume*	*Weight*	*Empty*	*Full*	*Empty*	*Full*
	(litre)	(bar)	(litre)	(kg)	(kg)	(kg)	(kg)	(kg)
Steel	12	200	2400	3.0	16.0	19.0	-1.2	-4.3
	15	200	3000	3.8	20.0	23.8	-1.4	-5.2
	16 (XS 130)	230	3680	4.7	19.5	23.9	-0.9	-5.3
	2x7	200	2800	3.5	19.5	23.0	-2.0	-5.6
	8	300	2400	3.0	13.0	16.0	-3.5	-6.5
	10	300	3000	3.8	17.0	20.8	-4.0	-7.8
	2x4	300	2400	3.0	15.0	18.0	-4.0	-7.0
	2x6	300	3600	4.6	21.0	25.6	-5.0	-9.6
Aluminium	9 (AL 63)	203	1826	2.3	12.2	13.5	+1.8	-0.5
	11 (AL 80)	203	2247	2.8	14.4	17.2	+1.8	-1.1
	13 (AL100)	203	2584	3.2	17.1	20.3	+1.4	-1.7

Filling Tanks

Tanks should only be filled with air from diving air compressors or with other breathing gases using gas blending techniques. Both these services should be provided by reliable suppliers such as dive shops. Breathing industrial compressed gases can be lethal because the high pressure increases the effect of any impurities in them.

Special precautions need to be taken with gases other than air:

- oxygen in high concentrations is a major cause of fire and rust.
- oxygen should be very carefully transferred from one tank to another and only ever stored in tanks that are certified and labelled for oxygen use.
- gas mixtures containing proportions of oxygen other than 21% could be extremely dangerous to divers who are unaware of the proportion of oxygen in them. All cylinders should be labelled with their composition.

Contaminated air at depth can be fatal. Common contaminants are: carbon monoxide a by-product of combustion, carbon dioxide a product of metabolism, oil and lubricants from the compressor.

Keeping the cylinder slightly pressurised at all times reduces the possibility of contaminating the inside of the cylinder with corrosive agents, such as sea water, or toxic material, such as oils, poisonous gases, fungi or bacteria.

The blast caused by a sudden release of the gas pressure inside a diving cylinder makes them very dangerous if mismanaged. The greatest risk of explosion exists at filling time and comes from thinning of the walls of the pressure vessel due to corrosion. Another cause of failure is damage or corrosion of the threads and neck of the cylinder where the pillar valve is screwed in. Aluminium cylinders have been observed occasionally to fail explosively, fragmenting the cylinder wall. Steel cylinders usually remain mostly intact, and tend to fail at the neck.

Manufacture and Testing

Most countries require tanks to be checked on a regular basis, see gas cylinder. This usually consists of an internal visual inspection and a hydrostatic test.

- In the United States, a visual inspection is NOT required by the USA DOT every year though they do require a hydrostatic every five years. The visual inspection requirement is a diving industry standard based on observations made during a review by the National Underwater Accident Data Centre.
- In European Union countries a visual inspection is required every 2.5 years, and a hydrostatic every five years.

- In Norway a hydrostatic (including a visual inspection) is required 3 years after production date, then every 2 years.
- Legislation in Australia requires that cylinders are hydrostatically tested every twelve months, regardless.
- In South Africa a hydrostatic test is required every 4 years, and visual inspection every year. Eddy current testing of neck threads must be done according to the manufacturer's recommendations.

A hydrostatic test involves pressurising the cylinder to its test pressure and measuring its volume before and after the test. A permanent increase in volume above the tolerated level means the cylinder fails the test and should be permanently removed from service.

When a cylinder is manufactured, its specification, including Working Pressure, Test Pressure, Data of Manufacture, Capacity and Weight are stamped on the cylinder.

After a cylinder passes the test, the test date, (or the test expiry date in some countries such as Germany), is punched into the shoulder of the tank for easy verification at fill time. Note: this is a European requirement. There is an international standard for the stamp format

Most compressor operators check these details before filling the cylinder and may refuse to fill non-standard or out-of-test cylinders. Note: this is a European requirement, a requirement of the USA DOT, and a South African requirement.

Safety

Before any cylinder is filled, verification of testing dates and a visual examination for external damage and corrosion are required by law in some jurisdictions, and are prudent even if not legally required at other places.

Before use the user should verify the contents of the cylinder and check the function of the cylinder valve. Pressure and gas mixture are critical information for the diver, and the valve should open freely without sticking or leaks from the spindle seals. Sniffing air bled from a cylinder may also reveal unpleasant surprises better left on land than discovered in the water.

Cylinders should not be left standing unattended unless secured so that they can not fall in reasonable foreseeable circumstances as an impact could damage the cylinder valve mechanism, and cocievably fracture the valve at the neck threads. This is more likely with taper thread valves, and when it happens the energy of the compressed gas is released within a second, and can accelerate the cylinder to speeds which can cause severe injury or damage to the surroundings.

A neatly assembled setup, with regulators, gauges, and delicate computers butterflied inside the BCD, or clipped where they will not be walked on, and stowed under the boat bench or secured to a rack, is the practice of a competent diver.

As the scuba set is a life support system, one should not touch a fellow diver's gear, even to move it, without their knowledge and approval.

Full cylinders should not be exposed to temperatures above 65°C and cylinders should not be filled to pressures greater than the developed pressure appropriate to the certified working pressure of the cylinder except by a test station performing a hydrostatic test.

Cylinders should be clearly labelled with their current contents. A generic "Nitrox" or "Trimix" label will alert the user that the contents may not be air, and must be analysed before use. In some parts of the world a label is required specifically indicating that the contents are air, and in other places a colour code without additional labels indicates by default that the contents are air.

Cases of lateral epicondylitis are also reported from the handling of diving cylinders.

Gas cylinder Colour Coding and Labelling

European Union

In the European Union gas cylinders may be colour coded according to EN 1098-3. The "shoulder" is the top of the cylinder close to the pillar valve. For mixed gases, the colours can be either bands or "quarters".

- Air has either a white (RAL 9010) top and black (RAL 9005) band on the shoulder, or white (RAL 9010) and black (RAL 9005) "quartered" shoulders.
- Heliox has either a white (RAL 9010) top and brown (RAL 8008) band on the shoulder, or white (RAL 9010) and brown (RAL 8008) "quartered" shoulders.
- Nitrox, like Air, has either a white (RAL 9010) top and black (RAL 9005) band on the shoulder, or white (RAL 9010) and black (RAL 9005) "quartered" shoulders.
- Pure oxygen has a white shoulder (RAL 9010).
- Pure helium has a brown shoulder (RAL 9008).
- Trimix has a white, black and brown segmented shoulder.

Note: As of the end of 2006, the quartered parts is obsolete, and new cylinders are now with the band, and the old system is repainted.

In the European Union breathing gas cylinders must be labelled with their contents. The label should state the type of breathing gas contained by the cylinder.

South Africa

Scuba cylinders are required to comply with the colours and markings specified in SANS 10019:2006.

- Cylinder colour is Golden yellow with a French grey shoulder.
- Cylinders containing gases other than air or medical oxygen must have a transparent adhesive label stuck on below the shoulder with the word NITROX or TRIMIX in green and the composition of the gas listed.
- Cylinders containing medical oxygen must be black with a white shoulder.

Worldwide

In many recreational diving settings where air and nitrox are the widely used gases, nitrox cylinders are colour-coded with a green stripe on yellow bottom. The normal colour of aluminium diving cylinders is their natural silver. Steel diving cylinders are often painted, to reduce corrosion, mainly yellow or white to increase visibility. In some industrial cylinder identification colour tables, yellow shoulders means chlorine and more generally within Europe it refers to cylinders with Toxic and/or Corrosive contents; but this is of no significance in SCUBA since gas fittings would not be compatible.

Cylinders that are subject to gas blending with pure oxygen also need an "oxygen service certificate" label indicating they have been prepared for use in an oxygen-rich environment.

Diving Chamber

A diving chamber has two main functions:

- as a simpler form of submersible vessel to take divers underwater and to provide a temporary base and retrieval system in the depths;
- as a land or ship-based hyperbaric chamber to artificially reproduce the hyperbaric conditions under the sea (pressures above normal atmospheric pressure) for diving-related and non-diving medical applications such as hyperbaric medicine).

Basic Types of Diving Chamber

There are two basic types of submersible diving chamber differentiated by the way in which the pressure in the diving chamber is produced and controlled.

Open Diving Bell

The historically older open diving chamber, open diving bell or wet bell is in effect a large diving bell, utilising an open bottom, the equivalent of a moon pool, to equalise internal air pressure and external water pressure automatically without the need, necessarily, to measure and control it. An air compressor or bottled compressed air is required to maintain the volume of the air as it becomes compressed with increasing depth, or to make up for oxygen depleted by the occupants' breathing and for carbon dioxide removed from exhaled air by a carbon dioxide scrubber system. This type of diving chamber can only be used underwater, as the internal air pressure is directly proportional to the depth underwater and raising or lowering the chamber is the only way to adjust the pressure.

Hyperbaric Chamber

A sealable diving chamber, closed bell or dry bell is a pressure vessel with hatches large enough for people to enter and exit, and an compressed breathing gas supply to raise the internal air pressure. This type is called a hyperbaric chamber whether used underwater or at the water surface or on land to produce underwater pressures, though some use submersible chamber to mean those used underwater and hyperbaric chamber to mean those used out of water. There are two related terms which reflect particular usages rather than technically-different types:

- Decompression chamber, a hyperbaric chamber used by surface-supplied divers to make their surface decompression stops
- Recompression chamber, a hyperbaric chamber used to treat or prevent decompression sickness.

Early decompression (recompression) chamber in the park at Broome, Western Australia. The chamber was used to treat decompression sickness in Japanese pearl divers who were, generally, of small build. The chamber is now located indoors in the Broome Historical Society Museum.

When used underwater there are two ways to prevent water flooding in when the submersible hyperbaric chamber's hatch is opened. The hatch could open into a moon poolchamber, and then its internal pressure must first be equalised to that of the moon pool chamber. More commonly the hatch opens into an underwater airlock, in which case the main chamber's pressure can stay constant, while it is the airlock pressure which shifts. This common design is called a lock-out chamber, and is used in submarines, submersibles, and underwater habitats as well as diving chambers. Another arrangement utilises a dry airlock between a sealable hyperbaric compartment and an open diving bell compartment (so that effectively the whole structure is a mixture of the two types of diving chamber). When used underwater all types of diving chamber are attached to a diving support vessel by a strongcable for raising and lowering and an umbilical cable delivering compressed air, power, and communications, and all need weights attached or built in to overcome their buoyancy. The greatest depth reached using a cable-suspended chamber is about 1500 m; beyond this the cable becomes unmanageable.

Related Equipment

In addition to the diving bell and hyperbaric chamber, related diving equipment includes the following.

- Underwater habitat: consists of compartments operating under the same principles as diving bells and diving chambers, but fixed to the sea floor for long-term use.
- Submersibles and submarines differ in being able to move under their own power. The interiors are usually maintained at surface pressure, but some examples include air locks and internal hyperbaric chambers.
- There is also other deep diving equipment which has atmospheric internal pressure, including:
- Bathysphere: name given to an experimental deep-sea diving chamber of the 1920s and 1930s.
- Benthoscope: a successor to the bathysphere built to go to greater depths.
- Bathyscaphe: a self-propelled submersible vessel able to adjust its own buoyancy for exploring extreme depths.

Diving Chambers in Use Underwater

As well as transporting divers, a diving chamber carries tools and equipment, breathing gas cylinders to replenish scuba tanks, and communications and emergency equipment. It provides a temporary dry air environment during extended dives for rest, eating meals, carrying out tasks which can't be done underwater, and for emergencies. Diving chambers also act as an underwater base for surface supplied diving operations, with the divers' umbilicals (air supply, etc) attached to the diving chamber rather than to the diving support vessel.

Use of Open Diving Bell Type

Diving bells and open diving chambers of the same principle were more common in the past owing to their simplicity, since they do not necessarily need to monitor, control and mechanically adjust the internal pressure. Secondly since internal air pressure and external water pressure on the bell wall are almost balanced, the chamber does not have to be as strong as a pressurised diving chamber (dry bell). (Actually if h is the distance between a point on the side of the bell and the air/water interface at the bottom, the air pressure at that point is higher than the water on the other side by a water head pressure equivalent to h, but this is a small and constant amount, and is not a structural problem).

A wet diving bell or open diving chamber must be raised slowly to the surface with decompression stops appropriate to the dive profile so that the occupants can avoid decompression sickness. This may take hours, and so limits its use.

Use of Hyperbaric Chambers

Submersible hyperbaric chambers can be brought to the surface without delay to allow divers to decompress since they can maintain the same pressure at which the divers were working. The divers can stay in the chamber on the support vessel to decompress. This flexibility makes them safer to use and more useful in an accident or emergency, including problems affecting the dive support vessel, such as sudden bad weather. They are used to support saturation diving for which the decompression times are very long. A diving chamber based on a pressure vessel is more expensive to construct since it has to withstand very high pressure differentials. These may be both crushing pressures when the chamber is lowered into the sea and the internal pressure is kept less than ambient water pressure, or it may be an outwards pressure when it is out of the water and its internal pressure is set the same as water pressure at a certain depth.

Hyperbaric chambers also require more sophisticated systems to set and control internal gas pressure. However modern manufacturing techniques and control systems have reduced the cost and this type of diving chamber is now more common than the older dive bell type.

Hyperbaric lifeboats are specialised diving chambers or submersibles able to retrieve divers or occupants of diving chambers or underwater habitats in an emergency and to keep them in the required decompression phase. They have airlocks for underwater entry or to form a watertight seal with hatches on the target structure to effect a dry transfer of personnel. Rescuing occupants of submarines or submersibles with internal air pressure of one atmosphere requires being able to withstand the huge pressure differential to effect a dry transfer, and has the advantage of not requiring decompression measures on returning to the surface.

Diving Chambers in Use on Vessels and on Land

Hyperbaric chambers are also used on land and at the ocean surface

- to treat divers for decompression sickness (recompression chambers)
- to take surface supplied divers who have been brought up from underwater through their decompression stops (decompression chambers)
- to train and test divers to adapt to hyperbaric conditions and decompression routines
- to treat people using raised oxygen pressure in hyperbaric oxygen therapy, HBOT
- In saturation diving life support systems
- in scientific research requiring elevated gas pressures.

Hyperbaric chambers designed only for use out of water do not have to resist inward crushing forces, only outward expansion forces. Those for medical applications typically only operate up to two or three atmospheres, while those for diving applications may have to go to six atmospheres and above.

Lightweight portable hyperbaric chambers which can be lifted by helicopter are used by commercial diving operators and rescue services to carry one or more divers requiring hospitalisation.

Decompression Chamber

A decompression chamber is a pressure vessel used in surface supplied diving to allow the divers to complete their decompression stops at the end of a dive on the surface rather than underwater. This eliminates many of the risks of long decompressions underwater, in cold or dangerous conditions.

History

The decompression chamber was invented in 1916 by the Italian engineer Alberto Gianni.

Recompression Chamber

Figure: Recompression chamber

A recompression chamber is a pressure vessel used to treat divers suffering from certain diving disorders such as decompression sickness.

Hyperbaric Treatment Chamber

A hyperbaric oxygen therapy chamber is used in a hospital or sporting context to treat patients whose condition might benefit from hyperbaric oxygen treatment, including divers.

Hyperbaric chambers capable of admitting more than one patient (multiplace) and an inside attendant have advantages for the treatment of decompression sickness and are preferred by the U.S. Coast Guard. Divers with serious complications or injuries may be attended to in this manner during recompression. In addition, multiplace chambers are generally capable of greater depth of recompression, should the need arise. For less seriously ill divers, the more common monoplace chamber is often used. A modified treatment regimen is then used.

- Treatment is ordered by the treating physician (medical diving officer), and is usually in accordance with the U.S. Navy Diving Manual (U.S Navy Diving Tables). Other treatment tables have been developed, including the Catalina Tables, and others, including proprietary tables.
- Test of pressure. If the diagnosis of decompression illness is considered questionable, the diving officer may order a test of

pressure. This typically consists of a recompression to 60 feet seawater for up to 20 minutes. If the diver notes significant improvement in symptoms, or the operator can detect changes in the physical examination, a recompression algorithm is followed.

- Representative diving tables:
- U.S. Navy Table 6. This diving table consists of compression to the depth of 60 feet seawater with the patient on oxygen. The diver is later decompressed to 30 feet on oxygen, then slowly pushed to surface pressure. This table typically takes 4 hours 45 minutes. It may be extended further. It is the most common treatment for type 2 decompression illness.
- U.S. Navy Table 5. This diving table is similar to Table 6 above, but is shorter in duration. It may be used in divers with less severe complaints (type 1 decompression illness).
- U.S. Navy Table 9. This diving table consists of compression to 45 feet seawater with the patient on oxygen, with later decompression to surface pressure. This table is often used by monoplace hyperbaric chambers, or as a follow-up treatment in multiplace chambers.

Fractional Distillation

Fractional distillation is the separation of a mixture into its component parts, or fractions, such as in separating chemical compounds by their boiling point by heating them to a temperature at which several fractions of the compound will evaporate. It is a special type of distillation. Generally the component parts boil at less than 25 °C from each other under a pressure of one atmosphere (atm). If the difference in boiling points is greater than 25 °C, a simple distillation is used.

Laboratory Setup

Fractional distillation in a laboratory makes use of common laboratory glassware and apparatuses, typically including a Bunsen burner, a round-bottomed flask and a condenser, as well as the single-purpose fractionating column.

Apparatus

- heat source, such as a hot plate with a bath, and ideally with a magnetic stirrer.
- distilling flask, typically a round-bottom flask
- receiving flask, often also a round-bottom flask

- fractionating column
- distillation head
- thermometre and adapter if needed
- condenser, such as a Liebig condenser, Graham condenser or Allihn condenser
- vacuum adapter (not used in image to the right)
- boiling chips, also known as anti-bumping granules
- Standard laboratory glassware with ground glass joints, e.g. quickfit apparatus.

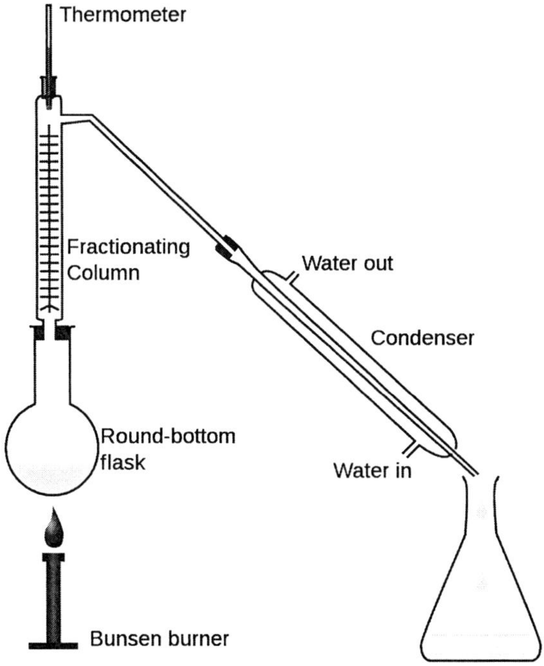

Figure: Fractional distillation apparatus using a Liebig condenser. A conical flask is used as a receiving flask. Here the distillation head and fractionating column are combined in one piece.

Discussion

As an example, consider the distillation of a mixture of water and ethanol. Ethanol boils at 78.4 °C while water boils at 100 °C. So, by heating the mixture, the most volatile component will concentrate to a greater degree in the vapour leaving the liquid. Some mixtures for mazeotropes, where the mixture boils at a lower temperature than either component. In this example, a mixture of 96% ethanol and 4%

water boils at 78.2 °C, being more volatile than pure ethanol. For this reason, ethanol cannot be completely purified by direct fractional distillation of ethanol-water mixtures.

The apparatus is assembled as in the diagram. (The diagram represents a batch apparatus, as opposed to a continuous apparatus.) The mixture is put into the round bottomed flask along with a fewanti-bumping granules (or a Teflon coated magnetic stirrer bar if using magnetic stirring), and the fractionating column is fitted into the top. The fractional distillation column is set up with the heat source at the bottom on the still pot. As the distance from the stillpot of the distillation column increases a heat gradient is formed in the column where it is coolest at top and hottest at the bottom. As the mixed vapour ascends the temperature gradient, some of the vapour condenses and revapourises along the temperature gradient. Each time the vapour condenses and vapourises, the composition of the more volatile liquid in the vapour increases. This distills the vapour along the length of the column, and eventually the vapour is composed solely of the more volatile liquid. The vapour condenses on the glass platforms, known as trays, inside the column, and runs back down into the liquid below, refluxing distillate. The efficiency in terms of the amount of heating and time required to get fractionation can be improved by insulating the outside of the column in an insulator such as wool, aluminium foil or preferably a vacuum jacket. The hottest tray is at the bottom and the coolest is at the top. At steady state conditions, the vapour and liquid on each tray are at equilibrium. Only the most volatile of the vapors stays in gaseous form all the way to the top. The vapour at the top of the column then passes into the condenser, which cools it down until it liquefies. The separation is more pure with the addition of more trays (to a practical limitation of heat, flow, etc.) The condensate that was initially very close to the azeotrope composition becomes gradually richer in water. The process continues until all the ethanol boils out of the mixture. This point can be recognised by the sharp rise in temperature shown on the thermometre.

Typically the example above now only reflects the theoretical way fractionation works. Normal laboratory fractionation columns will be simple glass tubes (often vacuum jacketed, and sometimes internally silvered) filled with a packing, often small glass helices of 4 to 7 mm diametre. Such a column can be calibrated by the distillation of a known mixture system to quantify the column in terms of number of theoretical plates. To improve fractionation the apparatus is set up to return condensate to the column by the use of some sort of reflux splitter (reflux wire, gago, Magnetic swinging bucket, etc.) - a typical careful fractionation would employ a reflux ratio of around 10:1 (10 parts returned condensate to 1 part condensate take off).

In laboratory distillation, several types of condensers are commonly found. The Liebig condenser is simply a straight tube within a water jacket, and is the simplest (and relatively least expensive) form of condenser. The Graham condenser is a spiral tube within a water jacket, and the Allihn condenser has a series of large and small constrictions on the inside tube, each increasing the surface area upon which the vapour constituents may condense.

Alternate set-ups may utilise a "cow" or "pig" which is connected to three or four receiving flasks. By turning the "cow" or "pig", the distillates can be channelled into the appropriate receiver. A Perkin triangle is versatile piece of apparatus that can also be used to collect distillation fractions which does not require a "cow" or "pig" adapter. A Perkin triangle is most often used where the distillates are air-sensitive or where the fractions distill and are collected under reduced pressure, but can be used for a simple and fractional distillation.

Vacuum distillation systems operate at reduced pressure, thereby lowering the boiling points of the materials. Note that the use of anti-bumping granules will not work at reduced pressures.

Industrial Distillation

Distillation is the most common form of separation technology used in petroleum refineries, petrochemical and chemical plants, natural gas processing and cryogenic air separation plants. Fractional distillation is also a crucial step in the processing of "not-from-concentrate" orange juice. In most cases, the distillation is operated at a continuous steady state. New feed is always being added to the distillation column and products are always being removed. Unless the process is disturbed due to changes in feed, heat, ambient temperature, or condensing, the amount of feed being added and the amount of product being removed are normally equal. This is known as continuous, steady-state fractional distillation.

Industrial distillation is typically performed in large, vertical cylindrical columns known as "distillation or fractionation towers" or "distillation columns" with diametres ranging from about 65 centimetres to 6 metres and heights ranging from about 6 metres to 60 metres or more. The distillation towers have liquid outlets at intervals up the column which allow for the withdrawal of different fractions or products having different boiling points or boiling ranges.

By increasing the temperature of the product inside the columns, the different hydrocarbons are separated. The "lightest" products (those with the lowest boiling point) exit from the top of the columns and the "heaviest" products (those with the highest boiling point) exit from the bottom of the column.

For example, fractional distillation is used in oil refineries to separate crude oil into useful substances (or fractions) having different hydrocarbons of different boiling points. The crude oil fractions with higher boiling points:

- have more carbon atoms
- have higher molecular weights
- are more branched chain alkanes
- are darker in colour
- are more viscous
- are more difficult to ignite and to burn

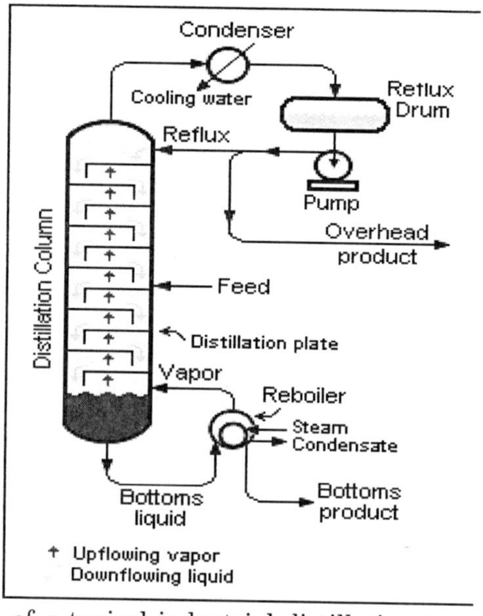

Figure : Diagram of a typical industrial distillation tower

Large-scale industrial towers use reflux to achieve a more complete separation of products. Reflux refers to the portion of the condensed overhead liquid product from a distillation or fractionation tower that is returned to the upper part of the tower as shown in the schematic diagram of a typical, large-scale industrial distillation tower. Inside the tower, the reflux liquid flowing downwards provides the cooling needed to condense the vapours flowing upwards, thereby increasing the effectiveness of the distillation tower. The more reflux is provided for a given number of theoretical plates, the better the tower's separation of lower boiling materials from higher boiling materials. Alternatively, the more reflux provided for a given desired separation, the fewer theoretical plates are required.

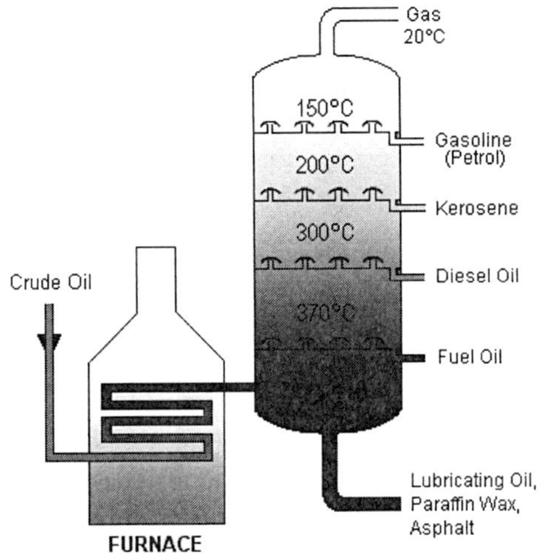

Figure : Crude oil is separated into fractions by fractional distillation. The fractions at the top of the fractionating column have lower boiling points than the fractions at the bottom. The heavy bottom fractions are often cracked into lighter, more useful products. All of the fractions are processed further in other refining units.

Fractional distillation is also used in air separation, producing liquid oxygen, liquid nitrogen, and highly concentrated argon. Distillation of chlorosilanes also enable the production of high-purity silicon for use as a semiconductor.

In industrial uses, sometimes a packing material is used in the column instead of trays, especially when low pressure drops across the column are required, as when operating under vacuum. This packing material can either be random dumped packing (1-3" wide) such as Raschig rings or structured sheet metal. Typical manufacturers are Koch, Sulzer and other companies. Liquids tend to wet the surface of the packing and the vapours pass across this wetted surface, where mass transfer takes place. Unlike conventional tray distillation in which every tray represents a separate point of vapour liquid equilibrium the vapour liquid equilibrium curve in a packed column is continuous. However, when modelling packed columns it is useful to compute a number of "theoretical plates" to denote the separation efficiency of the packed column with respect to more traditional trays. Differently shaped packings have different surface areas and void space between packings. Both of these factors affect packing performance.

Design of Industrial Distillation Columns

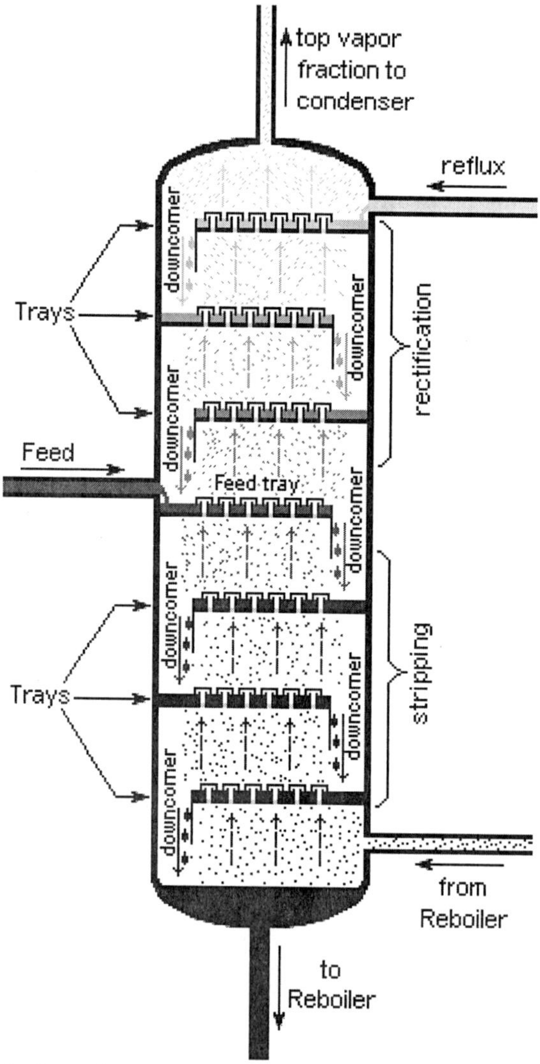

Figure : Chemical engineering schematic of typical bubble-cap trays in a distillation tower

Design and operation of a distillation column depends on the feed and desired products. Given a simple, binary component feed, analytical methods such as the McCabe-Thiele method or the Fenske equation can be used. For a multi-component feed, simulation models are used both for design and operation.

Moreover, the efficiencies of the vapour-liquid contact devices (referred to as plates or trays) used in distillation columns, as seen in Figure, are typically lower than that of a theoretical 100% efficient equilibrium stage. Hence, a distillation column needs more plates than the number of theoretical vapour-liquid equilibrium stages.

An indication of numbers: the separation of two compounds with relative volatility of 1.1 requires a minimum of 130 theoretical plates with a minimum reflux ratio of 20. With a relative volatility of 4, the required number of theoretical plates decreased to 9 with a reflux ratio of 0.66. In another source, a boiling point difference of 30 °C requires 12 theoretical plates and, for a difference of 3 °C, the number of plates increased to 1000.

The reflux ratio is the ratio of the amount of moles returned as refluxed liquid to the fractionating column and the amount of moles of final product, both per unit time.

Autoclave

An autoclave is an instrument used to sterilize equipment and supplies by subjecting them to high pressure saturated steam at 121 °C for around 15–20 minutes depending on the size of the load and the contents. It was invented by Charles Chamberland in 1879, although a precursor known as the steam digester was created by Denis Papin in 1679. The name comes from Greek auto-, ultimately meaning self, and Latin clavis meaning key — a self-locking device.

Uses

Autoclaves are widely used in microbiology, medicine, tattooing, body piercing, veterinary science, mycology, dentistry, chiropody and prosthetics fabrication. They vary in size and function depending on the media to be sterilized.

Typical loads include laboratory glassware, surgical instruments, medical waste, patient care utensils, animal cage bedding, and lysogeny broth.

A notable growing application of autoclaves is the pre-disposal treatment and sterilization of waste material, such as pathogenic hospital waste. Machines in this category largely operate under the same principles as conventional autoclaves in that they are able to neutralise potentially infectious agents by utilising pressurised steam and superheated water. A new generation of waste converters is capable of achieving the same effect without a pressure vessel to sterilize culture media, rubber material, gowns, dressing, gloves, etc. It is particularly useful for materials which cannot withstand the higher temperature of a hot air oven. For all-glass syringes, sterilizing in a hot air oven is a better method.

Autoclaves are also widely used to cure composites and in the vulcanisation of rubber. The high heat and pressure that autoclaves allow help to ensure that the best possible physical properties are repeatably attainable. The aerospace industry and sparmakers (for sailboats in particular) have autoclaves well over 50 feet long, some over 10 feet wide.

Air Removal

It is very important to ensure that all of the trapped air is removed from the autoclave before activation, as hot air is a very poor medium for achieving sterility. Steam at 134 °C can achieve in three minutes the same sterility that hot air at 160 °C takes two hours to achieve. Methods of achieving air removal include:

Downward displacement (or gravity type) - As steam enters the chamber, it fills the upper areas first as it is less dense than air. This compresses the air to the bottom, forcing it out through a drain which often contains a temperature-sensing device. Only when air evacuation is complete does the discharge stop. Flow is usually controlled by a steam trap or a solenoid valve, but bleed holes are sometimes used, often in conjunction with a solenoid valve. As the steam and air mix it is also possible to force out the mixture from locations in the chamber other than the bottom.

Steam pulsing - air dilution by using a series of steam pulses, in which the chamber is alternately pressurised and then depressurised to near atmospheric pressure.

Vacuum pumps - a vacuum pump sucks air or air/steam mixtures from the chamber.

Superatmospheric cycles - achieved with a vacuum pump. It starts with a vacuum followed by a steam pulse followed by a vacuum followed by a steam pulse. The number of pulses depends on the particular autoclave and cycle chosen.

Subatmospheric cycles - similar to the superatmospheric cycles, but chamber pressure never exceeds atmospheric pressure until they pressurise up to the sterilizing temperature.

Autoclaves in Medicine

Figure: Stovetop autoclaves - the simplest of autoclaves

A medical autoclave is a device that uses steam to sterilize equipment and other objects. This means that all bacteria, viruses, fungi, and spores are inactivated. However,prions, like those associated with Creutzfeldt-Jakob disease, may not be destroyed by autoclaving at the typical 134 °C for three minutes or 121 °C for 15 minutes. Also, some recently-discovered organisms, such as Strain 121 microbes, can survive at temperatures above 121 °C.

Autoclaves are found in many medical settings, laboratories, and other places that need to ensure the sterility of an object. Many procedures today employ single-use items rather than sterilizable, reusable items. This first happened with hypodermic needles, but today many surgical instruments (such as forceps, needle holders, and scalpel handles) are commonly single-use rather than reusable items (see waste autoclave).

Because damp heat is used, heat-labile products (such as some plastics) cannot be sterilized this way or they will melt. Paper and other products that may be damaged by steam must also be sterilized another way. In all autoclaves, items should always be separated to allow the steam to penetrate the load evenly.

Autoclaving is often used to sterilize medical waste prior to disposal in the standard municipal solid waste stream. This application has become more common as an alternative to incineration due to environmental and health concerns raised because of the combustion by-products emited by incinerators, especially from the small units which were commonly operated at individual hospitals. Incineration or a similar thermal oxidation process is still generally mandated for pathological waste and other very toxic and/or infectious medical waste.

Autoclave Quality Assurance

There are physical, chemical, and biological indicators that can be used to ensure that an autoclave reaches the correct temperature for the correct amount of time.

Chemical indicators on medical packaging and autoclave tapechange colour once the correct conditions have been met, indicating that the object inside the package, or under the tape, has been appropriately processed. Biological indicators contain spores of a heat-resistant bacterium, Geobacillus stearothermophilus. If the autoclave does not reach the right temperature, the spores will germinate when incubated and their metabolism will change the colour of a pH-sensitive chemical. Some physical indicators consist of an alloy designed to melt only after being subjected to a given temperature for the relevant holding time. If the alloy melts, the change will be visible.

Some computer-controlled autoclaves use an F_0 (F-nought) value to control the sterilization cycle. F_0values are set for the number of minutes of sterilization equivalent to 121 °C (250 °F) at 15 psi (100 kPa) above atmospheric pressure for 15 minutes. Since exact temperature control is difficult, the temperature is monitored, and the sterilization time adjusted accordingly.

Mining

Mining is the extraction of valuable minerals or other geological materials from the earth, from an orebody, veinor (coal) seam. The term also includes the removal of soil. Materials recovered by mining include base metals, precious metals, iron, uranium, coal, diamonds, limestone, oil shale, rock salt and potash. Any material that cannot be grown through agricultural processes, or created artificially in a laboratory or factory, is usually mined. Mining in a wider sense comprises extraction of any non-renewable resource (e.g., petroleum, natural gas, or even water).

Mining of stone and metal has been done since pre-historic times. Modern mining processes involve prospecting for ore bodies, analysis of the profit potential of a proposed mine, extraction of the desired materials and finally reclamation of the land to prepare it for other uses once the mine is closed.

The nature of mining processes creates a potential negative impact on the environment both during the mining operations and for years after the mine is closed. This impact has led to most of the world's nations adopting regulations to moderate the negative effects of mining operations. Safety has long been a concern as well, though modern practices have improved safety in mines significantly.

History

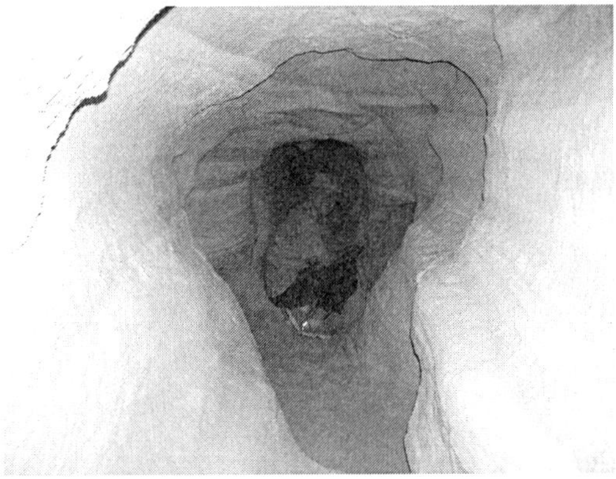

Figure: Chalcolithic copper mine in Timna Valley, Negev Desert, Israel.

Prehistoric Mining: Since the beginning of civilization, people have used stone, ceramics and later, metals found on or close to the Earth's surface.

These were used to manufacture early tools and weapons, for example, high quality flint found in northern France and southern England was used to create flint tools. Flint mines have been found in chalk areas where seams of the stone were followed underground by shafts and galleries. The mines at Grimes Graves are especially famous, and like most other flint mines, are Neolithic in origin (ca 4000 BC-ca 3000 BC). Other hard rocks mined or collected for axes included the greenstone of the Langdale axe industry based in the English Lake District.

The oldest known mine on archaeological record is the "Lion Cave" in Swaziland, which radiocarbon dating shows to be about 43,000 years old. At this site paleolithic humans mined hematite to make the red pigment ochre. Mines of a similar age in Hungary are believed to be sites where Neanderthals may have mined flint for weapons and tools.

Ancient Egypt

Ancient Egyptians mined malachite at Maadi. At first, Egyptians used the bright green malachite stones for ornamentations and pottery. Later, between 2613 and 2494 BC, large building projects required expeditions abroad to the area of Wadi Maghara in order "to secure minerals and other resources not available in Egypt itself." Quarries for turquoise and copper were also found at "Wadi Hamamat, Tura, Aswan and various other Nubian sites" on the Sinai Peninsula and at Timna.

Mining in Egypt occurred in the earliest dynasties. The gold mines of Nubia were among the largest and most extensive of any in Ancient Egypt, and are described by the Greek author Diodorus Siculus. He mentions that fire-setting was one method used to break down the hard rock holding the gold. One of the complexes is shown in one of earliest known maps. They crushed the ore and ground it to a fine powder before washing the powder for the gold dust.

Ancient Greece and Rome

Mining in Europe has a very long history, examples including the silver mines of Laurium, which helped support the Greek city state of Athens. However, it is the Romans who developed large scale mining methods, especially the use of large volumes of water brought to the minehead by numerous aqueducts. The water was used for a variety of purposes, including using it to remove overburden and rock debris, called hydraulic mining, as well as washing comminuted or crushed ores, and driving simple machinery.

The Romans used hydraulic mining methods on a large scale to prospect for the veins of ore, especially a now obsolete form of mining known as hushing. It involved building numerous aqueducts to supply water to the minehead where it was stored in large reservoirs and tanks. When a full tank was opened, the wave of water sluiced away the overburden to expose the bedrock underneath and any gold veins. The rock was then attacked by fire-setting to heat the rock, which would be quenched with a stream of water. The thermal shock cracked the rock, enabling it to be removed, aided by further streams of water from the overhead tanks. They used similar methods to work cassiterite deposits in Cornwall and lead ore in the Pennines.

The methods had been developed by the Romans in Spain in 25 AD to exploit large alluvial gold deposits, the largest site being at Las Medulas, where seven long aqueducts were built to tap local rivers and to sluice the deposits. Spain was one of the most important mining regions, but all regions of the Roman Empire were exploited. They used reverse overshot water-wheels for dewatering their deep mines such as those at Rio Tinto. In Great Britain the natives had mined minerals for millennia , but when the Romans came, the scale of the operations changed dramatically.

The Romans needed what Britain possessed, especially gold, silver, tin and lead. Roman techniques were not limited to surface mining. They followed the ore veins underground once opencast mining was no longer feasible. At Dolaucothi they stoped out the veins, and drove adits through barren rock to drain the stopes. The same adits were also used to ventilate the workings, especially important when fire-setting was used. At other parts of the site, they penetrated the water table and dewatered the mines using several kinds of machine, especially reverse overshot water-wheels. These were used extensively in the copper mines at Rio Tinto in Spain, where one sequence comprised 16 such wheels arranged in pairs, and lifting water about 80 feet (24 m). They were worked as treadmills with miners standing on the top slats. Many examples of such devices have been found in old Roman mines and some examples are now preserved in the British Museum and the National Museum of Wales.

Medieval Europe

Mining as an industry underwent dramatic changes in medieval Europe. The mining industry in the early Middle Ages was mainly focused on the extraction of copper and iron. Other precious metals were also used mainly for gilding or coinage. Initially, many metals were obtained through open-pit mining, and ore was primarily extracted from shallow depths, rather than though the digging of deep mine shafts. Around the 14th century, the demand for weapons, armour, stirrups, and horseshoes greatly increased the demand for iron. Medieval knights for example were often laden with up to 100 pounds of plate or chain link armour in addition to swords, lances and other weapons. The overwhelming dependency on iron for military purposes helped to spur increased iron production and extraction processes.

These new military applications coincided with a population explosion throughout Europe in the 11th-14th centuries which increased the demand for precious metals in order to fill a currency shortage. The silver crisis of 1465 occurred when the mines had all reached depths at which the shafts could no longer be pumped dry with the available technology. Although the increased use of bank notes and the use of credit during this period did decrease the dependence and value of precious metals, these forms of currency still remained vital to the story of mediaeval mining. Use of water power in the form of water mills was extensive; they were employed in crushing ore, raising ore from shafts and ventilating galleries by powering giant bellows. Black powder was first used in mining in Selmecbánya, Kingdom of Hungary in 1627. Black powder allowed blasting of rock and earth to loosen and reveal ore veins. Blasting was much faster than fire-setting and allowed the mining of previously impenetrable metals and ores. In 1762, the world's first mining academy was established in the same town.

The widespread adoption of agricultural innovations such as the iron plowshare, as well as the growing use of metal as a building material, was also a driving force in the tremendous growth of the iron industry during this period. Inventions like the arrastra were often used by the Spanish to pulverize ore after being mined. This device was powered by animals and used the same principles used for grainthreshing.

Much of our knowledge of medieval mining techniques comes from books such as Biringuccio's De la pirotechnia and probably most importantly from Georg Agricola's De re metallica (1556). These books detail many different mining methods used in German and Saxon mines. One of the prime issues confronting mediaeval miners (and one which Agricola explains in detail) was the removal of water from mining shafts. As miners dug deeper to access new veins, flooding became a very real obstacle. The mining industry became dramatically more efficient and prosperous with the invention of mechanical and animal driven pumps.

North and South America

There are ancient, prehistoric coppermines along Lake Superior, and metallic copper was still found there, near the surface, in colonial times. Indians availed themselves of this copper starting at least 5000 years ago," and copper tools, arrowheads, and other artifacts that were part of an extensive native trade network have been discovered. In addition, obsidian, flint, and other minerals were mined, worked, and traded.

While the early French explorers that encountered the sites made no use of the metals due to the difficulties in transporting it, the copper was eventually traded throughout the continent along major river routes. In Saskatchewan, Canada, there also are ancient quartz mines near Waddy Lake and surrounding regions.

Figure: Miners at the Tamarack Mine in Copper Country, Michigan, U.S. in 1905.

In the early colonial history of the Americas, "native gold and silver was quickly expropriated and sent back to Spain in fleets of gold- and silver-laden galleons" mostly from mines in Central and South America. Turquoise dated at 700 A.D. was mined in pre-Columbian America; in the Cerillos Mining District in New Mexico, estimates are that "about 15,000 tons of rock had been removed from Mt Chalchihuitl using stone tools before 1700." Mining in the United States became prevalent in the 19th century, and the General Mining Act of 1872 was passed to encourage mining of federal lands. As with the California Gold Rush in the mid 19th century, mining for minerals and precious metals, along with ranching, was a driving factor in the Westward Expansion to the Pacific coast. With the exploration of the West, mining camps were established and "expressed a distinctive spirit, an enduring legacy to the new nation;" Gold Rushers would experience the same problems as the Land Rushers of the transient West that preceded them. Aided by railroads, many travelled West for work opportunities in mining. Western cities such as Denver and Sacramento originated as mining towns. As new areas were explored, it was usually the gold (placer and then load) and then silver that were taken first, with other metals often waiting for railroads or canals. Gold does not require smelting, is easy to identify and is easily transported.

Mining Methods and Procedures

The process of mining from discovery of an ore body through extraction of minerals and finally to returning the land to its natural state consists of several distinct steps. The first is discovery of the ore body, which is carried out through prospecting or exploration to find and then define the extent, location and value of the ore body. This leads to a mathematical resource estimation to estimate the size and grade of the deposit.

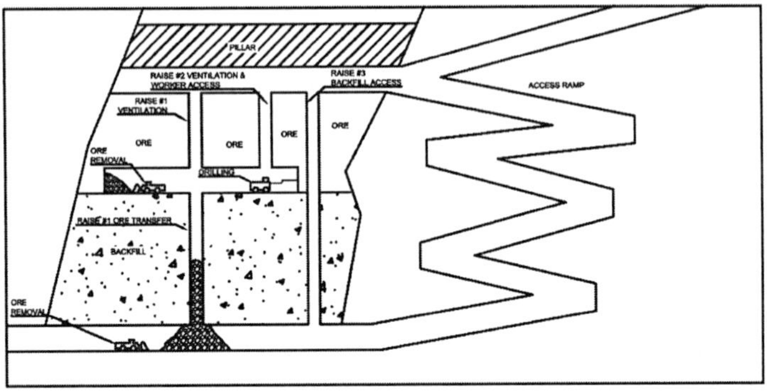

Figure: Schematic of a cut and fill mining operation in hard rock.

This estimation is used to conduct a pre-feasibility study to determine the theoretical economics of the ore deposit. This identifies, early on, whether further investment in estimation and engineering studies is warranted and identifies key risks and areas for further work. The next step is to conduct a feasibility study to evaluate the financial viability, technical and financial risks and robustness of the project.

This is when the mining company makes the decision to develop the mine or to walk away from the project. This includes mine planning to evaluate the economically recoverable portion of the deposit, the metallurgy and ore recoverability, marketability and payability of the ore concentrates, engineering concerns, milling and infrastructure costs, finance and equity requirements and an analysis of the proposed mine from the initial excavation all the way through to reclamation. The proportion of a deposit that is economically recoverable is dependent on the enrichment factor of the ore in the area.

Once the analysis determines a given ore body is worth recovering, development begins to create access to the ore body. The mine buildings and processing plants are built and any necessary equipment is obtained. The operation of the mine to recover the ore begins and continues as long as the company operating the mine finds it economical to do so. Once all the ore that the mine can produce profitably is recovered, reclamation begins to make the land used by the mine suitable for future use.

Mining Techniques

Mining techniques can be divided into two commonexcavation types: surface mining and subsurface (underground) mining. Surface mining is much more common, and produces, for example, 85% of minerals (excluding petroleum and natural gas) in the United States, including 98% of metallic ores.

Targets are divided into two general categories of materials: placer deposits, consisting of valuable minerals contained within river gravels, beach sands, and other unconsolidated materials; and lode deposits, where valuable minerals are found in veins, in layers, or in mineral grains generally distributed throughout a mass of actual rock. Both types of ore deposit, placer or lode, are mined by both surface and underground methods.

Processing of placer ore material consists of gravity-dependent methods of separation, such as sluiceboxes. Only minor shaking or washing may be necessary to disaggregate (unclump) the sands or gravels before processing. Processing of ore from a lode mine, whether it is a surface or subsurface mine, requires that the rock ore be crushed and pulverized before extraction of the valuable minerals begins. After lode ore is crushed, recovery of the valuable minerals is done by one, or a combination of several, mechanical and chemical techniques.

Some mining, including much of the rare earth elements and uranium mining, is done by less-common methods, such as in-situ leaching: this technique involves digging neither at the surface nor underground. The extraction of target minerals by this technique requires that they be soluble, e.g., potash, potassium chloride, sodium chloride, sodium sulfate, which dissolve in water. Some minerals, such as copper minerals and uranium oxide, require acid or carbonate solutions to dissolve.

Surface mining is done by removing (stripping) surface vegetation, dirt, and if necessary, layers of bedrock in order to reach buried ore deposits. Techniques of surface mining include; Open-pit mining which consists of recovery of materials from an open pit in the ground, quarrying or gathering building materials from an open pit mine, strip mining which consists of stripping surface layers off to reveal ore/seams underneath, and mountaintop removal, commonly associated with coal mining, which involves taking the top of a mountain off to reach ore deposits at depth. Most (but not all) placer deposits, because of their shallowly buried nature, are mined by surface methods. Landfill mining, finally, involves sites where landfills are excavated and processed. Subsurface mining consists of digging tunnels or shafts into the earth to reach buried ore deposits.

Ore, for processing, and waste rock, for disposal, are brought to the surface through the tunnels and shafts. Subsurface mining can be classified by the type of access shafts used, the extraction method or the technique used to reach the mineral deposit. Drift mining utilises horizontal access tunnels, slope mining uses diagonally sloping access shafts and shaft mining consists of vertical access shafts. Mining in hard and soft rock formations require different techniques.

Other methods include shrinkage stope mining which is mining upward creating a sloping underground room, long wall mining which is grinding a long ore surface underground and room and pillar which is removing ore from rooms while leaving pillars in place to support the roof of the room. Room and pillar mining often leads to retreat mining which is removing the pillars which support rooms, allowing the room to cave in, loosening more ore. Additional subsurface mining methods include hard rock mining which is mining of hard materials, bore hole mining, drift and fill mining, long hole slope mining, sub level caving and block caving

Machinery

Heavy machinery is needed in mining for exploration and development, to remove and stockpile overburden, to break and remove rocks of various hardness and toughness, to process the ore and for reclamation efforts after the mine is closed. Bulldozers, drills, explosives and trucks are all necessary for excavating the land. In the case of placer mining, unconsolidated gravel, or alluvium, is fed into machinery consisting of a hopper and a shaking screen or trommel which frees the desired minerals from the waste gravel. The minerals are then concentrated using sluices or jigs.

Large drills are used to sink shafts, excavate stopes and obtain samples for analysis. Trams are used to transport miners, minerals and waste. Lifts carry miners into and out of mines, as well as moving rock and ore out, and machinery in and out of underground mines. Huge trucks, shovels and cranes are employed in surface mining to move large quantities of overburden and ore. Processing plants can utilise large crushers, mills, reactors, roasters and other equipment to consolidate the mineral-rich material and extract the desired compounds and metals from the ore.

Extractive Metallurgy

The science of extractive metallurgy is a specialised area in the science of metallurgy that studies the extraction of valuable metals from their ores, especially through chemical or mechanical means. Mineral processing (or mineral dressing) is a specialised area in the science of metallurgy that studies the mechanical means of crushing, grinding, and washing that enable the separation (extractive metallurgy) of valuable metals or minerals from their gangue (waste material). Since most metals are present in ores as oxides or sulfides, the metal needs to be reduced to its metallic form. This can be accomplished through chemical means such as smelting or through electrolytic reduction, as in the case of aluminium. Geometallurgy combines the geologic sciences with extractive metallurgy and mining.

Mining Industry

Mining exists in many countries but Australia and Canada have a reputation for domestic mining expertise, and London is known as the capital of global "mining houses" such as Rio Tinto, BHP Billiton, and Anglo American PLC. The US mining industry is also large but it is dominated by the coal and nonmetal minerals, and the various regulations have worked to reduce the significance of mining in the United States. In 2007 the total market cap of mining companies was reported at US$962 billion, which compares to a total global market cap of publicly traded companies of about US$50 trillion in 2007.

While exploration and mining can sometimes be conducted by individual entrepreneurs or small business, most modern-day mines are large enterprises requiring large amounts of capital to establish. Consequently, the mining sector of the industry is dominated by large, often multinational companies, most of them publicly listed. See Mining Companies for a list. It can be argued that what is referred to as the 'mining industry' is actually two sectors, one specialising in exploration for new resources, the other specialising in mining those resources. The exploration sector is typically made up of individuals and small mineral resource companies ("juniors") dependent on venture capital. The mining sector is typically large and multinational companies sustained by mineral production from their mining operations. In addition to these two sectors, various other industries such as equipment manufacture, environmental testing and metallurgy analysis also rely on and support the mining industry throughout the world. Canadian stock exchanges have a particular focus on mining companies, particularly junior exploration companies through the

TSX Venture Exchange; Canadian companies raise capital on these exchanges and then invest the money in exploration globally. Some have argued that below juniors there exists a substantial sector of illegitimate companies primarily focused on manipulating stock prices.

Mining operations can be grouped into five major categories in terms of their respective resources. These are, oil and gas extraction, coal mining, metal ore mining, nonmetallic mineral mining and quarrying, and support activities for mining. Out of all these categories, oil and gas extraction remains one of the largest in terms of its global economic importance. Prospecting potential mining sites, a vital area of concern for the mining industry is now done using sophisticated new technologies such as seismic prospecting and remote-sensing satellites.

Corporate Classifications

Mining companies can be classified based on their size and financial capabilities:

- Major companies are considered to have an adjusted annual mining-related revenue of more than US$500 million, with the financial capability to develop a major mine on its own.
- Intermediate companies have at least $50 million in annual revenue but less than $500 million.
- Junior companies rely on equity financing as their principal means of funding exploration. Juniors are mainly pure exploration companies, but may also produce minimally, and do not have a revenue of US$50 million.

Safety

Safety has long been a controversial issue in the mining business especially with subsurface mining. While mining today is substantially safer than it was in the previous decades, mining accidents are often very high profile, such as the Quecreek Mine Rescue saving 9 trapped Pennsylvania coal miners in 2002. The Courrières mine disaster, Europe's worst mining accident, caused the death of 1,099 miners (including many children) in Northern France on 10 March 1906. It seems that this disaster was surpassed only by the Benxihu Colliery accident in Chinaon April 26, 1942, which killed 1,549 miners. Government figures indicate that 5,000 Chinese miners die in accidents each year, while other reports have suggested a figure as high as 20,000. Mining ventilation is a significant safety concern for many miners. Poor ventilation of the mines causes exposure to harmful gases, heat and dust inside subsurface mines. These can cause harmful physiological effects, including death. The concentration of methane and other airborne contaminants underground can generally be controlled by dilution (ventilation), capture before entering the host air stream (methane drainage), or isolation (seals and stoppings).

Ignited methane gas is a common source of explosions in coal mines, or, the more violent coal dust explosions. Gases in mines can also poison the workers or displace the oxygen in the mine, causing asphyxiation. For this reason, the MHSA requires that workers have gas detection equipment in groups of miners. It must be able to detect common gases, such as CO, O_2, H_2S, CH_4 and % Lower Explosive Limit. Regulation requires that all production stops if there is a concentrate of 1.4% of flamable gas present. Additionally, further regulation is being requested for more gas detection as newer technology such as nanotechnology is introduced.

High temperatures and humidity may result in heat-related illnesses, including heat stroke which can be fatal. Dusts can cause lung problems, including silicosis, asbestosis and pneumoconiosis (also known as miners lung or black lung disease). A ventilation system is set up to force a stream of air through the working areas of the mine. The air circulation necessary for the effective ventilation of a mine is generated by one or more large mine fans, usually located above ground. Air flows in one direction only, making circuits through the mine such that each main work area constantly receives a supply of fresh air. Watering down in coal mines also helps to keep dust levels down, this is achieved using sprays on the machine and a scrubber fan that filters the dust filled water trapping the dust.

Miners utilise equipment strong enough to break through extremely hard layers of the Earth's crust. This equipment, combined with the closed workspace that underground miners work in, can cause hearing loss. For example, a roof bolter (commonly used by mine roof bolter operators) can reach sound power levels of up to 115 dB. Combined with the reverberant effects of underground mines, a miner without proper hearing protection is at a high risk for hearing loss.

Since mining entails removing dirt and rock from its natural location creating large empty pits, rooms and tunnels, cave-ins are a major concern within mines. Modern techniques for timbering and bracing walls and ceilings within subsurface mines have reduced the number of fatalities due to cave-ins, but accidents still occur. The presence of heavy equipment in confined spaces also poses a risk to miners, and despite modern improvements to safety practices, mining remains dangerous throughout the world.

Abandoned Mines

Figure: Abandoned mine in Nevada.

There are upwards of 560,000 abandoned mines on public and privately owned lands in the United States alone. Abandoned mines may be dangerous to anyone who attempts to explore them without proper knowledge and safety training. Old mines are often dangerous and can contain deadly gases. Standing water in mines from seepage or infiltration poses a significant hazard as the water can hide deep pits and trap gases below the water. Additionally, since weather may have eroded the earth and rock surrounding it, the entrance to an old mine in particular can be very dangerous. Old mine workings, caves, etc. are commonly hazardous simply due to the lack of oxygen in the air, a condition in mines known as blackdamp.

Records

As of 2008, the deepest mine in the world is TauTona in Carletonville, South Africa at 3.9 kilometres, replacing the neighbouring Savuka Mine in the North West Province of South Africa at 3,774 metres. East Rand Mine in Boksburg, South Africa briefly held the record at 3,585 metres, and the first mine declared the deepest in the world was also TauTona when it was at 3,581 metres. The deepest mine in Europe is Bergwerk Saar in Saarland, Germany at 1,750 metres. The second deepest mine in Europe is Pyhäsalmi Mine in Pyhäjärvi, Finland at 1,444 metres. The third deepest mine in Europe is Boulby Mine England at 1,400 metres (shaft depth 1,100 metres).

The deepest open pit mine in the world is Bingham Canyon Mine in Bingham Canyon, Utah, United States at over 1,200 metres. The largest and second deepest open pit copper mine in the world is Chuquicamata in Chuquicamata, Chile at 900 metres, 940,600 tons of copper and 17,700 tons of molybdenum produced annually.

The deepest open pit mine with respect to sea level is Tagebau Hambach in Germany, where the ground of the pit is 293 metres below sea level.

The largest underground mine: El Teniente, in Rancagua, Chile, 2,400 kilometres of underground drifts, 418,000 tons of copper yearly. The deepest borehole in the world is Kola Superdeep Borehole at 12,262 metres. This, however, is not a matter of mining but rather related to scientific drilling.

Chapter 6

Metal Reserves and Recycling

The extraction of metals has been closely linked to economic growth. During the 20th century, the variety of metals uses in society grew rapidly. Today, the development of major nations, such as China and India, and advances in technologies, are fuelling ever more demand. The result is that metal mining activities are expanding, and more and more of the world's metal stocks are above ground in use, rather than below ground as unused reserves. An example is the in-use stock of copper. Between 1932 and 1999, copper in use in the USA rose from 73 g to 238 g per person.

Metals are inherently recyclable, so in principle, can be used over and over again, minimising the negative environmental impacts of mining and saving energy at the same time. For example, 95% of the energy used to make aluminium from bauxite ore is saved by using recycled material. However, levels of metals recycling are generally low. In 2010, the International Resource Panel, hosted by the United Nations Environment Programme (UNEP) published reports on metal stocks that exist within society and their recycling rates.

The report authors observed that the metal stocks in society can serve as huge mines above ground. However, they warned that the recycling rates of some rare metals used in applications such as mobile phones, battery packs for hybrid cars and fuel cells, are so low that unless future end-of-life recycling rates are dramatically stepped up these critical metals will become unavailable for use in modern technology.

Oil Refinery

An oil refinery or petroleum refinery is an industrial process plant where crude oil is processed and refined into more useful petroleum products, such as gasoline, diesel fuel, asphalt base, heating oil, kerosene, and liquefied petroleum gas. Oil refineries are typically large, sprawling industrial complexes with extensive piping running throughout, carrying streams of fluids between large chemical processing units. In many ways, oil refineries use much of the technology of, and can be thought of, as types of chemical plants. The crude oil feedstock has typically been processed by an oil production plant. There is usually an oil depot (tank farm) at or near an oil refinery for storage of bulk liquid products.

An oil refinery is considered an essential part of the downstream side of the petroleum industry.

Operation

Crude oil is separated into fractions by fractional distillation. The fractions at the top of the fractionating column have lower boiling points than the fractions at the bottom. The heavy bottom fractions are often cracked into lighter, more useful products. All of the fractions are processed further in other refining units.

Raw or unprocessed crude oil is not generally useful in industrial applications, although "light, sweet" (low viscosity, low sulfur) crude oil has been used directly as a burner fuel for steam vessel propulsion. The lighter elements, however, form explosive vapours in the fuel tanks and are therefore hazardous, especially in warships. Instead, the hundreds of different hydrocarbon molecules in crude oil are separated in a refinery into components which can be used as fuels, lubricants, and as feedstock in petrochemical processes that manufacture such products as plastics, detergents, solvents, elastomers and fibres such as nylon and polyesters.

Petroleum fossil fuels are burned in internal combustion engines to provide power for ships, automobiles, aircraft engines, lawn mowers, chainsaws, and other machines. Different boiling points allow the hydrocarbons to be separated by distillation. Since the lighter liquid products are in great demand for use in internal combustion engines, a modern refinery will convert heavy hydrocarbons and lighter gaseous elements into these higher value products.

Oil can be used in a variety of ways because it contains hydrocarbons of varying molecular masses, forms and lengths such as paraffins, aromatics, naphthenes (orcycloalkanes), alkenes, dienes, and alkynes. While the molecules in crude oil include different atoms such as sulfur and nitrogen, the hydrocarbons are the most common form of molecules, which are molecules of varying lengths and complexity made of hydrogen and carbonatoms, and a small number of oxygen atoms. The differences in the structure of these molecules account for their varying physical and chemical properties, and it is this variety that makes crude oil useful in a broad range of applications.

Once separated and purified of any contaminants and impurities, the fuel or lubricant can be sold without further processing. Smaller molecules such as isobutane and propylene or butylenes can be recombined to meet specific octane requirements by processes such as alkylation, or less commonly, dimerization.

Octane grade of gasoline can also be improved by catalytic reforming, which involves removing hydrogen from hydrocarbons producing compounds with higher octane ratings such as aromatics. Intermediate products such as gasoils can even be reprocessed to break a heavy, long-chained oil into a lighter short-chained one, by various forms of cracking such as fluid catalytic cracking, thermal cracking, and hydrocracking.

The final step in gasoline production is the blending of fuels with different octane ratings, vapour pressures, and other properties to meet product specifications.

Oil refineries are large scale plants, processing about a hundred thousand to several hundred thousand barrels of crude oil a day. Because of the high capacity, many of the units operate continuously, as opposed to processing in batches, at steady state or nearly steady state for months to years. The high capacity also makes process optimisation and advanced process control very desirable.

Major Products

Petroleum products are usually grouped into three categories: light distillates (LPG, gasoline, naphtha), middle distillates (kerosene, diesel), heavy distillates and residuum (heavy fuel oil, lubricating oils, wax, asphalt). This classification is based on the way crude oil is distilled and separated into fractions (called distillates and residuum) as in the above drawing.

- Liquified petroleum gas (LPG)
- Gasoline (also known as petrol)
- Naphtha
- Kerosene and related jet aircraft fuels
- Diesel fuel
- Fuel oils
- Lubricating oils
- Paraffin wax

- Asphalt and tar
- Petroleum coke

Oil refineries also produce various intermediate products such as hydrogen, light hydrocarbons, reformate and pyrolysis gasoline. These are not usually transported but instead are blended or processed further on-site. Chemical plants are thus often adjacent to oil refineries. For example, light hydrocarbons are steam-cracked in an ethylene plant, and the produced ethylene is polymerised to produce polyethene.

Common Process units Found in a Refinery

- Desalter unit washes out salt from the crude oil before it enters the atmospheric distillation unit.
- Atmospheric distillation unit distills crude oil into fractions. See Continuous distillation.
- Vacuum distillation unit further distills residual bottoms after atmospheric distillation.
- Naphtha hydrotreater unit uses hydrogen to desulfurize naphtha from atmospheric distillation. Must hydrotreat the naphtha before sending to a Catalytic Reformer unit.
- Catalytic reformer unit is used to convert the naphtha-boiling range molecules into higher octanereformate (reformer product). The reformate has higher content of aromatics and cyclic hydrocarbons). An important byproduct of a reformer is hydrogen released during the catalyst reaction. The hydrogen is used either in the hydrotreaters or the hydrocracker.
- Distillate hydrotreater unit desulfurizes distillates (such as diesel) after atmospheric distillation.
- Fluid catalytic cracker (FCC) unit upgrades heavier fractions into lighter, more valuable products.
- Hydrocracker unit uses hydrogen to upgrade heavier fractions into lighter, more valuable products.
- Visbreaking unit upgrades heavy residual oils by thermally cracking them into lighter, more valuable reduced viscosity products.
- Merox unit treats LPG, kerosene or jet fuel by oxidising mercaptans to organic disulfides.
- Coking units (delayed coking, fluid coker, and flexicoker) process very heavy residual oils into gasoline and diesel fuel, leaving petroleum coke as a residual product.

- Alkylation unit produces high-octane component for gasoline blending.
- Dimerization unit converts olefins into higher-octane gasoline blending components. For example, butenes can be dimerized into isooctene which may subsequently be hydrogenated to form isooctane. There are also other uses for dimerization.
- Isomerization unit converts linear molecules to higher-octane branched molecules for blending into gasoline or feed to alkylation units.
- Steam reforming unit produces hydrogen for the hydrotreaters or hydrocracker.
- Liquified gas storage units store propane and similar gaseous fuels at pressure sufficient to maintain them in liquid form. These are usually spherical vessels or bullets (horizontal vessels with rounded ends.
- Storage tanks store crude oil and finished products, usually cylindrical, with some sort of vapour emission control and surrounded by an earthen berm to contain spills.
- Slug catcher used when product (crude oil and gas) that comes from a pipeline with two-phase flow, has to be buffered at the entry of the units.
- Amine gas treater, Claus unit, and tail gas treatment convert hydrogen sulfide from hydrodesulfurization into elemental sulfur.
- Utility units such as cooling towers circulate cooling water, boiler plants generates steam, and instrument air systems include pneumatically operated control valves and an electrical substation.
- Wastewater collection and treating systems consist of API separators, dissolved air flotation (DAF) units and further treatment units such as an activated sludge biotreater to make water suitable for reuse or for disposal.
- Solvent refining units use solvent such as cresol or furfural to remove unwanted, mainly aromatics from lubricating oil stock or diesel stock.
- Solvent dewaxing units remove the heavy waxy constituents petrolatum from vacuum distillation products.

Flow Diagram of Typical Refinery

The image below is a schematic flow diagram of a typical oil refinery that depicts the various unitprocesses and the flow of intermediate product streams that occurs between the inlet crude oil feedstock and the final end products. The diagram depicts only one of the literally hundreds of different oil refinery configurations. The diagram also does not include any of the usual refinery facilities providing utilities such as steam, cooling water, and electric power as well as storage tanks for crude oil feedstock and for intermediate products and end products.

There are many process configurations other than that depicted above. For example, the vacuum distillation unit may also produce fractions that can be refined into endproducts such as: spindle oil used in the textile industry, light machinery oil, motor oil, and steam cylinder oil. As another example, the vacuum residue may be processed in a coker unit to produce petroleum coke.

Speciality end Products

These will blend various feedstocks, mix appropriate additives, provide short term storage, and prepare for bulk loading to trucks, barges, product ships, and railcars.

- Gaseous fuels such as propane, stored and shipped in liquid form under pressure in specialised railcars to distributors.
- Liquid fuels blending (producing automotive and aviation grades of gasoline, kerosene, various aviation turbine fuels, and diesel fuels, adding dyes, detergents, antiknock additives, oxygenates, and anti-fungal compounds as required). Shipped by barge, rail, and tanker ship. May be shipped regionally in dedicated pipelines to point consumers, particularly aviation jet fuel to major airports, or piped to distributors in multi-product pipelines using product separators called pipeline inspection gauges("pigs").
- Lubricants (produces light machine oils, motor oils, and greases, adding viscosity stabilisers as required), usually shipped in bulk to an offsite packaging plant.
- Wax (paraffin), used in the packaging of frozen foods, among others. May be shipped in bulk to a site to prepare as packaged blocks.
- Sulfur (or sulfuric acid), byproducts of sulfur removal from petroleum which may have up to a couple percent sulfur as

organic sulfur-containing compounds. Sulfur and sulfuric acid are useful industrial materials. Sulfuric acid is usually prepared and shipped as the acid precursor oleum.

- Bulk tar shipping for offsite unit packaging for use in tar-and-gravel roofing.
- Asphalt unit. Prepares bulk asphalt for shipment.
- Petroleum coke, used in speciality carbon products or as solid fuel.
- Petrochemicals or petrochemical feedstocks, which are often sent to petrochemical plants for further processing in a variety of ways. The petrochemicals may be olefins or their precursors, or various types of aromatic petrochemicals.

Safety and Environmental Concerns

The refining process releases numerous different chemicals into the atmosphere; consequently, there are substantial air pollution emissions and a notable odor normally accompanies the presence of a refinery. Aside from air pollution impacts there are also wastewater concerns, risks of industrial accidents such as fire and explosion, and noise health effects due to industrial noise.

The public has demanded that many governments place restrictions on contaminants that refineries release, and most refineries have installed the equipment needed to comply with the requirements of the pertinent environmental protection regulatory agencies. In the United States, there is strong pressure to prevent the development of new refineries, and no major refinery has been built in the country since Marathon's Garyville, Louisiana facility in 1976. However, many existing refineries have been expanded during that time. Environmental restrictions and pressure to prevent construction of new refineries may have also contributed to rising fuel prices in the United States. Additionally, many refineries (over 100 since the 1980s) have closed due to obsolescence and/or merger activity within the industry itself. This activity has been reported to Congress and in specialised studies not widely publicised.

Environmental and safety concerns mean that oil refineries are sometimes located some distance away from major urban areas. Nevertheless, there are many instances where refinery operations are close to populated areas and pose health risks such as in the Campo de Gibraltar, a CEPSA refinery near the towns of Gibraltar, Algeciras, La Linea, San Roqueand Los Barrios with a combined population of over 300,000 residents within a 5-mile (8.0 km) radius and the CEPSA refinery in Santa Cruz on the island of Tenerife, Spain which is sited in a densely populated city centre and next to the only two major evacuation routes in and out of the city. In California's Contra Costa County and Solano County, a shoreline necklace of refineries, built in the early 20th century before this area was populated, and associated chemical plants are adjacent to urban areas in Richmond, Martinez, Pacheco, Concord, Pittsburg, Vallejo and Benicia, with occasional accidental events that require "shelter in place" orders to the adjacent populations.

Corrosion Problems and Prevention

Petroleum refineries run as efficiently as possible to reduce costs. One major factor that decreases efficiency is corrosion of the metal components found throughout the process line of the hydrocarbon refining process. Corrosion causes the failure of parts in addition to dictating the cleaning schedule of the refinery, during which the entire production facility must be shut down and cleaned. The cost of corrosion in the petroleum industry has been estimated at US$3.7 billion.

Corrosion occurs in various forms in the refining process, such as pitting corrosion from water droplets, embrittlement from hydrogen, and stress corrosion cracking from sulfide attack. From a materials standpoint, carbon steel is used for upwards of 80% of refinery components, which is beneficial due to its low cost. Carbon steel is resistant to the most common forms of corrosion, particularly from hydrocarbon impurities at temperatures below 205 °C, but other corrosive chemicals and environments prevent its use everywhere. Common replacement materials are low alloy steels containing chromium and molybdenum, with stainless steels containing more chromium dealing with more corrosive environments. More expensive materials commonly used are nickel, titanium, and copper alloys. These are primarily saved for the most problematic areas where extremely high temperatures or very corrosive chemicals are present.

Corrosion is fought by a complex system of monitoring, preventative repairs and careful use of materials. Monitoring methods include both off-line checks taken during maintenance and on-line monitoring. Off-line checks measure corrosion after it has occurred, telling the engineer when equipment must be replaced based on the historical information he has collected. This is referred to as preventative management.

On-line systems are a more modern development, and are revolutionizing the way corrosion is approached. There are several types of on-line corrosion monitoring technologies such as linear polarisation resistance, electrochemical noise and electrical resistance. On-Line monitoring has generally had slow reporting rates in the past (minutes or hours) and been limited by process conditions and sources of error but newer technologies can report rates up to twice per minute with much higher accuracy (referred to as real-time monitoring). This allows process engineers to treat corrosion as another process variable that can be optimised in the system. Immediate responses to process changes allow the control of corrosion mechanisms, so they can be minimised while also maximising production output. In an ideal situation having on-line corrosion information that is accurate and real-time will allow conditions that cause high corrosion rates to be identified and reduced. This is known as predictive management.

Materials methods include selecting the proper material for the application. In areas of minimal corrosion, cheap materials are preferable, but when bad corrosion can occur, more expensive but longer lasting materials should be used. Other materials methods come in the form of protective barriers between corrosive substances and the equipment metals. These can be either a lining of refractory material such as standard Portland cement or other special acid-resistant cements that are shot onto the inner surface of the vessel. Also available are thin overlays of more expensive metals that protect cheaper metal against corrosion without requiring lots of material.

History

The first oil refinery in the world was build in 1851 at Bathgate, Scotland, by Scottish chemist James Young shortly followed by Ignacy £ukasiewicz near Jas3o, Poland from 1854 to 1856, but they were initially small as there was no real demand for refined fuel. As £ukasiewicz's kerosene lampgained popularity, the refining industry grew in the area.

The world's first large refinery opened at Cîmpina, Romania, in 1856-1857, with United States investment. After being taken over by Nazi Germany, the Cîmpina refineries were bombed in Operation Tidal Wave by the Allies during the Oil Campaign of World War II. Another early large refinery is Oljeön, Sweden (1875) (Swedish name means The Petroleum Isle), now preserved as a museum near Engelsberg Ironworks, a UNESCO World Heritage Site, and part of the Ekomuseum Bergslagen.

At one point, the refinery in Ras Tanura, Saudi Arabia owned by Saudi Aramco was claimed to be the largest oil refinery in the world. For most of the 20th century, the largest refinery was the Abadan Refinery in Iran. This refinery suffered extensive damage during the Iran-Iraq war. The world's largest refinery complex is the Jamnagar Refinery Complex, consisting of two refineries side by side operated by Reliance Industries Limited in Jamnagar, India with a combined production capacity of 1,240,000 barrels per day (197,000 m^3/d) (J-1 660,000 bbl/d (105,000 m^3/d), J-2 580,000 bbl/d (92,000 m^3/d). PDVSA's Paraguana refinery complex in Venezuela with a capacity of 956,000 bbl/d (152,000 m^3/d) and SK Energy's Ulsan in South Korea with 840,000 bbl/d (134,000 m^3/d) are the second and third largest, respectively.

Oil refining in the United States

In the 19th century, refineries in the U.S. processed crude oil primarily to recover the kerosene. There was no market for the more volatile fraction, including gasoline, which was considered waste and was often dumped directly into the nearest river. The invention of the automobile shifted the demand to gasoline and diesel, which remain the primary refined products today. Today, national and state legislation requires refineries to meet stringent air and water cleanliness standards. In fact, oil companies in the U.S. perceive obtaining a permit to build a modern refinery to be so difficult and costly that no new refineries have been built (though many have been expanded) in the U.S. since 1976. More than half the refineries that existed in 1981 are now closed due to low utilisation rates and accelerating mergers. As a result of these closures total US refinery capacity fell between 1981 to 1995, though the operating capacity stayed fairly constant in that time period at around 15,000,000 barrels per day (2,400,000 m^3/d). Increases in facility size and improvements in efficiencies have offset much of the lost physical capacity of the industry. In 1982 (the earliest data provided), the United States operate 301 refineries with a combined capacity of 17.9 million barrels (2,850,000 m^3) of crude oil each calendar day. In 2010, there were 149 operable U.S. refineries with a combined capacity of 17.6 million barrels (2,800,000 m^3) per calendar day.

In 2009 through 2010, as revenue streams in the oil business dried up and profitability of oil refineries fell due to lower demand for product and high reserves of supply preceding the economic recession, oil companies began to close or sell refineries. Due to EPA regulations, the costs associated with closing a refinery are very high, meaning that many former refineries are re-purposed.

Petrochemical

Figure: Petrochemical Plant in The Kingdom of Saudi Arabia

Petrochemicals are chemical products derived from petroleum. Some chemical compounds made from petroleum are also obtained from other fossil fuels, such as coalor natural gas, or renewable sources such as corn or sugar cane.

This article focuses on organic compounds that are not burned as fuel (see alsoPetroleum product).

Two petrochemical classes are olefins including ethylene and propylene, and aromatics including benzene, toluene, and xylene isomers. Oil refineries produce olefins and aromatics by fluid catalytic cracking of petroleum fractions. Chemical plants produce olefins by steam cracking of natural gas liquids like ethane and propane. Aromatics are produced by catalytic reforming of naphtha. Olefins and aromatics are the building-blocks for a wide range of materials such as solvents, detergents, and adhesives. Olefins are the basis for polymers and oligomers used in plastics, resins, fibres, elastomers, lubricants and gels.

Global ethylene and propylene production are ~110 million tonnes and ~65 million tonnes per annum, respectively. Aromatics production is ~70 million tonnes. The largest petrochemical industries are located in the USA and Western Europe; however, major growth in new production capacity is in the Middle East and Asia. There is substantial inter-regional petrochemical trade.

Primary petrochemicals are divided into three groups depending on their chemical structure:

- Olefins includes ethylene, propylene, and butadiene. Ethylene and propylene are important sources of industrial chemicals and plastics products. Butadiene is used in making synthetic rubber.
- Aromatics includes benzene, toluene, and xylenes. Benzene is a raw material for dyes and synthetic detergents, and benzene and toluene for isocyanates MDI and TDI used in making polyurethanes. Manufacturers use xylenes to produce plastics and synthetic fibres.
- Synthesis gas is a mixture of carbon monoxide and hydrogen used to make ammonia and methanol. Ammonia is used to make the fertilizer urea and methanol is used as a solvent and chemical intermediate.

The prefix "petro-" is an arbitrary abbreviation of the word "petroleum"; since "petro-" is Ancient Greek for "rock" and "oleum" means "oil". Therefore, the etymologically correct term would be "oleochemicals". However, the term oleochemical is used to describe chemicals derived from plant and animal fats.

List of Significant Petrochemicals and Their Derivatives

The following is a partial list of the major commercial petrochemicals and their derivatives:

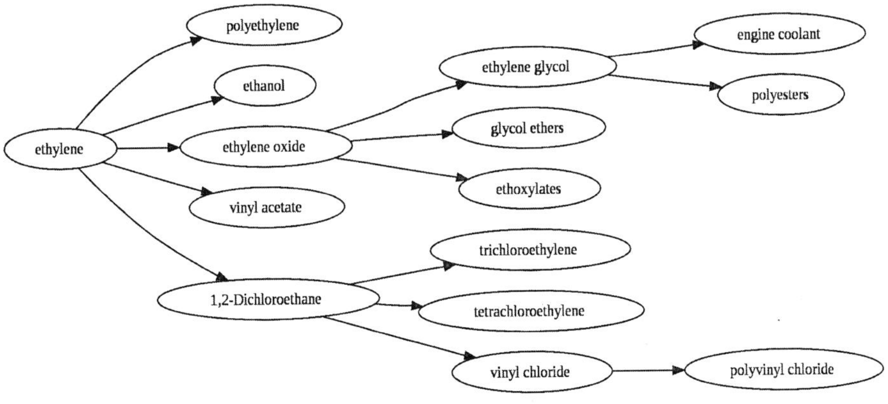

Chemicals produced from ethylene

- ethylene - the simplest olefin; used as a chemical feedstock and ripening hormone
- polyethylene - polymerised ethylene

- ethanol - via ethylene hydration (chemical reaction adding water) of ethylene
- ethylene oxide - via ethylene oxidation
- ethylene glycol - via ethylene oxide hydration
- engine coolant - ethylene glycol, water and inhibitor mixture
- polyesters - any of several polymers with ester linkages in the backbone chain
- glycol ethers - via glycol condensation
- ethoxylates
- vinyl acetate
- 1,2-dichloroethane
- trichloroethylene
- tetrachloroethylene - also called perchloroethylene; used as a dry cleaning solvent and degreaser
- vinyl chloride - monomer for polyvinyl chloride
- polyvinyl chloride (PVC) - type of plastic used for piping, tubing, other things

Chemicals produced from propylene

- propylene - used as a monomer and a chemical feedstock
- isopropyl alcohol - 2-propanol; often used as a solvent or rubbing alcohol
- acrylonitrile - useful as a monomer in forming Orlon, ABS
- polypropylene - polymerised propylene
- propylene oxide
- polyol - used in the production of polyurethanes
- propylene glycol - used in engine coolant and aircraft deicer fluid
- glycol ethers - from condensation of glycols
- acrylic acid
- acrylic polymers
- allyl chloride -
- epichlorohydrin - chloro-oxirane; used in epoxy resin formation
- epoxy resins - a type of polymerising glue from bisphenol A, epichlorohydrin, and someamine
- C4 hydrocarbons - a mixture consisting of butanes, butylenes and butadienes

- isomers of butylene - useful as monomers or co-monomers
- isobutylene - feed for making methyl tert-butyl ether (MTBE) or monomer for copolymerisation with a low percentage of isoprene to make butyl rubber
- 1,3-butadiene (or buta-1,3-diene) - a diene often used as a monomer or co-monomer for polymerisation to elastomers such as polybutadiene, styrene-butadiene rubber, or a plastic such as acrylonitrile-butadiene-styrene (ABS)
- synthetic rubbers - synthetic elastomers made of any one or more of several petrochemical (usually) monomers such as 1,3-butadiene, styrene, isobutylene, isoprene, chloroprene; elastomeric polymers are often made with a high percentage of conjugated diene monomers such as 1,3-butadiene, isoprene, or chloroprene
- higher olefins
- polyolefins such poly-alpha-olefins, which are used as lubricants
- alpha-olefins - used as monomers, co-monomers, and other chemical precursors. For example, a small amount of 1-hexene can be copolymerised with ethylene into a more flexible form of polyethylene.
- other higher olefins
- detergent alcohols

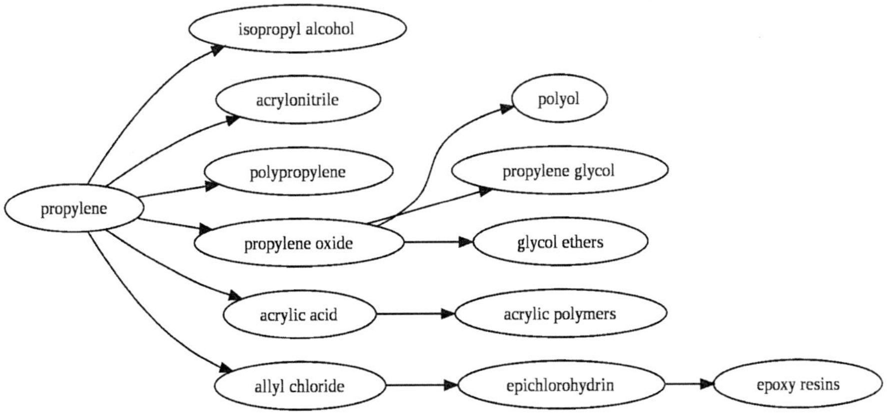

Chemicals produced from benzene

- benzene - the simplest aromatic hydrocarbon
- ethylbenzene - made from benzene and ethylene
- styrene made by dehydrogenation of ethylbenzene; used as a monomer

- polystyrenes - polymers with styrene as a monomer
- cumene - isopropylbenzene; a feedstock in the cumene process
- phenol - hydroxybenzene; often made by the cumene process
- acetone - dimethyl ketone; also often made by the cumene process
- bisphenol A - a type of "double" phenol used in polymerisation in epoxy resins and making a common type of polycarbonate
- epoxy resins - a type of polymerising glue from bisphenol A, epichlorohydrin, and someamine
- polycarbonate - a plastic polymer made from bisphenol A and phosgene (carbonyl dichloride)
- solvents - liquids used for dissolving materials; examples often made from petrochemicals include ethanol, isopropyl alcohol, acetone, benzene, toluene, xylenes
- cyclohexane - a 6-carbon aliphatic cyclic hydrocarbon sometimes used as a non-polar solvent
- adipic acid - a 6-carbon dicarboxylic acid, which can be a precursor used as a co-monomer together with a diamine to form an alternating copolymer form of nylon.
- nylons - types of polyamides, some are alternating copolymers formed from copolymerising dicarboxylic acid or derivatives with diamines
- caprolactam - a 6-carbon cyclic amide
- nylons - types of polyamides, some are from polymerising caprolactam
- nitrobenzene - can be made by single nitration of benzene
- aniline - aminobenzene
- methylene diphenyl diisocyanate (MDI) - used as a co-monomer with diols or polyols to form polyurethanes or with di- or polyamines to form polyureas
- polyurethanes
- alkylbenzene - a general type of aromatic hydrocarbon, which can be used as a presursor for a sulfonate surfactant (detergent)
- detergents - often include surfactants types such as alkylbenzenesulfonates and nonylphenol ethoxylates
- chlorobenzene

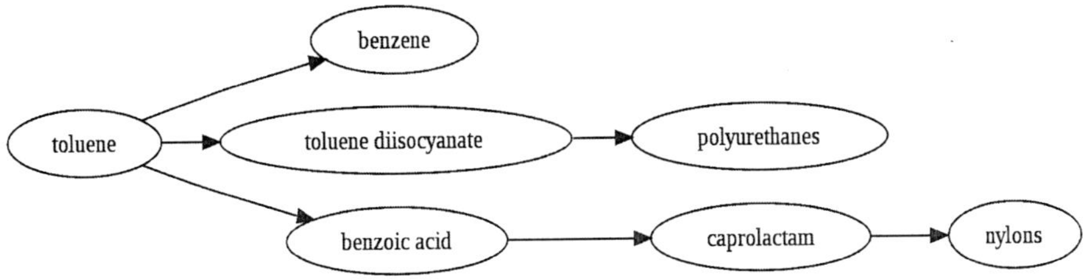

Chemicals produced from toluene

- toluene- methylbenzene; can be a solvent or precursor for other chemicals
- benzene
- toluene diisocyanate (TDI) - used as co-monomers with diols or polyols to form polyurethanes or with di- or polyamines to form polyureas
- polyurethanes - a polymer formed from diisocyanates and diols or polyols
- benzoic acid - carboxybenzene
- caprolactam
- nylon

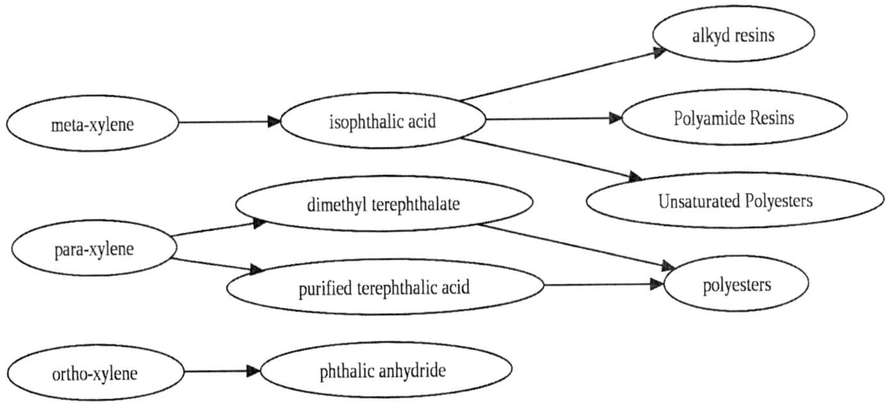

Chemicals produced from xylenes

- mixed xylenes - any of three dimethylbenzene isomers, could be a solvent but more often precursor chemicals
- ortho-xylene - both methyl groups can be oxidised to form (ortho-)phthalic acid
- phthalic anhydride
- para-xylene - both methyl groups can be oxidised to form terephthalic acid

- dimethyl terephthalate - can be copolymerised to form certain polyesters
- polyesters - although there can be many types, polyethylene terephthalate is made from petrochemical products and is very widely used.
- purified terephthalic acid - often copolymerised to form polyethylene terephthalate
- polyesters
- meta-xylene
- isophthalic acid
- alkyd resins
- Polyamide Resins
- Unsaturated Polyesters

Nuclear Reactor

Figure: Core of CROCUS, a small nuclear reactor used for research at the EPFL inSwitzerland

This article is a subarticle of Nuclear power.

A nuclear reactor is a device to initiate and control a sustained nuclear chain reaction. Most commonly they are used for generating electricity and for the propulsion of ships. Usually heat from nuclear fission is passed to a working fluid (water or gas), which runs through turbines that power either ship's propellers or generators. Some produce isotopes for medical and industrial use, and some are run only for research.

How it Works

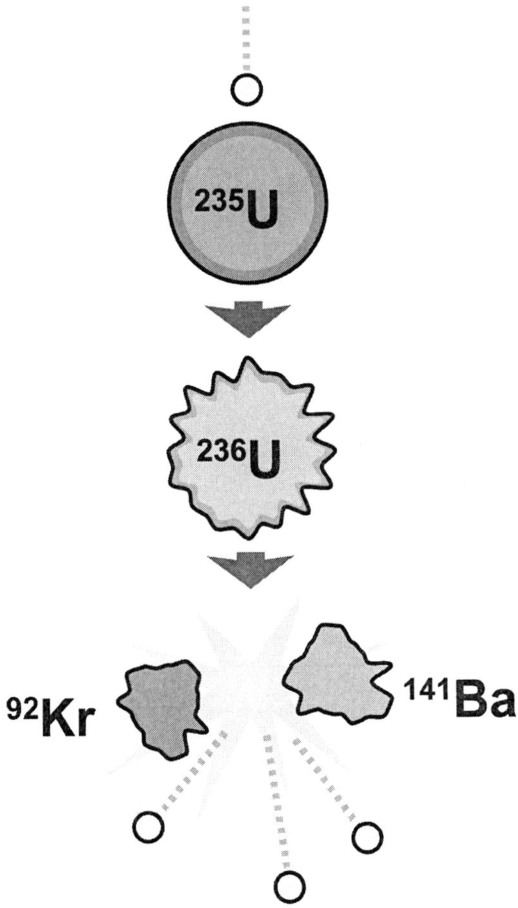

Figure: An induced nuclear fission event. A neutron is absorbed by the nucleus of a uranium-235 atom, which in turn splits into fast-moving lighter elements (fission products) and free neutrons. Though both reactors and nuclear weapons rely on nuclear chain reactions, the rate of reactions in a reactor is much slower than in a bomb.

Just as conventional power stations generate electricity by harnessing the thermal energy released from burning fossil fuels, nuclear reactors convert the thermal energy released from nuclear fission.

Fission

When a large fissile atomic nucleus such as uranium-235 or plutonium-239 absorbs a neutron, it may undergo nuclear fission. The heavy nucleus splits into two or more lighter nuclei (the fission products), releasing kinetic energy, gamma radiation and free neutrons. A portion of these neutrons may later be absorbed by other fissile atoms and trigger further fission events, which release more neutrons, and so on. This is known as a nuclear chain reaction.

This nuclear chain reaction can be controlled by using neutron poisons and neutron moderators to change the portion of neutrons that will go on to cause more fissions. Nuclear reactors generally have automatic and manual systems to shut the fission reaction down if unsafe conditions are detected.

Commonly used moderators include regular (light) water (75% of the world's reactors), solid graphite(20% of reactors) and heavy water (5% of reactors). Beryllium has also been used in some experimental types, and hydrocarbons have been suggested as another possibility.

Heat Generation

The reactor core generates heat in a number of ways:

- The kinetic energy of fission products is converted to thermal energy when these nuclei collide with nearby atoms.
- Some of the gamma rays produced during fission are absorbed by the reactor, their energy being converted to heat.
- Heat produced by the radioactive decay of fission products and materials that have been activated by neutron absorption. This decay heat source will remain for some time even after the reactor is shut down.

A kilogram of uranium-235 (U-235) converted via nuclear processes releases approximately three million times more energy than a kilogram of coal burned conventionally (7.2×10^{13} joules per kilogram of uranium-235 versus 2.4×10^{7} joules per kilogram of coal).

Cooling

A nuclear reactor coolant — usually water but sometimes a gas or a liquid metal or molten salt — is circulated past the reactor core to absorb the heat that it generates. The heat is carried away from the reactor and is then used to generate steam. Most reactor systems employ a cooling system that is physically separated from the water that will be boiled to produce pressurised steam for the turbines, like the pressurised water reactor. But in some reactors the water for the steam turbines is boiled directly by the reactor core, for example the boiling water reactor.

Reactivity Control

Main articles: Nuclear reactor control, Passive nuclear safety, Delayed neutron, Iodine pit, SCRAM, and Decay heat

The power output of the reactor is adjusted by controlling how many neutrons are able to create more fissions.

Control rods that are made of a neutron poison are used to absorb neutrons. Absorbing more neutrons in a control rod means that there are fewer neutrons available to cause fission, so pushing the control rod deeper into the reactor will reduce its power output, and extracting the control rod will increase it.

At the first level of control in all nuclear reactors, a process of delayed neutron emission by a number of neutron-rich fission isotopes is an important physical process. These delayed neutrons account for about 0.65% of the total neutrons produced in fission, with the remainder (termed "prompt neutrons") released immediately upon fission. The fission products which produce delayed neutrons have half lives for their decay by neutron emission that range from milliseconds to as long as several minutes. Keeping the reactor in the zone of chain-reactivity where delayed neutrons are necessary to achieve a critical mass state, allows time for mechanical devices or human operators to have time to control a chain reaction in "real time"; otherwise the time between achievement of criticality and nuclear meltdown as a result of an exponential power surge from the normal nuclear chain reaction, would be too short to allow for intervention.

In some reactors, the coolant also acts as a neutron moderator. A moderator increases the power of the reactor by causing the fast neutrons that are released from fission to lose energy and become thermal neutrons. Thermal neutrons are more likely than fast neutrons to cause fission, so more neutron moderation means more power output from the reactors. If the coolant is a moderator, then temperature changes can affect the density of the coolant/moderator and therefore change power output. A higher temperature coolant would be less dense, and therefore a less effective moderator.

In other reactors the coolant acts as a poison by absorbing neutrons in the same way that the control rods do. In these reactors power output can be increased by heating the coolant, which makes it a less dense poison. Nuclear reactors generally have automatic and manual systems to Scram the reactor in an emergency shut down. These systems insert large amounts of poison (often boron in the form of boric acid) into the reactor to shut the fission reaction down if unsafe conditions are detected or anticipated.

Most types of reactors are sensitive to a process variously known as xenon poisoning, or the iodine pit. Xenon-135 produced in the fission process acts as a "neutron poison" that absorbs neutrons and therefore tends to shut the reactor down. Xenon-135 accumulation can be controlled by keeping power levels high enough to destroy it as fast as it is produced. Fission also produces iodine-135, which in turn decays with a half life of under seven hours, to new xenon-135. When the reactor is shut down, iodine-135 continues to decay to xenon-135, making restarting the reactor more difficult for a day or two. This temporary state is the "iodine pit." If the reactor has sufficient extra reactivity capacity, it can be restarted. As the extra xenon-135 is transmuted to xenon-136 which is not a neutron poison, within a few hours the reactor experiences a "xenon burnoff (power) transient". Control rods must be further inserted to replace the neutron absorption of the lost xenon-135. Failure to properly follow such a procedure was a key step in the Chernobyl disaster.

Reactors used in nuclear marine propulsion (especially nuclear submarines) often cannot be run at continuous power around the clock in the same way that land-based power reactors are normally run, and in addition often need to have a very long core life without refuelling. For this reason many designs use highly enriched uranium but incorporate burnable neutron poison directly into the fuel rods. This allows the reactor to be constructed with a high excess of fissionable material, which is nevertheless made relatively more safe early in the reactor's fuel burn-cycle by the presence of the neutron-absorbing material which is later replaced by naturally produced long-lived neutron poisons (far longer-lived than xenon-135) which gradually accumulate over the fuel load's operating life.

Electrical Power Generation

The energy released in the fission process generates heat, some of which can be converted into usable energy. A common method of harnessing this thermal energy is to use it to boil water to produce pressurised steam which will then drive a steam turbine that generates electricity.

Early Reactors

The neutron was discovered in 1932. The concept of a nuclear chain reaction brought about by nuclear reactions mediated by neutrons, was first realised shortly thereafter, by Hungarian scientist Leó Szilárd, in 1933. He filed a patent for his idea of a simple nuclear reactor the following year while working at the Admiralty in London. However, Szilárd's idea did not incorporate the idea of nuclear fission as a neutron source, since that process was not yet discovered. Szilárd's ideas for nuclear reactors using neutron-mediated nuclear chain reactions in light elements proved unworkable.

Inspiration for a new type of reactor using uranium came from the discovery by Lise Meitner, Fritz Strassman and Otto Hahn in 1938 that bombardment of uranium with neutrons (provided by an alpha-on-beryllium fusion reaction, a "neutron howitzer") produced a barium residue, which they reasoned was created by the fissioning of the uranium nuclei. Subsequent studies in early 1939 (one of them by Szilárd and Fermi) revealed that several neutrons were also released during the fissioning, making available the opportunity for the nuclear chain reaction that Szilárd had envisioned six years previously.

On August 2, 1939 Albert Einstein signed a letter to President Franklin D. Roosevelt (written by Szilard) suggesting that the discovery of uranium's fission could lead to the development of "extremely powerful bombs of a new type", giving impetus to the study of reactors and fission. Szilárd and Einstein knew each other well and had worked together years previously, but Einstein had never thought about this possibility for nuclear energy until Szilard reported it to him, at the beginning of his quest to produce the Einstein-Szilard letter to alert the U.S. government.

Shortly after, Hitler's Germany invaded Poland in 1939, starting World War II in Europe. The U.S. was not yet officially at war, but in October, when the Einstein-Szilard letter was delivered to Roosevelt, he commented that the purpose of doing the research was to make sure "the Nazis don't blow us up." The U.S. nuclear project followed, although with some delay as there remained skepticism (some of it from Fermi) and also little action from the small number of officials in the government who were initially charged with moving the project forward.

The following year the U.S. Government received the Frisch–Peierls memorandum from the UK, which stated that the amount of uranium needed for a chain reaction was far lower than had previously been thought. The memorandum was a product of the MAUD Committee, which was working on the UK atomic bomb project, known as Tube Alloys, later to be subsumed within the Manhattan Project.

Eventually, the first artificial nuclear reactor, Chicago Pile-1, was constructed at the University of Chicago, by a team led by Enrico Fermi, in late 1942.

By this time, the program had been pressured for a year by U.S. entry into the war. The Chicago Pile achieved criticality on December 2, 1942 at 3:25 PM. The reactor support structure was made of wood, which supported a pile (hence the name) of graphite blocks, embedded in which was natural uranium-oxide 'pseudospheres' or 'briquettes'.

Soon after the Chicago Pile, the U.S. military developed a number of nuclear reactors for the Manhattan Project starting in 1943. The primary purpose for the largest reactors (located at the Hanford Site in Washington state), was the mass production of plutonium for nuclear weapons. Fermi and Szilard applied for a patent on reactors on 19 December 1944. Its issuance was delayed for 10 years because of wartime secrecy.

"World's first nuclear power plant" is the claim made by signs at the site of the EBR-I, which is now a museum near Arco, Idaho. This experimental LMFBR operated by the U.S. Atomic Energy Commission produced 0.8 kW in a test on December 20, 1951 and 100 kW (electrical) the following day, having a design output of 200 kW (electrical).

Besides the military uses of nuclear reactors, there were political reasons to pursue civilian use of atomic energy. U.S. President Dwight Eisenhower made his famous Atoms for Peace speech to the UN General Assembly on December 8, 1953. This diplomacy led to the dissemination of reactor technology to U.S. institutions and worldwide.

The first nuclear power plant built for civil purposes was the AM-1 Obninsk Nuclear Power Plant, launched on June 27, 1954 in the Soviet Union. It produced around 5 MW (electrical).

After World War II, the U.S. military sought other uses for nuclear reactor technology. Research by the Army and the Air Force never came to fruition; however, the U.S. Navy succeeded when they steamed the USS Nautilus (SSN-571) on nuclear power January 17, 1955.

The first commercial nuclear power station, Calder Hall in Sellafield, England was opened in 1956 with an initial capacity of 50 MW (later 200 MW).

The first portable nuclear reactor "Alco PM-2A" used to generate electrical power (2 MW) for Camp Century from 1960.

Components

Figure: The control room of NC State's Pulstar Nuclear Reactor.

The key components common to most types of nuclear power plants are:

- Nuclear fuel
- Nuclear reactor core
- Neutron moderator
- Neutron poison
- Coolant (often the Neutron Moderator and the Coolant are the same, usually both purified water)
- Control rods
- Reactor vessel
- Boiler feedwater pump
- Steam generators (not in BWRs)
- Steam turbine
- Electrical generator
- Condenser
- Cooling tower (not always required)

- Radwaste System (a section of the plant handling radioactive waste)
- Refuelling Floor
- Spent fuel pool
- Nuclear safety systems
  - Reactor Protective System (RPS)
  - Emergency Diesel Generators
  - Emergency Core Cooling Systems (ECCS)
  - Standby Liquid Control System (emergency boron injection, in BWRs only)
- Essential service water system (ESWS)
- Containment building
- Control room
- Emergency Operations Facility
- Nuclear training facility (usually contains a Control Room simulator)

Reactor Types

Figure: NC State's PULSTAR Reactor is a 1 MW pool-type research reactor with 4% enriched, pin-type fuel consisting of UO_2pellets in zircaloy cladding.

Classifications

Nuclear Reactors are classified by several methods; a brief outline of these classification methods is provided.

Classification by type of nuclear reaction:

- Nuclear fission. All commercial power reactors are based on nuclear fission. They generally use uranium and its product plutonium as nuclear fuel, though athorium fuel cycle is also possible. Fission reactors can be divided roughly into two classes, depending on the energy of the neutrons that sustain the fission chain reaction:
- Thermal reactors use slowed or thermal neutrons. Almost all current reactors are of this type. These contain neutron moderator materials that slow neutrons until their neutron temperature is thermalised, that is, until their kinetic energy approaches the average kinetic energy of the surrounding particles. Thermal neutrons have a far higher cross section (probability) of fissioning the fissile nuclei uranium-235, plutonium-239, and plutonium-241, and a relatively lower probability of neutron capture by uranium-238 (U-238) compared to the faster neutrons that originally result from fission, allowing use of low-enriched uranium or even natural uranium fuel. The moderator is often also the coolant, usually water under high pressure to increase the boiling point. These are surrounded by a reactor vessel, instrumentation to monitor and control the reactor, radiation shielding, and a containment building.
- Fast neutron reactors use fast neutrons to cause fission in their fuel. They do not have a neutron moderator, and use less-moderating coolants. Maintaining a chain reaction requires the fuel to be more highly enriched in fissile material (about 20% or more) due to the relatively lower probability of fission versus capture by U-238. Fast reactors have the potential to produce lesstransuranic waste because all actinides are fissionable with fast neutrons, but they are more difficult to build and more expensive to operate. Overall, fast reactors are less common than thermal reactors in most applications. Some early power stations were fast reactors, as are some Russian naval propulsion units. Construction of prototypes is continuing (see fast breederor generation IV reactors).
- Nuclear fusion. Fusion power is an experimental technology, generally with hydrogen as fuel. While not suitable for power

production, Farnsworth-Hirsch fusors are used to produce neutron radiation.

Classification by Moderator Material

Used by thermal reactors:

- Graphite moderated reactors
- Water moderated reactors
- Heavy water reactors
- Light water moderated reactors (LWRs). Light water reactors use ordinary water to moderate and cool the reactors. When at operating temperature, if the temperature of the water increases, its density drops, and fewer neutrons passing through it are slowed enough to trigger further reactions. That negative feedback stabilises the reaction rate. Graphite and heavy water reactors tend to be more thoroughly thermalised than light water reactors. Due to the extra thermalisation, these types can use natural uranium/unenriched fuel.
- Light element moderated reactors. These reactors are moderated by lithium or beryllium.
- Molten salt reactors (MSRs) are moderated by a light elements such as lithium or beryllium, which are constituents of the coolant/fuel matrix salts LiF and BeF_2.
- Liquid metal cooled reactors, such as one whose coolant is a mixture of Lead and Bismuth, may use BeO as a moderator.
- Organically moderated reactors (OMR) use biphenyl and terphenyl as moderator and coolant.

Classification by Coolant

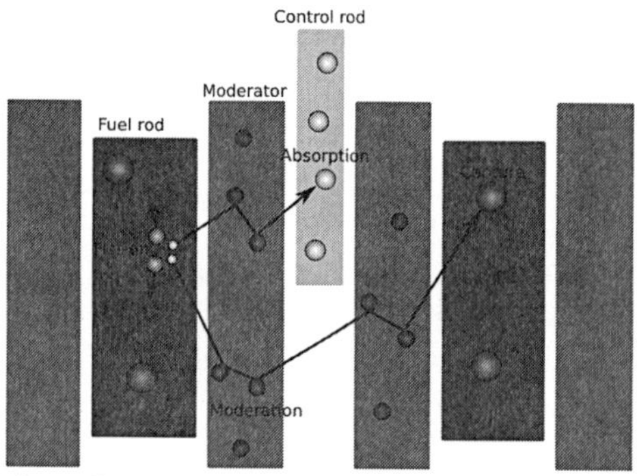

Figure: In thermal nuclear reactors (LWRs in specific), the coolant acts as a moderator that must slow down the neutrons before they can be efficiently absorbed by the fuel.

- Water cooled reactor. There are 104 operating reactors in the United States. Of these, 69 are pressurised water reactors (PWR), and 35 are boiling water reactors (BWR).

Pressurised water reactor (PWR)

- A primary characteristic of PWRs is a pressuriser, a specialised pressure vessel. Most commercial PWRs and naval reactors use pressurisers. During normal operation, a pressuriser is partially filled with water, and a steam bubble is maintained above it by heating the water with submerged heaters. During normal operation, the pressuriser is connected to the primary reactor pressure vessel (RPV) and the pressuriser "bubble" provides an expansion space for changes in water volume in the reactor. This arrangement also provides a means of pressure control for the reactor by increasing or decreasing the steam pressure in the pressuriser using the pressuriser heaters.
- Pressurised heavy water reactors are a subset of pressurised water reactors, sharing the use of a pressurised, isolated heat transport loop, but using heavy water as coolant and moderator for the greater neutron economies it offers.

Boiling Water Reactor (BWR)

- BWRs are characterised by boiling water around the fuel rods in the lower portion of a primary reactor pressure vessel. A boiling water reactor uses ^{235}U, enriched as uranium dioxide, as its fuel. The fuel is assembled into rods that are submerged in water and housed in a steel vessel. The nuclear fission causes the water to boil, generating steam. This steam flows through pipes into turbines. The turbines are driven by the steam, and this process generates electricity. During normal operation, pressure is controlled by the amount of steam flowing from the reactor pressure vessel to the turbine.
- Pool-type reactor
- Liquid metal cooled reactor. Since water is a moderator, it cannot be used as a coolant in a fast reactor. Liquid metal coolants have included sodium, NaK, lead, lead-bismuth eutectic, and in early reactors, mercury.
- Sodium-cooled fast reactor
- Lead-cooled fast reactor

- Gas cooled reactors are cooled by a circulating inert gas, often helium in high-temperature designs, while carbon dioxide has been used in past British and French nuclear power plants. Nitrogen has also been used. Utilisation of the heat varies, depending on the reactor. Some reactors run hot enough that the gas can directly power a gas turbine. Older designs usually run the gas through a heat exchanger to make steam for a steam turbine.
- Molten Salt Reactors (MSRs) are cooled by circulating a molten salt, typically a eutectic mixture of fluoride salts, such as FLiBe. In a typical MSR, the coolant is also used as a matrix in which the fissile material is dissolved.

Classification by Generation

- Generation I reactor
- Generation II reactor (most current nuclear power plants)
- Generation III reactor (evolutionary improvements of existing designs)
- Generation IV reactor (technologies still under development)

The "Gen IV"-term was dubbed by the United States Department of Energy (DOE) for developing new plant types in 2000. In 2003, the French Commissariat à l'Énergie Atomique (CEA) was the first to refer to Gen II types in Nucleonics Week;. First mentioning of Gen III was also in 2000 in conjunction with the launch of the Generation IV International Forum (GIF) plans.

Classification by Phase of Fuel

- Solid fuelled
- Fluid fuelled
- Aqueous homogeneous reactor
- Molten salt reactor
- Gas fuelled (theoretical)

Classification by Use

- Electricity
- Nuclear power plants
- Propulsion, see nuclear propulsion
- Nuclear marine propulsion
- Various proposed forms of rocket propulsion

- Other uses of heat
- Desalination
- Heat for domestic and industrial heating
- Hydrogen production for use in a hydrogen economy
- Production reactors for transmutation of elements
- Breeder reactors are capable of producing more fissile material than they consume during the fission chain reaction (by converting fertile U-238 to Pu-239, or Th-232 to U-233). Thus, a uranium breeder reactor, once running, can be re-fuelled with natural or even depleted uranium, and a thorium breeder reactor can be re-fuelled with thorium; however, an initial stock of fissile material is required.
- Creating various radioactive isotopes, such as americium for use in smoke detectors, and cobalt-60, molybdenum-99 and others, used for imaging and medical treatment.
- Production of materials for nuclear weapons such as weapons-grade plutonium
- Providing a source of neutron radiation (for example with the pulsed Godiva device) and positron radiation (e.g. neutron activation analysis and potassium-argon dating).
- Research reactor: Typically reactors used for research and training, materials testing, or the production of radioisotopes for medicine and industry. These are much smaller than power reactors or those propelling ships, and many are on university campuses. There are about 280 such reactors operating, in 56 countries. Some operate with high-enriched uranium fuel, and international efforts are underway to substitute low-enriched fuel.

Current Technologies

There are two types of nuclear power in current use:

- The Radioisotope thermoelectric generator produces heat through passive radioactive decay. Some radioisotope thermoelectric generators have been created to power space probes (for example, the Cassini probe), some lighthouses in the former Soviet Union, and some pacemakers. The heat output of these generators diminishes with time; the heat is converted to electricity utilising the thermoelectric effect.

- Nuclear fission reactors produce heat through a controlled nuclear chain reaction in a critical mass of fissile material. All current nuclear power plants are critical fission reactors, which are the focus of this article. The output of fission reactors is controllable. There are several subtypes of critical fission reactors, which can be classified as Generation I, Generation II and Generation III. All reactors will be compared to the Pressurised Water Reactor (PWR), as that is the standard modern reactor design.

Figure: Diablo Canyon — a PWR

Pressurised Water Reactors (PWR)

These reactors use a pressure vessel to contain the nuclear fuel, control rods, moderator, and coolant. They are cooled and moderated by high pressure liquid water.

Figure: Laguna Verde nuclear power plant — a BWR

The hot radioactive water that leaves the pressure vessel is looped through a steam generator, which in turn heats a secondary (non-radioactive) loop of water to steam that can run turbines.

They are the majority of current reactors, and are generally considered the safest and most reliable technology currently in large scale deployment. This is a thermal neutron reactor design, the newest of which are the VVER-1200, Advanced Pressurised Water Reactor and the European Pressurised Reactor. United States Naval reactors are of this type.

Boiling Water Reactors (BWR)

A BWR is like a PWR without the steam generator. A boiling water reactor is cooled and moderated by water like a PWR, but at a lower pressure, which allows the water to boil inside the pressure vessel producing the steam that runs the turbines. Unlike a PWR, there is no primary and secondary loop.

The thermal efficiency of these reactors can be higher, and they can be simpler, and even potentially more stable and safe. This is a thermal neutron reactor design, the newest of which are the Advanced Boiling Water Reactor and the Economic Simplified Boiling Water Reactor.

Figure: The CANDU Qinshan Nuclear Power Plant

Pressurised Heavy Water Reactor (PHWR)

A Canadian design (known as CANDU), these reactors are heavy-water-cooled and -moderated Pressurised-Water reactors. Instead of using a single large pressure vessel as in a PWR, the fuel is contained in hundreds of pressure tubes. These reactors are fuelled with natural uranium and are thermal neutron reactor designs. PHWRs can be refuelled while at full power, which makes them very efficient in their use of uranium (it allows for precise flux control in the core). CANDU PHWRs have been built in Canada, Argentina, China, India, Pakistan, Romania, and South Korea. India also operates a number of PHWRs, often termed 'CANDU-derivatives', built after the Government of Canada halted nuclear dealings with India following the 1974 Smiling Buddha nuclear weapon test.

Figure: The Ignalina Nuclear Power Plant — a RBMK type (closed 2009)

Reaktor Bolshoy Moschnosti Kanalniy (High Power Channel Reactor) (RBMK)

A Soviet design, built to produce plutonium as well as power. RBMKs are water cooled with a graphite moderator. RBMKs are in some respects similar to CANDU in that they are refuelable during power operation and employ a pressure tube design instead of a PWR-style pressure vessel.

However, unlike CANDU they are very unstable and large, making containment buildings for them expensive. A series of critical safety flaws have also been identified with the RBMK design, though some of these were corrected following the Chernobyl disaster. Their main attraction is their use of light water and un-enriched uranium. As of 2010, 11 remain open, mostly due to safety improvements and help from international safety agencies such as the DOE. Despite these safety improvements, RBMK reactors are still considered one of the most dangerous reactor designs in use. RBMK reactors were deployed only in the former Soviet Union.

Figure: The Magnox Sizewell A nuclear power station

Figure: The Torness nuclear power station — an AGR

Gas Cooled Reactor (GCR) and Advanced Gas Cooled Reactor (AGR)

These are generally graphite moderated and CO_2 cooled. They can have a high thermal efficiency compared with PWRs due to higher operating temperatures. There are a number of operating reactors of this design, mostly in the United Kingdom, where the concept was developed. Older designs (i.e. Magnox stations) are either shut down or will be in the near future. However, the AGCRs have an anticipated life of a further 10 to 20 years. This is a thermal neutron reactor design. Decommissioning costs can be high due to large volume of reactor core.

Liquid Metal Fast Breeder Reactor (LMFBR)

This is a reactor design that is cooled by liquid metal, totally unmoderated, and produces more fuel than it consumes. They are said to "breed" fuel, because they produce fissionable fuel during operation because of neutron capture. These reactors can function much like a PWR in terms of efficiency, and do not require much high pressure containment, as the liquid metal does not need to be kept at high pressure, even at very high temperatures. BN-350 and BN-600 in USSR and Superphénix in France were a reactor of this type, as was

Fermi-I in the United States. The Monju reactor in Japan suffered a sodium leak in 1995 and was restarted in May 2010. All of them use/used liquid sodium. These reactors are fast neutron, not thermal neutron designs. These reactors come in two types:

Figure: The Superphénix, one of the few FBRs

Lead Cooled

Using lead as the liquid metal provides excellent radiation shielding, and allows for operation at very high temperatures. Also, lead is (mostly) transparent to neutrons, so fewer neutrons are lost in the coolant, and the coolant does not become radioactive. Unlike sodium, lead is mostly inert, so there is less risk of explosion or accident, but such large quantities of lead may be problematic from toxicology and disposal points of view. Often a reactor of this type would use a lead-bismuth eutectic mixture. In this case, the bismuth would present some minor radiation problems, as it is not quite as transparent to neutrons, and can be transmuted to a radioactive isotope more readily than lead. The Russian Alfa class submarine uses a lead-bismuth-cooled fast reactor as its main power plant.

Sodium Cooled

Most LMFBRs are of this type. The sodium is relatively easy to obtain and work with, and it also manages to actually prevent corrosion on the various reactor parts immersed in it. However, sodium explodes violently when exposed to water, so care must be taken, but such explosions would not be vastly more violent than (for example) a leak of superheated fluid from a SCWR or PWR. EBR-I, the first reactor to have a core meltdown, was of this type.

Pebble Bed Reactors (PBR)

These use fuel molded into ceramic balls, and then circulate gas through the balls. The result is an efficient, low-maintenance, very safe reactor with inexpensive, standardised fuel. The prototype was the AVR.

Molten Salt Reactors

These dissolve the fuels in fluoride salts, or use fluoride salts for coolant. These have many safety features, high efficiency and a high power density suitable for vehicles. Notably, they have no high pressures or flammable components in the core. The prototype was the MSRE, which also used Thorium's fuel cycle to produce 0.1% of the radioactive waste of standard reactors.

Aqueous Homogeneous Reactor (AHR)

These reactors use soluble nuclear salts dissolved in water and mixed with a coolant and a neutron moderator.

Future and Developing Technologies

Advanced Reactors

More than a dozen advanced reactor designs are in various stages of development. Some are evolutionary from the PWR, BWR and PHWR designs above, some are more radical departures. The former include the Advanced Boiling Water Reactor (ABWR), two of which are now operating with others under construction, and the planned passively safe ESBWR and AP1000 units (see Nuclear Power 2010 Program).

- The Integral Fast Reactor (IFR) was built, tested and evaluated during the 1980s and then retired under the Clinton administration in the 1990s due to nuclear non-proliferation policies of the administration. Recycling spent fuel is the core of its design and it therefore produces only a fraction of the waste of current reactors.
- The Pebble Bed Reactor, a High Temperature Gas Cooled Reactor (HTGCR), is designed so high temperatures reduce power output by doppler broadening of the fuel's neutron cross-section. It uses ceramic fuels so its safe operating temperatures exceed the power-reduction temperature range. Most designs are cooled by inert helium. Helium is not subject to steam explosions, resists neutron absorption leading to radioactivity, and does not dissolve contaminants that can become radioactive. Typical designs have more layers (up to 7) of passive containment than light water reactors (usually 3). A unique feature that may aid safety is that the fuel-balls actually form

the core's mechanism, and are replaced one-by-one as they age. The design of the fuel makes fuel reprocessing expensive.

- The Small Sealed Transportable Autonomous Reactor (SSTAR) is being primarily researched and developed in the US, intended as a fast breeder reactor that is passively safe and could be remotely shut down in case the suspicion arises that it is being tampered with.
- The Clean And Environmentally Safe Advanced Reactor (CAESAR) is a nuclear reactor concept that uses steam as a moderator — this design is still in development.
- The Hydrogen Moderated Self-regulating Nuclear Power Module (HPM) is a reactor design emanating from the Los Alamos National Laboratory that uses uranium hydride as fuel.
- Subcritical reactors are designed to be safer and more stable, but pose a number of engineering and economic difficulties. One example is the Energy amplifier.
- Thorium based reactors. It is possible to convert Thorium-232 into U-233 in reactors specially designed for the purpose. In this way, thorium, which is more plentiful than uranium, can be used to breed U-233 nuclear fuel. U-233 is also believed to have favourable nuclear properties as compared to traditionally used U-235, including better neutron economy and lower production of long lived transuranic waste.
- Advanced Heavy Water Reactor (AHWR)— A proposed heavy water moderated nuclear power reactor that will be the next generation design of the PHWR type. Under development in the Bhabha Atomic Research Centre (BARC), India.
- KAMINI — A unique reactor using Uranium-233 isotope for fuel. Built in India by BARC and Indira Gandhi Centre for Atomic Research (IGCAR).
- India is also planning to build fast breeder reactors using the thorium – Uranium-233 fuel cycle. The FBTR (Fast Breeder Test Reactor) in operation at Kalpakkam (India) uses Plutonium as a fuel and liquid sodium as a coolant.

Generation IV Reactors

Generation IV reactors are a set of theoretical nuclear reactor designs currently being researched. These designs are generally not expected to be available for commercial construction before 2030.

Current reactors in operation around the world are generally considered second- or third-generation systems, with the first-generation systems having been retired some time ago. Research into these reactor types was officially started by the Generation IV International Forum (GIF) based on eight technology goals. The primary goals being to improve nuclear safety, improve proliferation resistance, minimise waste and natural resource utilisation, and to decrease the cost to build and run such plants.

- Gas cooled fast reactor
- Lead cooled fast reactor
- Molten salt reactor
- Sodium-cooled fast reactor
- Supercritical water reactor
- Very high temperature reactor

Generation V+ Reactors

Generation V reactors are designs which are theoretically possible, but which are not being actively considered or researched at present. Though such reactors could be built with current or near term technology, they trigger little interest for reasons of economics, practicality, or safety.

- Liquid Core reactor. A closed loop liquid core nuclear reactor, where the fissile material is molten uranium cooled by a working gas pumped in through holes in the base of the containment vessel.
- Gas core reactor. A closed loop version of the nuclear lightbulb rocket, where the fissile material is gaseous uranium-hexafluoride contained in a fused silica vessel. A working gas (such as hydrogen) would flow around this vessel and absorb the UV light produced by the reaction. In theory, using UF_6 as a working fuel directly (rather than as a stage to one, as is done now) would mean lower processing costs, and very small reactors. In practice, running a reactor at such high power densities would probably produce unmanageable neutron flux.
- Gas core EM reactor. As in the Gas Core reactor, but with photovoltaic arrays converting the UV light directly to electricity.
- Fission fragment reactor

Fusion Reactors

Controlled nuclear fusion could in principle be used in fusion power plants to produce power without the complexities of handling actinides, but significant scientific and technical obstacles remain. Several fusion reactors have been built, but as yet none has 'produced' more thermal energy than electrical energy consumed. Despite research having started in the 1950s, no commercial fusion reactor is expected before 2050. The ITER project is currently leading the effort to commercialize fusion power.

Nuclear Fuel Cycle

Thermal reactors generally depend on refined and enriched uranium. Some nuclear reactors can operate with a mixture of plutonium and uranium (see MOX). The process by which uranium ore is mined, processed, enriched, used, possibly reprocessed and disposed of is known as the nuclear fuel cycle.

Under 1% of the uranium found in nature is the easily fissionable U-235 isotope and as a result most reactor designs require enriched fuel. Enrichment involves increasing the percentage of U-235 and is usually done by means of gaseous diffusion or gas centrifuge. The enriched result is then converted into uranium dioxide powder, which is pressed and fired into pellet form. These pellets are stacked into tubes which are then sealed and called fuel rods. Many of these fuel rods are used in each nuclear reactor.

Most BWR and PWR commercial reactors use uranium enriched to about 4% U-235, and some commercial reactors with a high neutron economy do not require the fuel to be enriched at all (that is, they can use natural uranium). According to the International Atomic Energy Agency there are at least 100 research reactors in the world fuelled by highly enriched (weapons-grade/90% enrichment uranium). Theft risk of this fuel (potentially used in the production of a nuclear weapon) has led to campaigns advocating conversion of this type of reactor to low-enrichment uranium (which poses less threat of proliferation).

Fissile U-235 and non-fissile but fissionable and fertile U-238 are both used in the fission process. U-235 is fissionable by thermal (i.e. slow-moving) neutrons. A thermal neutron is one which is moving about the same speed as the atoms around it. Since all atoms vibrate proportionally to their absolute temperature, a thermal neutron has the best opportunity to fission U-235 when it is moving at this same vibrational speed. On the other hand, U-238 is more likely to capture a neutron when the neutron is moving very fast. This U-239 atom will soon decay into plutonium-239, which is another fuel. Pu-239 is a viable fuel and must be accounted for even when a highly enriched uranium fuel is used. Plutonium fissions will dominate the U-235 fissions in some reactors, especially after the initial loading of U-235 is spent. Plutonium is fissionable with both fast and thermal neutrons, which make it ideal for either nuclear reactors or nuclear bombs.

Most reactor designs in existence are thermal reactors and typically use water as a neutron moderator (moderator means that it slows down the neutron to a thermal speed) and as a coolant. But in a fast breeder reactor, some other kind of coolant is used which will not moderate or slow the neutrons down much. This enables fast neutrons to dominate, which can effectively be used to constantly replenish the fuel supply. By merely placing cheap unenriched uranium into such a core, the non-fissionable U-238 will be turned into Pu-239, "breeding" fuel.

Fuelling of Nuclear Reactors

The amount of energy in the reservoir of nuclear fuel is frequently expressed in terms of "full-power days," which is the number of 24-hour periods (days) a reactor is scheduled for operation at full power output for the generation of heat energy. The number of full-power days in a reactor's operating cycle (between refuelling outage times) is related to the amount of fissile uranium-235 (U-235) contained in the fuel assemblies at the beginning of the cycle. A higher percentage of U-235 in the core at the beginning of a cycle will permit the reactor to be run for a greater number of full-power days.

At the end of the operating cycle, the fuel in some of the assemblies is "spent" and is discharged and replaced with new (fresh) fuel assemblies, although in practice it is the buildup of reaction poisons in nuclear fuel that determines the lifetime of nuclear fuel in a reactor. Long before all possible fission has taken place, the buildup of long-lived neutron absorbing fission byproducts impedes the chain reaction. The fraction of the reactor's fuel core replaced during refuelling is typically one-fourth for a boiling-water reactor and one-third for a pressurised-water reactor. The disposition and storage of this spent fuel is one of the most challenging aspects of the operation of a commercial nuclear power plant. This nuclear waste is highly radioactive and its toxicity presents a danger for thousands of years. Not all reactors need to be shut down for refuelling; for example, pebble bed reactors, RBMK reactors, molten salt reactors, Magnox, AGR and CANDU reactors allow fuel to be shifted through the reactor while it is running. In a CANDU reactor, this also allows individual fuel elements to be situated within the reactor core that are best suited to the amount of U-235 in the fuel element.

The amount of energy extracted from nuclear fuel is called its burnup, which is expressed in terms of the heat energy produced per initial unit of fuel weight. Burn up is commonly expressed as megawatt days thermal per metric ton of initial heavy metal.

Safety

Nuclear safety covers the actions taken to prevent nuclear and radiation accidents or to limit their consequences. The nuclear power industry has improved the safety and performance of reactors, and has proposed new safer (but generally untested) reactor designs but there is no guarantee that the reactors will be designed, built and operated correctly. Mistakes do occur and the designers of reactors at Fukushima in Japan did not anticipate that a tsunami generated by an earthquake would disable the backup systems that were supposed to stabilise the reactor after the earthquake. According to UBS AG, the Fukushima I nuclear accidents have cast doubt on whether even an advanced economy like Japan can master nuclear safety. Catastrophic scenarios involving terrorist attacks are also conceivable. An interdisciplinary team from MIT have estimated that given the expected growth of nuclear power from 2005 – 2055, at least four serious nuclear accidents would be expected in that period.

Accidents

Some serious nuclear and radiation accidents have occurred. Nuclear power plant accidents include the Chernobyl disaster (1986), Fukushima Daiichi nuclear disaster (2011), the Three Mile Island accident (1979), and SL-1 accident (1961). Nuclear-powered submarinemishaps include the K-19 reactor accident (1961), the K-27 reactor accident (1968), and the K-431 reactor accident (1985).

Nuclear reactors have been launched into Earth orbit at least 34 times. A number of incidents connected with the unmanned nuclear-reactor-powered Soviet RORSAT radar satellite program resulted in spent nuclear fuel re-entering the Earth's atmosphere from orbit.

Spacecraft

A spacecraft or spaceship is a craft, vehicle, vessel or machine designed for spaceflight. Spacecraft are used for a variety of purposes, including communications, earth observation, meteorology, navigation, planetary exploration and transportation of humans and cargo.

Figure: More than 100 Russian Soyuz manned spacecraft (TMA version shown) have flown since 1967, originally for a Soviet manned lunar program, but currently supporting the International Space Station.

Figure: The US Space Shuttle flew 135 times from 1981 to 2011, supporting Spacelab, Mir, and ISS. (Columbia's first launch shown)

On a suborbital spaceflight, a spacecraft enters space and then returns to the surface, without having gone into an orbit. For orbital spaceflights, spacecraft enter closed orbits around the Earth or around other celestial bodies. Spacecraft used for human spaceflight carry people on board as crew or passengers from start or on orbit (space stations) only, while those used for robotic space missions operate either autonomously or telerobotically. Robotic spacecraft used to support scientific research are space probes. Robotic spacecraft that remain in orbit around a planetary body are artificial satellites. Only a handful of interstellar probes, such as Pioneer 10 and 11, Voyager 1 and 2, and New Horizons, are currently on trajectories that leave our Solar System. Orbital spacecraft may be recoverable or not. By method of reentry to Earth they may be divided in non-winged space capsules and winged spaceplanes.

Currently, only a few nations have spaceflight technology: Russia (Russian Federal Space Agency), the United States (NASA , the US Air Force, SpaceX (a U.S private aerospace company), the European Union (European Space Agency), the People's Republic of China (China National Space Administration), Japan (Japan Aerospace Exploration Agency), and India (Indian Space Research Organisation). As of 2011, only the US, Russia and China have demonstrated human spaceflight capability.

Spacecraft (spaceship) and space travel are common themes in works of science fiction.

History

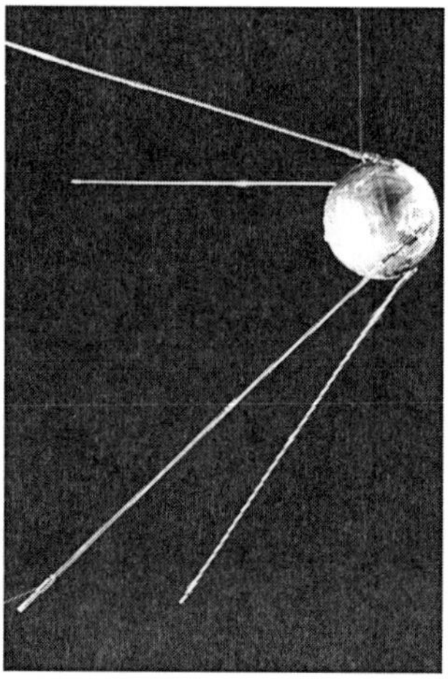

Figure: The first spacecraft, Sputnik 1.

The first artificial satellite. It was launched into an elliptical low Earth orbit by the Soviet Union on 4 October 1957. The launch ushered in new political, military, technological, and scientific developments. While the Sputnik launch was a single event, it marked the start of the Space Age. Apart from its value as a technological first, Sputnik also helped to identify the upper atmospheric layer's density, through measuring the satellite's orbital changes. It also provided data on radio-signal distribution in the ionosphere. Pressurised nitrogen in the satellite's body provided the first opportunity for meteoroid detection. If a meteoroid penetrated the satellite's outer hull, it would be detected by the temperature data sent back to Earth. Sputnik 1 was launched during the International

Geophysical Year from Site No.1/5, at the 5th Tyuratam range, in Kazakh SSR (now at the Baikonur Cosmodrome). The satellite travelled at 29,000 kilometres (18,000 mi) per hour, taking 96.2 minutes to complete an orbit, and emitted radio signals at 20.005 and 40.002 MHz

While Sputnik 1 was the first spacecraft to orbit the Earth, other man-made objects had previously reached an altitude of 100 km, which is the height required by the international organisation Fédération Aéronautique Internationale to count as a spaceflight. This altitude is called the Kármán line. In particular, in the 1940s there were several test launches of the V-2 rocket, some of which reached altitudes well over 100 km.

Past and Present Spacecraft

Manned Spacecraft

As of 2011, only three nations have flown manned spacecraft: USSR/Russia, USA, and China. India, Japan, Europe/ESA, Iran, DPRK, Denmark, and Romania have plans for manned spacecraft (for manned suborbital rockets).

The first manned spacecraft was Vostok 1, which carried Soviet cosmonaut Yuri Gagarin into space in 1961, and completed a full Earth orbit. There were five other manned missions which used a Vostok spacecraft. The second manned spacecraft was named Freedom 7, and it performed a suborbital spaceflight in 1961 carrying American as tronautAlan Shepard to an altitude of just over 187 kilometres (116 mi). There were five other manned missions using Mercury spacecraft.

Other Soviet manned spacecraft include the Voskhod, Soyuz, unflown as manned Zond/L1, L3, TKS, and the Salyut and Mirmanned space stations. Other American manned spacecraft include the Gemini Spacecraft, Apollo Spacecraft, the Skylabspace station, and the Space Shuttle with undetached European Spacelab and private US Spacehab space stations-modules. China developed unflown Shuguang and currently using Shenzhou (the first manned mission was in 2003).

All of these recovery manned spacecraft were space capsules. The International Space Station, manned since November 2000, is a joint venture between Russia, the United States, and several other countries.

Spaceplanes

Some reusable vehicles have been designed only for manned spaceflight, and these are often called spaceplanes. The first example of such was the North American X-15 spaceplane, which conducted two manned flights which reached a height over 100 km in the 1960s. The first reusable spacecraft, the X-15, was air-launched on a suborbital trajectory on July 19, 1963.

The first partially reusable orbital spacecraft, winged non-capsule, the Space Shuttle, was launched by the USA on the 20th anniversary of Yuri Gagarin's flight, on April 12, 1981. During the Shuttle era, six orbiters were built, all of which have flown in the atmosphere and five of which have flown in space. The Enterprise was used only for approach and landing tests, launching from the back of a Boeing 747 SCA and gliding to deadstick landings at Edwards AFB, California. The first Space Shuttle to fly into space was the Columbia, followed by the Challenger, Discovery, Atlantis, and Endeavour. The Endeavour was built to replace the Challenger when it was lost in January 1986. The Columbia broke up during reentry in February 2003.

The first automatic partially reusable spacecraft was the Buran (Snowstorm), launched by the USSR on November 15, 1988, although it made only one flight. This spaceplane was designed for a crew and strongly resembled the U.S. Space Shuttle, although its drop-off boosters used liquid propellants and its main engines were located at the base of what would be the external tank in the American Shuttle. Lack of funding, complicated by the dissolution of the USSR, prevented any further flights of Buran. The Space Shuttle has since been modified to allow for autonomous re-entry in case of necessity.

Per the Vision for Space Exploration, the Space Shuttle was retired in 2011 due mainly to its old age and high cost of program reaching over a billion dollars per flight. The Shuttle's human transport role is to be replaced by the partially reusable Crew Exploration Vehicle (CEV) no later than 2014. The Shuttle's heavy cargo transport role is to be replaced by expendable rockets such as the Evolved Expendable Launch Vehicle (EELV) or a Shuttle Derived Launch Vehicle.

Scaled Composites' SpaceShipOne was a reusable suborbital spaceplane that carried pilots Mike Melvill and Brian Binnie on consecutive flights in 2004 to win the Ansari X Prize. The Spaceship Company will build its successor SpaceShipTwo. A fleet of SpaceShipTwos operated by Virgin Galactic should begin reusable private spaceflight carrying paying passengers in 2011.

XCOR Aerospace also plans to initiate a suborbital commercial spaceflight service with the Lynx rocketplane in 2012 through a partnership with RocketShip Tours. First test flights are planned for 2011.

Spacecraft Under Development

Manned

- (USA-NASA) Orion Multi-Purpose Crew Vehicle – capsule
- (Russia-RKA) Prospective Piloted Transport System(PPTS) – capsule
- (USA-SpaceX) Dragon – capsule
- (USA-Boeing) CST-100 – capsule
- (USA-Sierra Nevada Corporation) Dream Chaser – suborbital spaceplane
- (USA-Blue Origin) New Shepard – VTVL capsule
- (India-ISRO) Orbital Vehicle – capsule
- (Europe-ESA) Advanced Crew Transportation System – capsule
- (China-CNSA) Shenzhou – capsule
- (Iranian Space Agency) Orbital Vehicle – capsule
- (USA-XCOR) Lynx rocketplane – suborbital spaceplane

Unmanned

- SpaceX Dragon – cargo delivery to the ISS
- Orbital Sciences Cygnus – cargo delivery to the ISS
- CNES Mars Netlander
- James Webb Space Telescope (delayed)
- ESA Darwin probe
- Mars Science Laboratory rover
- Shenzhou spacecraft Cargo
- Terrestrial Planet Finder cancelled probe
- System F6—a DARPA Fractionated Spacecraft demonstrator
- Reaction Engines Limited Skylon (spacecraft)

Subsystems

A spacecraft system comprises various subsystems, dependent upon mission profile. Spacecraft subsystems comprise the spacecraft "bus" and may include: attitude determination and control (variously called ADAC, ADC or ACS), guidance, navigation and control (GNC or GN&C), communications (Comms), command and data handling (CDH or C&DH), power (EPS), thermal control (TCS), propulsion, and structures. Attached to the bus are typically payloads.

Life Support

Spacecraft intended for human spaceflight must also include a life support system for the crew.

Figure: Reaction control system thrusters on the nose of the U.S. Space Shuttle

Attitude Control

A Spacecraft needs an attitude control subsystem to be correctly oriented in space and respond to external torques and forces properly. The attitude control subsystem consists of sensors and actuators, together with controlling algorithms. The attitude control subsystem permits proper pointing for the science objective, sun pointing for power to the solar arrays and earth-pointing for communications.

GNC

Guidance refers to the calculation of the commands (usually done by the CDH subsystem) needed to steer the spacecraft where it is desired to be. Navigation means determining a spacecraft's orbital elements or position. Control means adjusting the path of the spacecraft to meet mission requirements. On some missions, GNC and Attitude Control are combined into one subsystem of the spacecraft.

Chapter 7

Command and Data Handling

The CDH subsystem receives commands from the communications subsystem, performs validation and decoding of the commands, and distributes the commands to the appropriate spacecraft subsystems and components. The CDH also receives housekeeping data and science data from the other spacecraft subsystems and components, and packages the data for storage on a data recorder or transmission to the ground via the communications subsystem. Other functions of the CDH include maintaining the spacecraft clock and state-of-health monitoring.

On-Board Data Handling

Power

Spacecraft need an electrical power generation and distribution subsystem for powering the various spacecraft subsystems. For spacecraft near the Sun, solar panels are frequently used to generate electrical power. Spacecraft designed to operate in more distant locations, for example Jupiter, might employ a Radioisotope Thermoelectric Generator (RTG) to generate electrical power. Electrical power is sent through power conditioning equipment before it passes through a power distribution unit over an electrical bus to other spacecraft components. Batteries are typically connected to the bus via a battery charge regulator, and the batteries are used to provide electrical power during periods when primary power is not available, for example when a Low Earth Orbit (LEO) spacecraft is eclipsed by the Earth.

Thermal Control

Spacecraft must be engineered to withstand transit through the Earth's atmosphere and the space environment. They must operate in a vacuum with temperatures potentially ranging across hundreds of degrees Celsius as well as (if subject to reentry) in the presence of plasmas. Material requirements are such that either high melting temperature, low density materials such as beryllium and reinforced carbon-carbon or (possibly due to the lower thickness requirements despite its high density) tungsten or ablative carbon/carbon composites are used. Depending on mission profile, spacecraft may also need to operate on the surface of another planetary body. The thermal control subsystem can be passive, dependent on the selection of materials with specific radiative properties. Active thermal control makes use of electrical heaters and certain actuators such as louvers to control temperature ranges of equipments within specific ranges.

Figure: A launch vehicle, like this Proton rocket, is typically used to bring a spacecraft to orbit.

Propulsion

Spacecraft may or may not have a propulsion subsystem, depending upon whether or not the mission profile calls for propulsion. The Swift spacecraft is an example of a spacecraft that does not have a propulsion subsystem. Typically though, LEO spacecraft (for example Terra (EOS AM-1) include a propulsion subsystem for altitude adjustments (called drag make-up maneuvers) and inclination adjustment maneuvers. A propulsion system is also needed for spacecraft that perform momentum management maneuvers. Components of a conventional propulsion subsystem include fuel, tankage, valves, pipes, and thrusters. The TCS interfaces with the propulsion subsystem by monitoring the temperature of those components, and by preheating tanks and thrusters in preparation for a spacecraft maneuver.

Structures

Spacecraft must be engineered to withstand launch loads imparted by the launch vehicle, and must have a point of attachment for all the other subsystems. Depending upon mission profile, the structural subsystem might need to withstand loads imparted by entry into the atmosphere of another planetary body, and landing on the surface of another planetary body.

Payload

The payload is dependent upon the mission of the spacecraft, and is typically regarded as the part of the spacecraft "that pays the bills". Typical payloads could include scientific instruments (cameras, telescopes, or particle detectors, for example), cargo, or a human crew.

Ground Segment

The ground segment, though not technically part of the spacecraft, is vital to the operation of the spacecraft. Typical components of a ground segment in use during normal operations include a mission operations facility where the flight operations team conducts the operations of the spacecraft, a data processing and storage facility, ground stations to radiate signals to and receive signals from the spacecraft, and a voice and data communications network to connect all mission elements.

Launch Vehicle

The launch vehicle propels the spacecraft from the Earth's surface, through the atmosphere, and into an orbit, the exact orbit being dependent upon mission configuration. The launch vehicle may be expendable or reusable.

Submarine

A submarine is a watercraft capable of independent operation below the surface of the water. It differs from a submersible, which has more limited underwater capability. The term submarine most commonly refers to a large crewed autonomous vessel; however, historically or colloquially, submarine can also refer to medium sized or smaller vessels (midget submarines, wet subs), remotely operated vehicles or robots.

Figure: A Japan Maritime Self-Defence Force Oyashio-class submarine in 2006

The word submarine was originally an adjective meaning "under the sea"; consequently other uses such as "submarine engineering" or "submarine cable" may not actually refer at all to the vessel. Submarine was in fact shortened from the proper term, "submarine boat", and is often further shortened to "sub" when the word is employed informally. Submarines should always be referred to as "boats" rather than as "ships", regardless of their size. The English term U-boat for a German submarine comes from the German word for submarine, U-Boot, itself an abbreviation for Unterseeboot ("undersea boat").

Although experimental submarines had been built before, submarine design took off during the 19th century, and they were adopted by several different navies. Submarines were first widely used during World War I (1914–1918) and now feature in many large navies. Military usage includes attacking enemy surface ships or submarines, aircraft carrier protection, blockade running, ballistic missile submarines as part of a nuclear strike force, reconnaissance, conventional land attack (for example using a cruise missile), and covert insertion of special forces. Civilian uses for submarines include marine science, salvage, exploration and facility inspection/ maintenance. Submarines can also be modified to perform more specialised functions such as search-and-rescue missions or undersea cable repair. Submarines are also used in tourism, and for undersea archaeology.

Most large submarines comprise a cylindrical body with hemispherical (and/or conical) ends and a vertical structure, usually located amidships, which houses communications and sensing devices as well as periscopes. In modern submarines this structure is the "sail" in American usage, and "fin" in European usage. A "conning tower" was a feature of earlier designs: a separate pressure hull above the main body of the boat that allowed the use of shorter periscopes. There is a propeller (or pump jet) at the rear and various hydrodynamic control fins as well as ballast tanks. Smaller, deep diving and speciality submarines may deviate significantly from this traditional layout.

Submarines have one of the largest ranges of capabilities in any vessel, ranging from small autonomous examples to one- or two-person vessels operating for a few hours, to vessels which can remain submerged for 6 months such as the Russian Typhoon class - the biggest submarines ever built and in use. Submarines can work at greater depths than are survivable or practical for human divers. Modern deep diving submarines are derived from the bathyscaphe, which in turn was an evolution of the diving bell.

History of Submarines

Early history of submarines and the first submersibles: The first submersible with reliable information on its construction was built in 1620 by Cornelius Jacobszoon Drebbel, a Dutchman in the service of James I of England. It was created to the standards of the design outlined by English mathematician William Bourne. It was propelled by means of oars. The precise nature of the submarine type is a matter of some controversy; some claim that it was merely a bell towed by a boat. Two improved types were tested in the Thames between 1620 and 1624. In 2002 a two-person version of Bourne's design was built for the BBCTV programme Building the Impossible by Mark Edwards, and successfully rowed under water at Dorney Lake, Eton.

Though the first submersible vehicles were tools for exploring under water, it did not take long for inventors to recognise their military potential. The strategic advantages of submarines were set out by Bishop John Wilkins of Chester, England, in Mathematicall Magick in 1648:

1. Tis private: a man may thus go to any coast in the world invisibly, without discovery or prevented in his journey.

2. Tis safe, from the uncertainty of Tides, and the violence of Tempests, which do never move the sea above five or six paces deep. From Pirates and Robbers which do so infest other voyages; from ice and great frost, which do so much endanger the passages towards the Poles.
3. It may be of great advantages against a Navy of enemies, who by this may be undermined in the water and blown up.
4. It may be of special use for the relief of any place besieged by water, to convey unto them invisible supplies; and so likewise for the surprisal of any place that is accessible by water.
5. It may be of unspeakable benefit for submarine experiments.

First Military Submarines

The first military submarine was Turtle (1775), a hand-powered acorn-shaped device designed by the American David Bushnell to accommodate a single person. It was the first verified submarine capable of independent underwater operation and movement, and the first to use screws for propulsion. During the American Revolutionary War, Turtle (operated by Sgt. Ezra Lee, Continental Army) tried and failed to sink the British warship HMS Eagle, flagship of the blockaders in New York harbor on September 7, 1776.

In 1800, France built a human-powered submarine designed by American Robert Fulton, the Nautilus. The French eventually gave up on the experiment in 1804, as did the British when they later considered Fulton's submarine design.

During the War of 1812, in 1814, Silas Halsey lost his life while using a submarine in an unsuccessful attack on a British warship stationed in New London harbor.

The Submarino Hipopótamo was the first submarine in South America built and tested in Ecuador on September 18, 1837. It was designed by Jose Rodriguez Lavandera, who successfully crossed the Guayas River in Guayaquil accompanied by Jose Quevedo. Rodriguez Lavandera had enrolled in the Ecuadorian Navy in 1823, becoming a Lieutenant by 1830. The Hipopotamo crossed the Guayas on two more occasions, but it was then abandoned because of lack of funding and interest from the government. Today, few engravings and a scale model of the original design is preserved by the Maritime Museum of the Ecuadorian Navy.

In 1851, a Bavarian artillery corporal, Wilhelm Bauer, took a submarine designed by him called the Brandtaucher (incendiary-diver), which sank on its first test dive in Kiel Harbour—but its three crewmen managed to escape, after flooding the vessel, which allowed the inside pressure to equalise. This submarine was built by August Howaldt and powered by a treadwheel. The submarine was rediscovered during a dredging operation 1887, and was raised sixteen years later. The vessel is on display in a museum in Dresden.

The submarine Flach was commissioned in 1865 by the Chilean government during the war of Chile and Peru against Spain (1864–1866). It was built by the German engineer Karl Flach. The submarine sank during tests in Valparaiso bay on May 3, 1866, with the entire eleven-man crew.

Submarines in the American Civil War

Figure: The 1862 Alligator, first submarine of the United States Navy. It was designed by the French engineer, Brutus de Villeroi.

During the American Civil War both sides successfully built working submarines. The Confederate States of America submarines were all designed to attack the Union blockade of Southern ports. Two operational unnamed Confederate submarines were spotted during the latter half of 1861, one in the James River in Virginia and another in New Orleans. The United States Navy was first interested in submarines as a way to clear obstacles. Interest in attack submarines began at least by May 1861, when French engineer Brutus de Villeroi tested an early submarine design in Philadelphia harbor in what may have been a effort to attract the Navy's attention. Most Confederate submarines were built under the auspices of the Confederate Secret

Service rather than the Confederate Navy, with only three being well known and documented. Others, both Confederate and Union, are known to have existed but their names and designs have escaped the historical record. In all, evidence indicates that a combined total of over twenty operational submarines were built by both sides during the conflict.

First launched on May 1, 1862, the Villeroi-designed Alligator was the first U.S. Navy submarine. It was also the first to feature compressed air for an air supply and an air filtration system. Initially propelled by sixteen hand-powered paddles protruding from the sides, it was converted after six months to a screw propeller powered by a hand crank. At 47 feet (14.3 m) long and about 4 feet (1.2 m) in diametre, the Alligator was the largest of the documented American Civil War submarines. It was lost in a storm off Cape Hatteras on April 1, 1863 while under tow to its first combat deployment at Charleston.

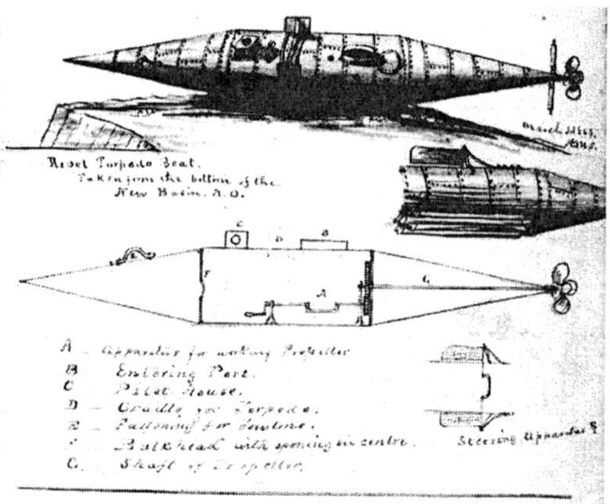

Figure: The Confederate Pioneer, as drawn by Ensign David M. Stauffer of the Mississippi Squadron in 1865.

The first documented Confederate submarine was the New Orleans-built Pioneer. It was 30 feet (9 m) long. This submarine sank a target schooner using a towed mine during tests on Lake Pontchartrain in February 1862. It was never used in combat, having been scuttled by Confederate forces before New Orleans was captured by the Union in April 1862. The Bayou St. John Confederate Submarine is another Louisiana-built submarine that is contemporaneous with the Pioneer, although no history is known. It is now on display at the Louisiana State Museum. The second documented Confederate submarine was the American Diver, also known as the Pioneer II, built in Mobile, Alabama. It was initially designed to be propelled by an electric motor but this proved to be too weak. A steam engine was installed next, but also proved to be insufficient. Finally, a hand-cranked propeller was installed. The Diver was 36 feet (11 m) long with a 3-foot (0.91 m) beam. It was lost during a storm while under tow during trials in Mobile Bay in late February 1863.

Figure: Confederate H.L. Hunley

The third documented Confederate submarine was H. L. Hunley, named for one of its financiers, Horace Lawson Hunley. Also built in Mobile, she was launched in July 1863. She was 39.5 feet (12.0 m) long with a 3.83-foot (1.17 m) beam. Hunley was designed to be propelled by a hand-crank from the beginning and could accommodate eight men to turn the hand-crank, as opposed to four men for Diver, and was armed with a spar torpedo.

The submarine had to approach an enemy vessel, attach the explosive with a barb, move away, and then detonate it. Hunley proved to be hazardous to operate, and had no air supply other than what was contained in the hull. On three occasions, she sank. On August 29, 1863, five out of the nine crew members drowned during a trial run. The second incident occurred on October 15, 1863, when all eight people on board, including Hunley, drowned during a diving exercise.

Then, on February 17, 1864, the salvaged and renovated vessel sank USS Housatonic off Charleston Harbor. Soon after signaling success, the submarine sank due to unknown cause; again the entire eight-man crew drowned. Submarines did not have a major impact on the outcome of the war, but did portend their future importance to, and increased interest in their use in, naval warfare. The location of Hunley was unknown until 1995, and she was raised in 2000. The sinking of Housatonic by Hunley was the first successful submarine attack on a warship.

Mechanically Powered Submarines, Late 19th Century

Figure: The Nordenfelt designed, Ottoman submarine Abdülhamid

The first submarine not relying on human power for propulsion was the French Plongeur (Diver), launched in 1863, and using compressed air at 180 psi (1241 kPa).

The first combustion-powered submarine was Ictineo II, designed in Catalonia, Spain by Narcís Monturiol. Originally launched in 1864 as human-powered, propelled by 16, it was converted to peroxide propulsion and steam in 1867. The 14 m (46 ft)-long craft was designed for a crew of two, could dive to 30 m (98 ft), and demonstrated dives of two hours. On the surface it ran on a steam engine, but underwater such an engine would quickly consume the submarine's oxygen, so Monturiol invented an air-independent propulsion system. While the air-independent power system drove the screw, the chemical process driving it also released oxygen into the hull for the crew and an auxiliary steam engine. Monturiol's fully functional, double hulled vessels also solved pressure and buoyancy control problems that had be deviled earlier designs. In 1870, the French writer Jules Verne, inspired by the recent efforts of Monturiol and of his own navy, published the science fiction classic 20,000 Leagues under the Sea, which concerns the adventures of a maverick inventor of the Nautilus, a submarine more advanced than any at the time. An international success, the story encouraged inventors around the world to work towards making such a vehicle a reality.

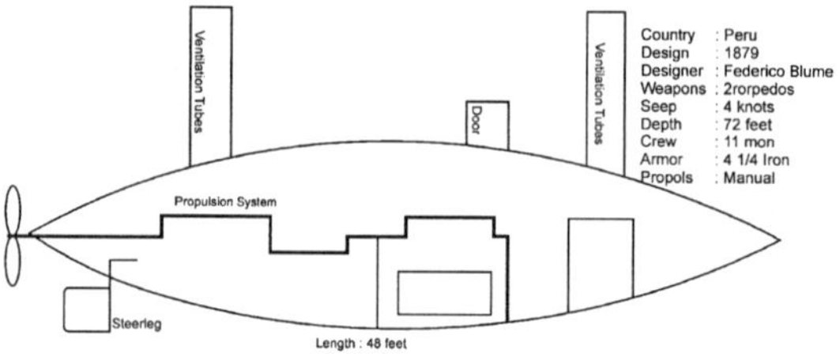

The Submarine Bull

Figure: Design of the Peruvian Toro, the first fully functional submarine built in Latin America.

In 1879, the Peruvian government, during the War of the Pacific, commissioned and built the fully operational submarine Toro Submarino. It never saw military action before being scuttled by the Peruvians after their defeat in the war to prevent its capture by the Chileans.

The first submarine to be mass-produced was human-powered. It was the submarine of the Polish inventor Stefan Drzewiecki—50 units were built in 1881 for the Russian government. In 1884 the same inventor built an electric-powered submarine.

Discussions between the English clergyman and inventor George Garrett and the industrially and commercially adept Swede Thorsten Nordenfelt led to a series of steam-powered submarines. The first was the Nordenfelt I, a 56 tonne, 19.5 metre (64 ft) vessel similar to Garret's ill-fated Resurgam (1879), with a range of 240 kilometres (150 mi, 130 nm), armed with a single torpedo, in 1885. Like Resurgam, Nordenfelt I operated on the surface by steam, then shut down its engine to dive. While submerged the submarine released pressure generated when the engine was running on the surface to provide propulsion for some distance underwater. Greece, fearful of the return of the Ottomans, purchased it. Nordenfelt then built Nordenfelt II (Abdülhamid) in 1886 and Nordenfelt III (Abdülmecid) in 1887, a pair of 30 metre (100 ft) submarines with twin torpedo tubes, for the Ottoman navy. Abdülhamid became the first submarine in history to fire a torpedo submerged. Nordenfelt's efforts culminated in 1887 with Nordenfelt IV which had twin motors and twin torpedoes. It was sold to the Russians, but proved unstable, ran aground, and was scrapped.

Figure: Peral submarine hull, Cartagena

Two submarines, both launched in September 1888, marked the maturing of naval submarine technology.

One was the Peral Submarine, launched by the Spanish Navy. It had two torpedoes, new air systems, hull shape, propeller, and cruciform external controls anticipating much later designs. Peral was the first all-electrical powered submarine. After two years of trials the project was scrapped by naval officialdom that cited concerns over the short range permitted by its batteries.

The other was the Gymnote, launched by the French Navy. Gymnote was also an electrically powered and fully functional military submarine. It completed over 2,000 successful dives using a 204-cell battery. Although she was scrapped for her limited range her side hydroplanes became the standard for future submarine designs.

Many more designs were built at this time by various inventors, but submarines were not put into service by navies until 1900.

End of the 19th Century to the Russo-Japanese War

The turn of the 20th century marked a pivotal time in the development of submarines, with a number of important technologies making their debut, as well as the widespread adoption and fielding of submarines by a number of nations. Diesel electric propulsion would become the dominant power system and equipment such as the periscope would become standardised. Large numbers of experiments were done by countries on effective tactics and weapons for submarines, all of which would culminate in them making a large impact on the coming World War I.

Figure: USS Plunger, launched in 1902

In 1896, the Irish-American inventor John Philip Holland designed submarines that, for the first time, made use of internal combustion engines on the surface and electric battery power submerged. Fenian Ram, in 1881, was world's first successful submarine. Holland VI was launched on May 17, 1897 at Navy Lt. Lewis Nixon's Crescent Shipyard of Elizabeth, New Jersey. On April 11, 1900 the United States Navy purchased the revolutionary Holland VI and renamed it USS Holland (SS-1), America's first commissioned submarine. (John P. Holland's company, the Holland Torpedo Boat Company/Electric Boat Company became General Dynamics' Cold Warprogeny and is the builder of technologically advanced submarines today).

Commissioned in June 1900, the French steam and electric Narval introduced the classic double-hull design, with a pressure hull inside the outer shell. These 200-ton ships had a range of over 100 miles (160 km) underwater. The French submarine Aigrette in 1904 further improved the concept by using a diesel rather than a gasoline engine for surface power. Large numbers of these submarines were built, with seventy-six completed before 1914.

Submarines During the Russo-Japanese War

The first mechanically powered series of submarines to be put into service by navies, which included Great Britain, Japan, Russia, and the United States, were the Holland submersibles built by Irish designer John Philip Holland in 1900. Several of each of them were retained in both the Imperial Russian and Japanese Navies during the Russo-Japanese War in 1904-1905.

The Imperial Japanese Navy (IJN) started their submarine service with five Holland Type VII submarines purchased from the Electric Boat Company in 1904. The five vessels were delivered in sections, arriving in Japan on 14 June 1904. After re-assembly, the five Hollands were ready for combat operations in August 1905, but the Russo-Japanese War was nearing its end by that date, and no IJN submarines would see action in that war.

The first submarines built in Japan were constructed by Kawasaki beginning in 1904. The Kaigun Holland Type #6 and #7 were each launched on 28 September but a year apart, in 1905 and 1906 respectively. Both submarines were modified versions of the original imported Hollands.

However, while the original vessels had each displaced over a 100 tons submerged, and were approximately 67' long and 11' wide; the Kawasaki boats displaced only 63/95 tons submerged, and measured 73'/84' by 7' respectively for the number 6 & 7 submarines. The Kawasaki machines had increased horse-power by 1/2, and reduced fuel consumption by 1/4, but could only launch one 18" torpedo and carried 14 men, while the Hollands could fire two 18" torpedoes and operate with only 13 crewmen. The Kaigun Holland #6 submarine has been preserved as a memorial at Kure, Japan. The Imperial Russian

Navy (IRN) preferred the German constructed submersibles built by the Germaniawerft shipyards out of Kiel. In 1903 Germany successfully completed its first fully functional engine-powered submarine, the Forelle (Trout). This vessel was sold to Russia in 1904 and shipped via the Trans-Siberian Railway to the combat zone during the Russo-Japanese War. In 1901 two IRN Lieutenants, Kolbasieff and Kuteinoff designed and built the electric submarine Piotr Koschka which was operated by bicycle pedals, but no other versions were built. During the final weeks of the Port Arthur siege in 1904, the IRN attempted to place the Piotr Koschka into operation, her bicycle pedals having been replaced by an automobile engine. But the attempt to deploy the submarine into the Port Arthur battle was unsuccessful.

A prototype version of the Plunger-class or A-class submarines, the Fulton, was developed at Nixon's Crescent Shipyard for the United States Navy before the construction of the A-class submarines there in 1901. A naval architect and shipbuilder from the United Kingdom, Arthur Leopold Busch, superintended the development of these first submarines for Holland's company. However the Fulton was never purchased by the U.S. Navy and was eventually sold to the Imperial Russian Navy during the Russo-Japanese War of 1904-1905. Two other A-class vessels were built on the West Coast of (USA) at Mare Island Naval Shipyard/Union Iron Works circa 1901. In 1902, Holland received a patent for his persistent pursuit to perfect the underwater naval craft. By this time, Holland was no longer in control of the day-to-day operations at Electric Boat, as others were now at the helm of the company he once founded. The acumen of business were now in control of these operations as Holland was forced to step down. His resignation from the company was to be effective as of April 1904.

Due to the blockade at Port Arthur, Russia sent the remainder of their submarines to Vladivostok, where by 1 January 1905 there were seven boats, enough to create the world's first "operational submarine fleet." The new submarine fleet sent out its first patrol on 14 February, usually lasting for about 24 hours. The first confrontation with Japanese warships occurred on 29 April 1905 when the IRN sub Som was fired upon by IJN torpedo boats, but then withdrew.

In 1904, the Imperial Russian Navy ordered several more submersibles from the Kiel shipyard, submarines from the Karp class. One sample of which was modified and improved, and commissioned into the Imperial German Navy in 1906 as its first U-Boat, the U-1. U-1 was retired from service in 1919, and is currently preserved and on display in the Deutsches Museum in Munich.

Submarines During World War I

Figure: The German submarine U-9, which sank three British cruisers in less than an hour in September 1914

Military submarines first made a significant impact in World War I. Forces such as the U-boats of Germany saw action in the First Battle of the Atlantic, and were responsible for the sinking of Lusitania, which was sunk as a result of unrestricted submarine warfare and is often cited among the reasons for the entry of the United States into the war.

In August 1914, a flotilla of ten U-boats sailed from their base in Heligoland to attack Royal Navy warships in the North Sea in the first submarine war patrol in history. Their aim was to sink capital ships of the British Grand Fleet, and so reduce the Grand Fleet's numerical superiority over the German High Seas Fleet. With much depending more on luck than strategy, the first sortie was not a success. Only one attack was carried out, when U-15 fired a torpedo (which missed) at HMS Monarch, while two of the ten U-boats were lost. The U-9 had better luck. On 22 September 1914 while patrolling the Broad Fourteens, a region of the southern North Sea,U-9 found a squadron of three obsolescent British Cressy-class armoured cruisers (HMS Aboukir, HMS Hogue, and HMS Cressy), which were assigned to prevent German surface vessels from entering the eastern end of the English Channel. She fired all six of her torpedoes, reloading while submerged, and sank all three in less than an hour.

Figure: German U-boat U 14

The U-boats' ability to function as practical war machines relied on new tactics, their numbers, and submarine technologies such as combination diesel-electric power system developed in the preceding years. More submersibles than true submarines, U-boats operated primarily on the surface using regular engines, submerging occasionally to attack under battery power. They were roughly triangular in cross-section, with a distinct keel to control rolling while surfaced, and a distinct bow. During World War I more than 5,000 Allied ships were sunk by U-boats.

Interwar Developments

Various new submarine designs were developed during the interwar years. Among the most notable ones were submarine aircraft carriers, equipped with a waterproof hangar and steam catapult to launch and recover one or more small seaplanes. The submarine and its plane could then act as a reconnaissance unit ahead of the fleet, an essential role at a time when radar still did not exist. The first example was the British HMS M2, followed by the French Surcouf, and numerous aircraft-carrying submarines in the Imperial Japanese Navy.

Submarines During World War II

Germany

Figure: German submarine U-175 on the surface.

Germany had the largest submarine fleet during World War II. Due to the Treaty of Versailles limiting the surface navy, the rebuilding of the German surface forces had only begun in earnest a year before the outbreak of World War II. Expecting to be able to defeat the Royal Navy through underwater warfare, the German High Command pursued commerce raiding and immediately stopped all construction on capital surface ships save the nearly completed Bismarck-class battleships and two cruisers, switching its resources to submarines, which could be built more quickly. Though it took most of 1940 to expand the production facilities and get the mass production started, more than a thousand submarines were built by the end of the war.

During World War II, Germany utilised submarines to devastating effect in the Second Battle of the Atlantic, attempting to cut Britain's supply routes by sinking more merchant ships than Britain could replace. (Shipping was vital to supply Britain's population with food, industry with raw material, and armed forces with fuel and armaments.)

While U-boats destroyed a significant number of ships, the strategy ultimately failed. Although the U-boats had been updated in the interwar years, the major innovation was improved communications, encrypted using the famous Enigma cipher machine. This allowed for mass-attack tactics (Rudeltaktik, commonly known as "wolfpack"), but was also ultimately the U-boats' downfall.

After putting to sea, U-boats operated mostly on their own, trying to find convoys in areas assigned to them by the High Command. If a convoy was found, the submarine did not attack immediately, but shadowed to guide other submarines in the area. These then attacked more or less simultaneously, preferably at night while surfaced, which offered a speed advantage over the escorting corvettes and denied the Allies the ability to use ASDIC, which was unable to detect surfaced submarines.

From September 1939 to the beginning of 1943, the Ubootwaffe ("U-boat force") scored unprecedented success with these tactics, but were too few to have any decisive success. By the spring of 1943, German U-boat construction was at full capacity, but this was more than nullified by increased numbers of convoy escorts and aircraft, as well as technical advances like radar and sonar. High Frequency Direction Finding (HF/DF, known as Huff-Duff) and Ultra allowed the Allies to route convoys around wolfpacks when they detected radio transmissions from trailing boats. The results were devastating: from March to July of that year, over 130 U-boats were lost, 41 in May alone. Concurrent Allied losses dropped dramatically, from 750,000 tons in March to only 188,000 in July. Although the Second battle of the Atlantic would continue to the last day of the war, the U-boat arm was unable to stem the tide of personnel and supplies, paving the way for Operation Torch, Operation Husky, and ultimately, D-Day. Winston Churchill wrote the U-boat "peril" was the only thing to ever give him cause to doubt eventual Allied victory.

By the end of the war, almost 3,000 Allied ships (175 warships, 2,825 merchantmen) were sunk by U-boats. Of the 40,000 men in the U-boat service, 28,000 (70%) lost their lives.

Japan

Figure: The Imperial Japanese Navy's I-400-class submarine, the largest submarine type of WWII

During World War II, the IJN operated the most varied fleet of submarines of any navy; including Kaiten crewed torpedoes, midget submarines (Ko-hyoteki and Kairyu), medium-range submarines, purpose-built supply submarines and long-range fleet submarines. They also had submarines with the highest submerged speeds during World War II (I-201-class submarines) and submarines that could carry multiple aircraft (I-400-class submarine). They were also equipped with one of the most advanced torpedoes of the conflict, the oxygen-propelled Type 95.

Nevertheless, despite their technical prowess, Japan had chosen to utilise its submarines for fleet warfare, and consequently were relatively unsuccessful, as warships were fast, maneuverable and well-defended compared to merchant ships. In 1942, a Japanese submarine sank one aircraft carrier, damaged one battleship, and damaged one destroyer (which sank later) from one torpedo salvo; and during the Battle of Midway were able to deliver the coup de grace to another fleet aircraft carrier, again, sinking another destroyer, for another multiple score from one salvo. But with the lack of fuel oil and air supremacy, Imperial submarines were not able to sustain those kind of results afterwards. By the end of the war, submarines were instead often relegated to transport supplies to island garrisons.

United States

After the attack on Pearl Harbor, many of the U.S. Navy's front-line Pacific Fleet surface ships were destroyed or severely damaged. The submarines survived the attack and carried the war to the enemy. Lacking support vessels, the submarines were asked to independently hunt and destroy Japanese ships and submarines. They did so very effectively and without the assistance of other supporting ships.

During World War II the submarine force was the most effective anti-ship and anti-submarine weapon in the entire American arsenal. Submarines, though only about 2 percent of the U.S. Navy, destroyed over 30 percent of the Japanese Navy, including 8 aircraft carriers, 1 battleship and 11 cruisers. U.S. submarines also destroyed over 60 percent of the Japanese merchant fleet, crippling Japan's ability to supply its military forces and industrial war effort. Allied submarines in the Pacific War destroyed more Japanese shipping, than all other weapons combined. This feat was considerably aided by the Imperial Japanese Navy's failure to provide adequate escort forces for the nation's merchant fleet.

Of note, whereas Japan had the finest submarine torpedoes of the war, the U.S. Navy had the worst: for example, the U.S. Mark 14 torpedo typically ran ten feet too deep and was tipped with a Mk VIexploder, with both magnetic influence and contact features, neither reliable. The faulty depth control mechanism of the Mark 14 was corrected in August 1942, but field trials for the exploders were not ordered until mid-1943, when tests in Hawaii and Australia confirmed the flaws. In addition, the Mark 14 sometimes suffered circular runs, which sank at least one U.S. submarine, Tullibee. Fully operational Mark 14 torpedoes were not put into service until September 1943. The Mark 15 torpedo used by U.S. surface combatants had the same Mk VI exploder and was not fixed until late 1943. One attempt to correct the problems resulted in a wakeless, electric torpedo (the Mark 18) being placed in submarine service; Tang was lost to a circular run by one of these torpedoes. Given the prevalence of circular runs, there were probably other losses among boats which simply disappeared.

During World War II, 314 submarines served in the United States Navy, of which nearly 260 were deployed to the Pacific. On December 7, 1941, 111 boats were in commission; 203 submarines from the Gato, Balao, and Tench classes were commissioned during the war. During the war, 52 US submarines were lost to all causes, with 48 directly due to hostilities; 3,505 sailors were lost, the highest percentage killed in action of any US service arm in World War II. U.S. submarines sank 1,560 enemy vessels, a total tonnage of 5.3 million tons (55% of the total sunk), including 8 aircraft carriers, a battleship, three heavy cruisers, and over 200 other warships. In addition, the Japanese merchant marine lost 16,200 sailors killed and 53,400 wounded, of some 122,000 at the start of the war, due to submarines.

United Kingdom

The Royal Navy Submarine Service was primarily used to enforce the classic British blockade role. It therefore chiefly operated in inshore waters and tended to only surface by night. Its major operating areas were around Norway, the Mediterranean (against the Axis supply routes to North Africa), and in the Far East. Royal Navy submarines operating out of Trincomalee and Australia were a constant threat to Japanese shipping passing through the Malacca Straits.

In the war British submarines sank 2 million tons of enemy shipping and 57 major warships, the latter including 35 submarines. Among these is the only documented instance of a submarine sinking another submarine while both were submerged. This occurred when HMS Venturer engaged the U864; the Venturer crew manually computed a successful firing solution against a three-dimensionally manoeveuring target using techniques which became the basis of modern torpedo computer targeting systems. Seventy-four British submarines were lost, the majority, 42, in the Mediterranean.

Snorkel

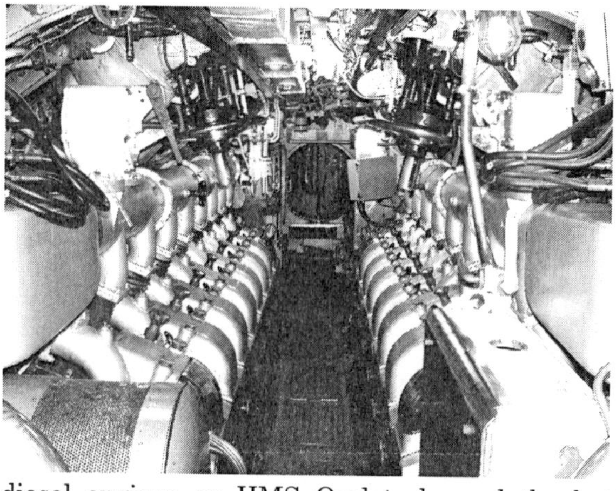

Figure: The diesel engines on HMS Ocelot charged the batteries located beneath the decking.

Diesel-electric submarines need air to run their diesel engines, and so carried very large batteries for submerged operation. The need to recharge the batteries from the diesel engines limited the endurance of the submarine while submerged and required it to surface regularly for extended periods, during which it was especially vulnerable to detection and attack. The snorkel, a prewar Dutch invention, was used to allow German submarines to run their diesel engines whilst running just under the surface, drawing air through a tube from the surface.

The German Navy also experimented with engines that would use hydrogen peroxide to allow diesel fuel to be used while submerged, but technical difficulties were great. The Allies experimented with a variety of detection systems, including chemical sensors to "smell" the exhaust of submarines.

Cold-war diesel-electric submarines, such as the Oberon class, used batteries to power their electric motors in order to run silently. They recharged the batteries using the diesel engines without ever surfacing.

Modern Military Submarines

Figure: HMAS Rankin, a Collins-class submarineat periscope depth

The first launch of a cruise missile (SSM-N-8 Regulus) from a submarine occurred in July 1953 from the deck of USSTunny, a World War II fleet boat modified to carry this missile with a nuclear warhead. Tunny and her sister boat Barbero were the United States's first nuclear deterrent patrol submarines. They were joined in 1958 by two purpose built Regulus submarines, Grayback, Growler, and, later, by the nuclear powered Halibut.

In the 1950s, nuclear power partially replaced diesel-electric propulsion. Equipment was also developed to extract oxygen from sea water. These two innovations gave submarines the ability to remain submerged for weeks or months, and enabled previously impossible voyages such as USS Nautilus' crossing of the North pole beneath the Arctic ice cap in 1958 and the USS Triton's submerged circumnavigation of the world in 1960. Most of the naval submarines built since that time in the United States and the Soviet Union/Russia have been powered by nuclear reactors. The limiting factors in submerged endurance for these vessels are food supply and crew morale in the space-limited submarine.

In 1959–1960, the first ballistic missile submarines were put into service by both the United States (George Washington class) and the Soviet Union (Hotel class) as part of the Cold War nuclear deterrent strategy.

While the greater endurance and performance from nuclear reactors makes nuclear submarines better for long-distance missions or the protection of a carrier battle group, their reactor cooling pumps have traditionally made them noisier, and thus easier to detect, than conventional diesel-electric submarines. Diesel-electrics have continued to be produced by both nuclear and non-nuclear powers as they lack this limitation, except when required to run the diesel engine to recharge the ship's battery. Recent technological advances in sound damping, noise isolation, and cancellation have made nuclear subs quieter and substantially eroded this disadvantage. Though far less capable regarding speed and weapons payload, conventional submarines are also cheaper to build. The introduction of air-independent propulsion boats, conventional diesel-electric submarines with some kind of auxiliary air-independent electricity generator, have led to increased sales of such types of submarines.

Figure: Nuclear powered Los Angeles-class submarines form the backbone of the United States submarine fleet.

During the Cold War, the United States and the Soviet Union maintained large submarine fleets that engaged in cat-and-mouse games. The Soviet Union suffered the loss of at least four submarines during this period: K-129 was lost in 1968 (which the CIA attempted to retrieve from the ocean floor with the Howard Hughes -designed ship Glomar Explorer), K-8 in 1970, K-219 in 1986, and Komsomolets in 1989 (which held a depth record among military submarines—1000 m). Many other Soviet subs, such as K-19 (the first Soviet nuclear submarine, and the first Soviet sub to reach the North Pole) were badly damaged by fire or radiation leaks. The US lost two nuclear submarines during this time: USS Thresher due to equipment failure during a test dive while at its operational limit, and USS Scorpion due to unknown causes.

During the Indo-Pakistani War of 1971, the Pakistan Navy's Hangor sank the Indian frigate INS Khukri. This was the first kill by a submarine since World War II, and the only one until the United Kingdom employed nuclear-powered submarines against Argentina in 1982 during the Falklands War. The Argentine cruiser General Belgrano was sunk by HMS Conqueror (the first sinking by a nuclear-powered submarine in war). The PNS Ghazi, a Tench-class submarine on loan to Pakistan from the US, was sunk in the Indo-Pakistani War. It was the first submarine casualty since World War II during war time.

More recently, Russia has had three high profile submarine accidents. The Kursk went down with all hands in 2000; the K-159 sank while being towed to a scrapyard in 2003, with nine lives lost; and the Nerpa had an accident with the fire-extinguishing system resulting in twenty deaths in late 2008.

India launched its first locally built nuclear-powered submarine, the INS Arihant, on July 26, 2009.

A North Korean submarine's torpedo allegedly sank the South Korean navy ship ROKS Cheonan on 26 March 2010.

Technology

Submersion and Trimming

All surface ships, as well as surfaced submarines, are in a positively buoyant condition, weighing less than the volume of water they would displace if fully submerged. To submerge hydrostatically, a ship must have negative buoyancy, either by increasing its own weight or decreasing its displacement of water. To control their weight, submarines have ballast tanks, which can be filled with outside water or pressurised air.

For general submersion or surfacing, submarines use the forward and aft tanks, called Main Ballast Tanks or MBTs, which are filled with water to submerge, or filled with air to surface. Under submerged conditions, MBTs generally remain flooded, which simplifies their design, and on many submarines these tanks are a section of interhull space. For more precise and quick control of depth, submarines use smaller Depth Control Tanks or DCTs, also called hard tanks, due to their ability to withstand higher pressure. The amount of water in depth control tanks can be controlled either to reflect changes in outside conditions or change depth. Depth control tanks can be located either near the submarine's centre of gravity, or separated along the submarine body to prevent affecting trim.

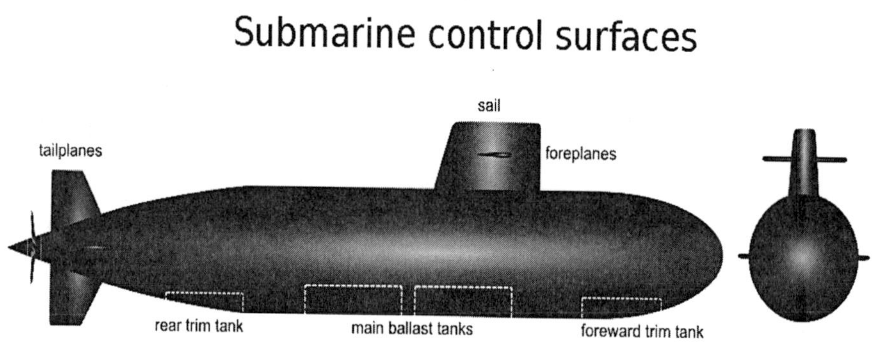

When submerged, the water pressure on submarine's hull can reach 4 MPa (580 psi) for steel submarines and up to 10 MPa (1,500 psi) for titanium submarines like Komsomolets, while interior pressure remains relatively unchanged. This difference results in hull compression, which decreases displacement. Water density also increases with depth, as the salinity and pressure are higher, but this incompletely compensates for hull compression, so buoyancy decreases as depth increases. A submerged submarine is in an unstable equilibrium, having a tendency to either fall or float to the surface. Keeping a constant depth requires continual operation of either the depth control tanks or control surfaces.

Submarines in a neutral buoyancy condition are not intrinsically trim-stable. To maintain desired trim, submarines use forward and aft trim tanks. Pumps can move water between these, changing weight distribution, creating a moment pointing the sub up or down. A similar system is sometimes used to maintain stability.

The hydrostatic effect of variable ballast tanks is not the only way to control the submarine underwater. Hydrodynamic maneuvering is done by several surfaces, which can be moved to create hydrodynamic forces when a submarine moves at sufficient speed. The stern planes, located near the propeller and normally horizontal, serve the same purpose as the trim tanks, controlling the trim, and are commonly used, while other control surfaces may not be present on many submarines. The fairwater planes on the sail and/or bow planes on the main body, both also horizontal, are closer to the centre of gravity, and are used to control depth with less effect on the trim.

When a submarine performs an emergency surfacing, all depth and trim methods are used simultaneously, together with propelling the boat upwards. Such surfacing is very quick, so the sub may even partially jump out of the water, potentially damaging submarine systems.

Submarine Hull

Figure: The US Navy Los Angeles-class attack submarine USS Greeneville in dry dock, showing typical cigar-shaped hull.

Modern submarines are cigar-shaped. This design, visible in early submarines (see below) is sometimes called a "teardrop hull". It reduces the hydrodynamic drag when submerged, but decreases the sea-keeping capabilities and increases drag while surfaced. Since the limitations of the propulsion systems of early submarines forced them to operate surfaced most of the time, their hull designs were a compromise. Because of the slow submerged speeds of those subs, usually well below 10 kt (18 km/h), the increased drag for underwater travel was acceptable. Late in World War II, when technology allowed faster and longer submerged operation and increased aircraft surveillance forced submarines to stay submerged, hull designs became teardrop shaped again to reduce drag and noise. On modern military submarines the outer hull is covered with a layer of sound-absorbing rubber, or anechoic plating, to reduce detection.

The occupied pressure hulls of deep diving submarines such as DSV Alvin are spherical instead of cylindrical. This allows a more even distribution of stress at the great depth. A titanium frame is usually affixed to the pressure hull, providing attachment for ballast and trim systems, scientific instrumentation, battery packs, syntactic flotation foam, and lighting.

A raised tower on top of a submarine accommodates the periscope and electronics masts, which can include radio, radar, electronic warfare, and other systems including the snorkel mast. In many early classes of submarines (see history), the control room, or "conn", was located inside this tower, which was known as the "conning tower". Since then, the conn has been located within the hull of the submarine, and the tower is now called the "sail". The conn is distinct from the "bridge", a small open platform in the top of the sail, used for observation during surface operation.

"Bathtubs" are related to conning towers but are used on smaller submarines. The bathtub is a metal cylinder surrounding the hatch that prevents waves from breaking directly into the cabin. It is needed because surfaced submarines have limited freeboard, that is, they lie low in the water. Bathtubs help prevent swamping the vessel.

Single and Double Hulls

Modern submarines and submersibles, as well as the oldest ones, usually have a single hull. Large submarines generally have an additional hull or hull sections outside. This external hull, which actually forms the shape of submarine, is called the outer hull (casing in the Royal Navy) or light hull, as it does not have to withstand a pressure difference. Inside the outer hull there is a strong hull, or pressure hull, which withstands sea pressure and has normal atmospheric pressure inside.

As early as World War I, it was realised that the optimal shape for withstanding pressure conflicted with the optimal shape for seakeeping and minimal drag, and construction difficulties further complicated the problem. This was solved either by a compromise shape, or by using two hulls; internal for holding pressure, and external for optimal shape. Until the end of World War II, most submarines had an additional partial cover on the top, bow and stern, built of thinner metal, which was flooded when submerged. Germany went further with the Type XXI, a general predecessor of modern submarines, in which the pressure hull was fully enclosed inside the light hull, but optimised for submerged navigation, unlike earlier designs that were optimised for surface operation.

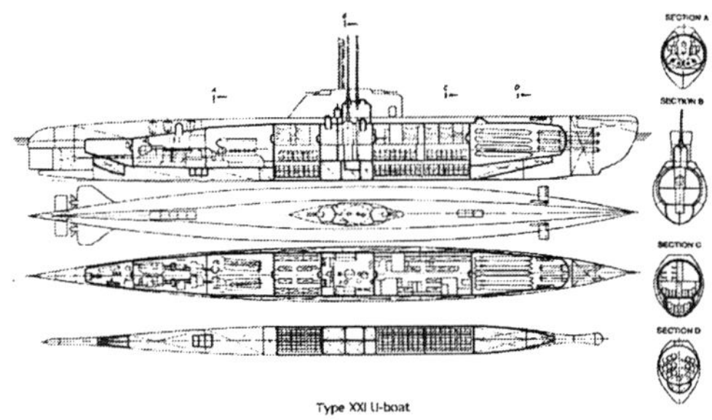

Figure: Type XXI U-Boat, late WWII, with pressure hull almost fully enclosed inside the light hull

After World War II, approaches split. The Soviet Union changed its designs, basing them on German developments. All post–World War II heavy Soviet and Russian submarines are built with a double hull structure. American and most other Western submarines switched to a primarily single-hull approach. They still have light hull sections in the bow and stern, which house main ballast tanks and provide a hydrodynamically optimised shape, but the main cylindrical hull section has only a single plating layer. The double hulls are being considered for future submarines in the United States to improve payload capacity, stealth and range.

Pressure Hull

The pressure hull is generally constructed of thick high strength steel with a complex structure and high strength reserve, and is separated with watertight bulkheads into several compartments. There are also examples of more than two hulls in a submarine, like the Typhoon class, which has two main pressure hulls and three smaller ones for control room, torpedoes and steering gear, with the missile launch system between the main hulls. The dive depth cannot be increased easily. Simply making the hull thicker increases the weight and requires reduction of onboard equipment weight, ultimately resulting in a bathyscaphe. This is acceptable for civilian research submersibles, but not military submarines.

WWI submarines had hulls of carbon steel, with a 100-metre (330 ft) maximum depth. During WWII, high-strength alloyed steel was introduced, allowing 200-metre (660 ft) depths. High-strength alloy steel remains the primary material for submarines today, with 250–400-metre (820–1,300 ft) depths, which cannot be exceeded on a military submarine without design compromises. To exceed that limit, a few submarines were built with titanium hulls. Titanium can be stronger than steel, lighter, and is not ferromagnetic, important for stealth. Titanium submarines were built by the Soviet Union, which developed specialised high-strength alloys. It has produced several types of titanium submarines. Titanium alloys allow a major increase in depth, but other systems need to be redesigned to cope, so test depth was limited to 1,000 metres (3,300 ft) for the Soviet submarine Komsomolets, the deepest-diving combat submarine. An Alfa-class submarine may have successfully operated at 1,300 metres (4,300 ft), though continuous operation at such depths would produce excessive stress on many submarine systems. Titanium does not flex as readily as steel, and may become brittle during many dive cycles. Despite its benefits, the high cost of titanium construction led to the abandonment of titanium submarine construction as the Cold War ended. Deep diving civilian submarines have used thick acrylic pressure hulls.

The deepest Deep Submergence Vehicle (DSV) to date is Trieste. On October 5, 1959 Triestedeparted San Diego for Guam aboard the freighter Santa Maria to participate in Project Nekton, a series of very deep dives in the Mariana Trench. On January 23, 1960, Trieste reached the ocean floor in the Challenger Deep (the deepest southern part of the Mariana Trench), carrying Jacques Piccard (son of Auguste) and Lieutenant Don Walsh, USN.

This was the first time a vessel, manned or unmanned, had reached the deepest point in the Earth's oceans. The onboard systems indicated a depth of 11,521 metres (37,799 ft), although this was later revised to 10,916 metres (35,814 ft) and more accurate measurements made in 1995 have found the Challenger Deep to be slightly shallower, at 10,911 metres (35,797 ft).

Building a pressure hull is difficult, as it must withstand pressures at its required diving depth. When the hull is perfectly round in cross-section, the pressure is evenly distributed, and causes only hull compression. If the shape is not perfect, the hull is bent, with several points heavily strained. Inevitable minor deviations are resisted by stiffener rings, but even a one inch (25 mm) deviation from roundness results in over 30 percent decrease of maximal hydrostatic load and consequently dive depth. The hull must therefore be constructed with high precision. All hull parts must be welded without defects, and all joints are checked multiple times with different methods, contributing to the high cost of modern submarines. (For example, each Virginia-class attack submarine costs US$2.6 billion, over US$200,000 per ton of displacement.)

Propulsion

Originally, submarines were human propelled. The first mechanically driven submarine was the 1863 French Plongeur, which used compressed air for propulsion. Anaerobic propulsion was first employed by the Spanish Ictineo II in 1864, which used a solution of zinc, manganese dioxide, and potassium chlorate to generate sufficient heat to power a steam engine, while also providing oxygen for the crew. A similar system was not employed again until 1940 when the German Navy tested a hydrogen peroxide-based system, the Walter turbine, on the experimental V-80 submarine and later on the naval U-791 and type XVII submarines.

Until the advent of nuclear marine propulsion, most 20th century submarines used batteries for running underwater and gasoline (petrol) or diesel engines on the surface, and for battery recharging. Early submarines used gasoline, but this quickly gave way to kerosene (paraffin), then diesel, because of reduced flammability. Diesel-electric became the standard means of propulsion.

The diesel or gasoline engine and the electric motor, separated by clutches, were initially on the same shaft driving the propeller. This allowed the engine to drive the electric motor as a generator to recharge the batteries and also propel the submarine. The clutch between the motor and the engine would be disengaged when the submarine dived, so that the motor could drive the propeller. The motor could have multiple armatures on the shaft, which could be electrically coupled in series for slow speed and in parallel for high speed. (These connections were called “group down” and “group up”, respectively.)

Electric Transmission

Diesel-electric

Early submarines used a direct mechanical connection between the engine and propeller, switching between diesel engines for surface running, and electric motors for submerged propulsion.

In 1928 the United States Navy’s Bureau of Engineering proposed a diesel-electric transmission; instead of driving the propeller directly while running on the surface, the submarine's diesel would drive a generator which could either charge the submarine's batteries or drive the electric motor. This meant that motor speed was independent of the diesel engine's speed, and the diesel could run at an optimum and non-critical speed, while one or more of the diesel engines could be shut down for maintenance while the submarine continued to run using battery power. The concept was pioneered in 1929 in the S-class submarines S-3, S-6, and S-7 to test the concept. No other navy adopted the system before 1945, apart from the Royal Navy's U-class submarines, though some submarines of the Imperial Japanese Navy used separate diesel generators for low speed running.

Other advantages of such an arrangement were that a submarine could travel slowly with the engines at full power to recharge the batteries quickly, reducing time on the surface or on snorkel. It was then possible to insulate the noisy diesel engines from the pressure hull, making the submarine quieter. Additionally, diesel-electric transmissions were more compact.

Figure: German Type XXI submarines, also known as "Elektroboote", were the first submarines designed to operate submerged for extended periods

Air-independent Propulsion

During the Second World War, German Type XXI submarines were designed to carry hydrogen peroxide for long-term, fast air-independent propulsion, but were ultimately built with very large batteries instead. At the end of the War, the British and Russians experimented with hydrogen peroxide/kerosene (paraffin) engines which could be used surfaced and submerged. The results were not encouraging; although the Russians deployed a class of submarines with this engine type (codenamed Quebec by NATO), they were considered unsuccessful.

Figure: German Type 212 submarine with AIP propulsion in dock at HDW/ Kiel

Today several navies use air-independent propulsion. Notably Sweden uses Stirling technology on the Gotland-class and Södermanland-class submarines. The Stirling engine is heated by burning diesel fuel with liquid oxygen from cryogenic tanks. A newer development in air-independent propulsion is hydrogen fuel cells, first used on the German Type 212 submarine, with nine 34 kW or two 120 kW cells and soon to be used in the new Spanish S-80 class submarines.

Nuclear Power

Figure: HMS Astute is amongst the most advanced nuclear submarines in the world.

Steam power was resurrected in the 1950s with a nuclear-powered steam turbine driving a generator. By eliminating the need for atmospheric oxygen, the length of time that a modern submarine could remain submerged was limited only by its food stores, as breathing air was recycled and fresh water distilled from seawater. Nuclear-powered submarines have a relatively small battery and diesel engine/generator powerplant for emergency use if the reactors must be shut down.

Nuclear power is now used in all large submarines, but due to the high cost and large size of nuclear reactors, smaller submarines still use diesel-electric propulsion. The ratio of larger to smaller submarines depends on strategic needs. The US Navy, French Navy, and the British Royal Navy operate only nuclear submarines, which is explained by the need for distant operations. Other major operators rely on a mix of nuclear submarines for strategic purposes and diesel-electric submarines for defence. Most fleets have no nuclear submarines, due to the limited availability of nuclear power and submarine technology.

Diesel-electric submarines have a stealth advantage over their nuclear counterparts. Nuclear submarines generate noise from coolant pumps and turbo-machinery needed to operate the reactor, even at low power levels. Some nuclear submarines such as the American Ohio class can operate with their reactor coolant pumps secured, making them quieter than electric subs. A conventional submarine operating on batteries is almost completely silent, the only noise coming from the shaft bearings, propeller, and flow noise around the hull, all of which stops when the sub hovers in mid water to listen, leaving only the noise from crew activity. Commercial submarines usually rely only on batteries, since they never operate independently of a mother ship.

Several serious nuclear and radiation accidents have involved nuclear submarine mishaps. The Soviet submarine K-19 reactor accident in 1961 resulted in 8 deaths and more than 30 other people were over-exposed to radiation. The Soviet submarine K-27 reactor accident in 1968 resulted in 9 fatalities and 83 other injuries. The Soviet submarine K-431 accident in 1985 resulted in 10 fatalities and 49 other people suffered radiation injuries.

Alternative Propulsion

Oil-fired steam turbines powered the British K-class submarines, built during the first World War and later, to give them the surface speed to keep up with the battle fleet. The K-class subs were not very successful, however. Toward the end of the 20th century, some submarines, such as the British Vanguard class, began to be fitted with pump-jet propulsors instead of propellers. Although these are heavier, more expensive, and less efficient than a propeller, they are significantly quieter, giving an important tactical advantage.

Magnetohydro dynamic drive (MHD) was portrayed as the operating principle behind the titular submarine's nearly silent propulsion system in the film adaptation of The Hunt for Red October. However, in the novel, the Red October did not use MHD. Although experimental surface ships have used this system, speeds have been below expectations. In addition, the drive system can induce bubble formation, compromising stealth, and the low efficiency requires high powered reactors. These factors make it unlikely for military usage.

Armament

The success of the submarine is inextricably linked to the development of the torpedo, invented by Robert Whitehead in 1866. His invention is essentially the same now as it was 140 years ago. Only with self propelled torpedoes could the submarine make the leap from novelty to a weapon of war. Until the perfection of the guided torpedo, multiple "straight-running" torpedoes were required to attack a target. With at most 20 to 25 torpedoes stored on board, the number of attacks was limited. To increase combat endurance most World War I submarines functioned as submersible gunboats, using their deck guns against unarmed targets, and diving to escape and engage enemy warships. The importance of guns encouraged the development of the unsuccessful Submarine Cruiser such as the French Surcouf and the Royal Navy's X1 and M-classsubmarines. With the arrival of ASW aircraft, guns became more for defence than attack. A more practical method of increasing combat endurance was the external torpedo tube, loaded only in port.

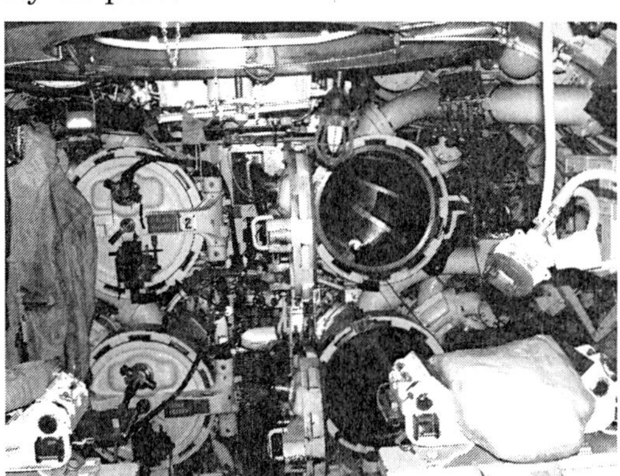

Figure: The forward torpedo tubes in HMSOcelot

The ability of submarines to approach enemy harbours covertly led to their use as minelayers. Minelaying submarines of World War I and World War II were specially built for that purpose. Modern submarine-laid mines, such as the British Mark 6 Sea Urchin, are designed to be deployed by a submarine's torpedo tubes.

After World War II, both the US and the USSR experimented with submarine launched cruise missiles such as the SSM-N-8 Regulus and P-5 Pyatyorka. Such missiles required the submarine to surface to fire its missiles. They were the forerunners of modern submarine launched cruise missiles, which can be fired from the torpedo tubes of submerged submarines, for example the US BGM-109 Tomahawk and Russian RPK-2 Viyuga and versions of surface to surface anti-ship missiles such as the Exocet and Harpoon, encapsulated for submarine launch. Ballistic missiles can also be fired from a submarine's torpedo tubes, for example missiles such as the anti-submarine SUBROC. With internal volume as limited as ever and the desire to carry heavier warloads, the idea of the external launch tube was revived, usually for encapsulated missiles, with such tubes being placed between the internal pressure and outer streamlined hulls.

The strategic mission of the SSM-N-8 and the P-5 were taken up by submarine-launched ballistic missile beginning with the US Navy's Polaris missile, and subsequently the Poseidon and Trident missiles.

Germany is working on the short-range IDAS (missile) which is launched from a torpedo tube and can be used against ASW helicopters as well as surface ships and coastal targets.

Sensors

A submarine will have a variety of sensors determined by its missions. Modern military submarines rely almost entirely on a suite of passive and active sonars to find their prey. Active sonar relies on an audible "ping" to generate echoes to reveal objects around the submarine. Active systems are rarely used, as doing so reveals the sub's presence. Passive sonar is a set of sensitive hydrophones set into the hull or trailed in a towed array, generally several hundred feet long. The towed array is the mainstay of NATO submarine detection systems, as it reduces the flow noise heard by operators. Hull mounted sonar is employed to back up the towed array, and in confined waters where a towed array could be fouled by obstacles.

Submarines also carry radar equipment for detection of surface ships and aircraft. Sub captains are more likely to use radar detection gear rather than active radar to detect targets, as radar can be detected far beyond its own return range, revealing the submarine. Periscopes are rarely used, except for position fixes and to verify a contact's identity. Civilian submarines, such as the DSV Alvin or the Russian Mir submersibles, rely on small active sonar sets and viewing ports to navigate. Sunlight does not penetrate below about 300 feet (91 m) underwater, so high intensity lights are used to illuminate the viewing area.

Navigation

Figure: The larger search periscope, and the smaller, less detectable attack periscope on HMS Ocelot

Early submarines had few navigation aids, but modern subs have a variety of navigation systems. Modern military submarines use an inertial guidance system for navigation while submerged, but drift error unavoidably builds up over time. To counter this, the Global Positioning System will occasionally be used to obtain an accurate position. The periscope - a retractable tube with prisms allowing a view to the surface - is only used occasionally in modern submarines, since the range of visibility is short. The Virginia-class submarines and Astute-class submarineshave photonics masts rather than hull-penetrating optical periscopes. These masts must still be hoisted above the surface, and employ electronic sensors for visible light, infrared, laser range-finding, and electromagnetic surveillance. One benefit to hoisting the mast above the surface is that while the mast is above the water the entire sub is still below the water and is much harder to detect visibly or by radar.

Communication

Military submarines have several systems for communicating with distant command centres or other ships. One is VLF (Very Low

Frequency) radio, which can reach a submarine either on the surface or submerged to a fairly shallow depth, usually less than 250 feet (76 m). ELF (Extremely Low Frequency) can reach a submarine at much greater depths, but has a very low bandwidth and are generally used to call a submerged sub to a shallower depth where VLF signals can reach. A submarine also has the option of floating a long, buoyant wire antenna to a shallower depth, allowing VLF transmissions to be made by a deeply submerged boat.

By extending a radio mast, a submarine can also use a "burst transmission" technique. A burst transmission takes only a fraction of a second, minimising a submarine's risk of detection. To communicate with other submarines, a system known as Gertrude is used. Gertrude is basically asonar telephone. Voice communication from one submarine is transmitted by low power speakers into the water, where it is detected by passive sonars on the receiving submarine. The range of this system is probably very short, and using it radiates sound into the water, which can be heard by the enemy. Civilian submarines can use similar, albeit less powerful systems to communicate with support ships or other submersibles in the area.

Crew

A typical nuclear submarine has a crew of over 80. Non-nuclear boats typically have fewer than half as many. The conditions on a submarine can be difficult because crew members must work in isolation for long periods of time, without family contact. Submarines normally maintain radio silence to avoid detection. Operating a submarine is dangerous, even in peacetime, and many submarines have been lost in accidents.

Women as Part of Crew

Most navies prohibited women from serving on submarines, even after they had been permitted to serve on surface warships. The Royal Norwegian Navy became the first navy to allow female crew on its submarines in 1985. The Royal Danish Navy allowed female submariners in 1988. Others followed suit including the Swedish Navy (1989), the Royal Australian Navy (1998), the German Navy (2001) and the Canadian Navy (2002). In 1995, Solveig Krey of the Royal Norwegian Navy became the first female officer to assume command on a military submarine, HNoMS Kobben.

The British Royal Navy does not yet permit women to serve on its submarines because of "medical concerns for the safety of the fetus and hence its mother" due to the potentially compromised air quality on board submarines. However, on 8 December 2011, Defence secretary Philip Hammond announced that the ban was to be lifted from 2013. Similar dangers to the pregnant woman and her fetus barred females from submarine service in Sweden 1983, when all other positions were made available for them in the Swedish Navy. Today, pregnant women are still not allowed to serve on submarines in Sweden. However, the policymakers thought that it was discriminatory with a general ban and demanded that females should be tried on their individual merits and have their suitability evaluated and compared to other candidates. Further, they noted that a female complying with such high demands is unlikely to become pregnant.

Women have served on U.S. Navy surface ships since 1993, and as of 2011-2012 will begin serving on submarines for the first time. Until presently, the Navy only allowed three exceptions for women being on board military submarines: female civilian technicians for a few days at most, women midshipmenon an overnight during summer training for both Navy ROTC and Naval Academy, and family members for one-day dependent cruises. In 2009, senior officials, including then-Secretary of the Navy Ray Mabus, Joint Chief of Staff Admiral Michael Mullen, and Chief of Naval Operations Admiral Gary Roughead, began the process of finding a way to implement females onboard submarines. In 2011, the first classes of female submarine officers graduated from Naval Submarine School's Submarine Officer Basic Course (SOBC) at the Naval Submarine Base New London. Additionally, more senior ranking and experienced female supply officers from the surface warfare speciality will attend SOBC as well, and proceed to fleet Ballistic Missile (SSBN) and Guided Missile (SSGN) submarines along with the new female submarine line officers beginning in late 2011/early 2012.

Both the U.S. and British navies operate nuclear-powered submarines which deploy for periods of six months or longer, whereas the other navies that do permit women to serve on submarines operate conventionally-powered submarines, which deploy for much shorter periods, usually only for one or two months. Prior the recent change by the U.S., no nation using nuclear submarines permitted women to serve onboard them.

Life Support Systems

With nuclear power, submarines can remain submerged for months at a time. Diesel submarines must periodically resurface or snorkel to recharge their batteries. Most modern military submarines generate breathing oxygen by electrolysis of water. Atmosphere control equipment includes a CO_2 scrubber, which uses an amine absorbent to remove the gas from air and diffuse it into waste pumped overboard. A machine that uses a catalyst to convert carbon monoxide into carbon dioxide (removed by the CO_2 scrubber) and bonds hydrogen produced from the ship's storage battery with oxygen in the atmosphere to produce water, is also used. An atmosphere monitoring system samples the air from different areas of the ship for nitrogen, oxygen, hydrogen, R-12 and R-114 refrigerants, carbon dioxide, carbon monoxide, and other gases. Poisonous gases are removed, and oxygen is replenished by use of an oxygen bank located in a main ballast tank. Some heavier submarines have two oxygen bleed stations (forward and aft). The oxygen in the air is sometimes kept a few percent less than atmospheric concentration to reduce fire danger.

Fresh water is produced by either an evaporator or a reverse osmosis unit. The primary use for fresh water is to provide feed water for the reactor and steam propulsion plants. It is also available for showers, sinks, cooking and cleaning once propulsion plant needs have been met. Seawater is used to flush toilets, and the resulting "black water" is stored in a sanitary tank until it is blown overboard using pressurised air or pumped overboard by using a special sanitary pump. The method for blowing sanitaries overboard is difficult to operate, and the German Type VIIC boat U-1206 was lost with casualties because of a mistake with the toilet. Water from showers and sinks is stored separately in "grey water" tanks, which are pumped overboard using the drain pump.

Trash on modern large submarines is usually disposed of using a tube called a Trash Disposal Unit (TDU), where it is compacted into a galvanized steel can. At the bottom of the TDU is a large ball valve. An ice plug is set on top of the ball valve to protect it, the cans atop the ice plug. The top breech door is shut, and the TDU is flooded and equalised with sea pressure, the ball valve is opened and the cans fall out assisted by scrap iron weights in the cans. The TDU is also flushed with seawater to ensure it is completely empty and the ball valve is clear before shutting the valve.

Pneumatics

Pneumatics is a branch of technology, which deals with the study and application of use of pressurised gas to effect mechanical motion. Pneumatic systems are extensively used in industry, where factories are commonly plumbed with compressed air or compressed inert gases. This is because a centrally located and electrically powered compressor that powers cylinders and other pneumatic devices through solenoid valves is often able to provide motive power in a cheaper, safer, more flexible, and more reliable way than a large number of electric motors and actuators.

Pneumatics also has applications in dentistry, construction, mining, and other areas.

Examples of Pneumatic Systems and Components

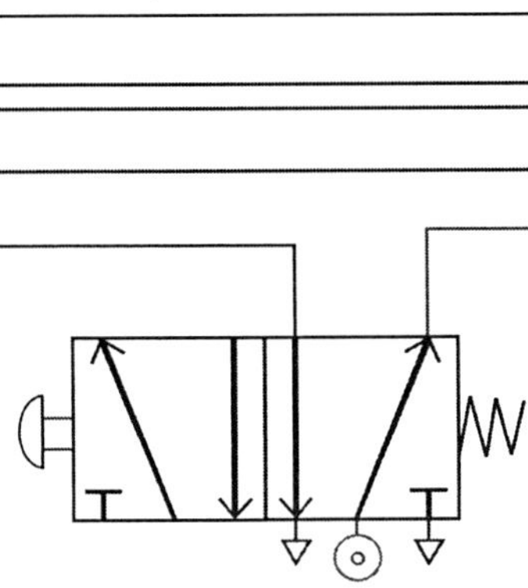

Pneumatic circuit.

- Air brakes on buses and trucks
- Air brakes, on trains
- Air compressors
- Air engines for pneumatically powered vehicles
- Barostat systems used in Neurogastroenterology and for researching electricity
- Cable jetting, a way to install cables in ducts
- Compressed-air engine and compressed-air vehicles
- Gas-operated reloading
- Holman Projector, a pneumatic anti-aircraft weapon
- Inflatable structures
- Lego pneumatics can be used to build pneumatic models
- Pipe organs:
- Electro-pneumatic action
- Tubular-pneumatic action
- Pneumatic actuator
- Pneumatic air guns
- Pneumatic cylinder

- Pneumatic Launchers, a type of spud gun
- Pneumatic mail systems
- Pneumatic motor
- Pneumatic tire
- Pneumatic tools:
- Jackhammer used by road workers
- Pneumatic nailgun
- Pressure regulator
- Pressure sensor
- Pressure switch
- Vacuum pump

Gases Used in Pneumatic Systems

Pneumatic systems in fixed installations such as factories use compressed air because a sustainable supply can be made by compressing atmospheric air. The air usually has moisture removed and a small quantity of oil added at the compressor, to avoid corrosion of mechanical components and to lubricate them. Factory-plumbed, pneumatic-power users need not worry about poisonous leakages as the gas is commonly just air. Smaller or stand-alone systems can use other compressed gases which are an asphyxiation hazard, such as nitrogen - often referred to as OFN (oxygen-free nitrogen), when supplied in cylinders. Any compressed gas other than air is an asphyxiation hazard - including nitrogen, which makes up 77% of air. Compressed oxygen (approx. 23% of air) would not asphyxiate, but it would be an extreme fire hazard, so is never used in pneumatically powered devices.

Portable pneumatic tools and small vehicles such as Robot Wars machines and other hobbyist applications are often powered by compressed carbon dioxide because containers designed to hold it such as soda stream canisters and fire extinguishers are readily available, and the phase change between liquid and gas makes it possible to obtain a larger volume of compressed gas from a lighter container than compressed air would allow. Carbon dioxide is an asphyxiant and can also be a freezing hazard when vented inappropriately.

Comparison to Hydraulics

Both pneumatics and hydraulics are applications of fluid power. Pneumatics uses an easily compressible gas such as air or a suitable pure gas, while hydraulics uses relatively incompressible liquid media such as oil. Most industrial pneumatic applications use pressures of about 80 to 100pounds per square inch (550 to 690 kPa). Hydraulics applications commonly use from 1,000 to 5,000 psi (6.9 to 34 MPa), but specialised applications may exceed 10,000 psi (69 MPa).

Advantages of Pneumatics

- Simplicity of Design And Control
  - Machines are easily designed using standard cylinders & other components. Machines operate by simple ON - OFF type control.
- Reliability
  - Pneumatic systems tend to have long operating lives and require very little maintenance.
  - Because gas is compressible, the equipment is less likely to be damaged by shock. The gas in pneumatics absorbs excessive force, whereas the fluid of hydraulics directly transfers force.
- Storage
  - Compressed gas can be stored, allowing the use of machines when electrical power is lost.
- Safety
  - Very low chance of fire (compared to hydraulic oil).
  - Machines can be designed to be overload safe.

Advantages of Hydraulics

- Liquid (as a gas is also a 'fluid') does not absorb any of the supplied energy.
- Capable of moving much higher loads and providing much higher forces due to the incompressibility.
- The hydraulic working fluid is basically incompressible, leading to a minimum of spring action. When hydraulic fluid flow is stopped, the slightest motion of the load releases the pressure on the load; there is no need to "bleed off" pressurised air to release the pressure on the load.

Pneumatic Logic

Pneumatic logic systems (sometimes called air logic control) are often used to control industrial processes, consisting of primary logic units such as:

- And Units
- Or Units

- 'Relay or Booster' Units
- Latching Units
- 'Timer' Units
- Sorteberg relay
- Fluidics amplifiers with no moving parts other than the air itself

Pneumatic logic is a reliable and functional control method for industrial processes. In recent years, these systems have largely been replaced by electrical control systems, due to the smaller size and lower cost of electrical components. Pneumatic devices are still used in processes where compressed air is the only energy source available or upgrade cost, safety, and other considerations outweigh the advantage of modern digital control.

Hydraulics

Hydraulics is a topic in applied science and engineering dealing with the mechanical properties of liquids. Fluid mechanics provides the theoretical foundation for hydraulics, which focuses on the engineering uses of fluid properties. In fluid power, hydraulics is used for the generation, control, and transmission of power by the use of pressurised liquids. Hydraulic topics range through most science and engineering disciplines, and cover concepts such as pipe flow, dam design, fluidics and fluid control circuitry, pumps, turbines, hydropower, computational fluid dynamics, flow measurement, river channel behaviour and erosion.

Free surface hydraulics is the branch of hydraulics dealing with free surface flow, such as occurring in rivers, canals, lakes, estuaries and seas. Its sub-field open channel flow studies the flow in openchannels. The word "hydraulics" originates from the Greek word hydraulikos which in turn originates from hydor, Greek for water and aulos, meaning pipe.

Railway Air Brake

An air brake is a conveyance braking system actuated by compressed air. Modern trains rely upon a fail-safe air brake system that is based upon a design patented by George Westinghouse on March 5, 1872. The Westinghouse Air Brake Company (WABCO) was subsequently organised to manufacture and sell Westinghouse's invention. In various forms, it has been nearly universally adopted. The Westinghouse system uses air pressure to charge air reservoirs (tanks) on each car. Full air pressure signals each car to release the brakes. A reduction or loss of air pressure signals each car to apply its brakes, using the compressed air in its reservoirs.

Prior to the introduction of air brakes, stopping a train was a difficult business. In the early days when trains consisted of one or two cars and speeds were low, the engine driver could stop the train by reversing the steam flow to the cylinders, causing the locomotive to act as a brake. However, as trains got longer, heavier and faster, and started to operate in mountainous regions, it became necessary to fit each car with brakes, as the locomotive was no longer capable of bringing the train to a halt in a reasonable distance.

The introduction of brakes to railcars necessitated the employment of additional crew members called brakemen, whose job it was to move from car to car and apply or release the brakes when signaled to do so by the engineer with a series of whistle blasts. Occasionally, whistle signals were not heard, incorrectly given or incorrectly interpreted, and derailments or collisions would occur because trains were not stopped in time. Brakes were manually applied and released by turning a large brake wheel located at one end of each car. The brake wheel pulled on the car's brake rigging and clamped the brake shoes against the wheels. As considerable force was required to overcome the friction in the brake rigging, the brakeman used a stout piece of wood called a "club" to assist him in turning the brake wheel.

The job of a passenger train brakeman wasn't too difficult, as he was not exposed to the weather and could conveniently move from car to car through the vestibules, which is where the brake wheel was (and still is, in many cases) located. Also, passenger trains were not as heavy or lengthy as their freight counterparts, which eased the task of operating the brakes. A brakeman's job on a freight train was far more difficult, as he was exposed to the elements and was responsible for many more cars. To set the brakes on a boxcar (UIC: covered wagon) the brakeman had to climb to the roof ("coon the buggy" in railroad slang) and walk a narrow catwalk to reach the brake wheel while the car was swaying and pitching beneath his feet. There was nothing to grasp other than the brake wheel itself, and getting to the next car often required jumping. Needless to say, a freight brakeman's job was extremely dangerous, and many were maimed or killed in falls from moving trains.

Complicating matters, the manually operated brakes had limited effectiveness and controlling a train's speed in mountainous terrain was a dicey affair. Occasionally, the brakemen simply could not set enough brakes to a degree where they were able to reduce speed while descending a grade, which usually resulted in a runaway—followed by a disastrous wreck. When adopted, the Westinghouse system had a major effect on railroad safety. Reliable braking was assured, reducing the frequent accidents that plagued the industry. Brakemen were no longer required to risk life and limb to stop a train, and with the engineer now in control of the brakes, misunderstood whistle signals were eliminated. As a result, longer and heavier trains could be safely run at higher speeds.

During his lifetime, Westinghouse made many improvements to his invention. The United States Congress passed the Safety Appliance Act in 1893 making the use of some automatic brake system mandatory. By 1905, over 2,000,000 freight, passenger, mail, baggage and express railroad cars and 89,000 locomotives in the United States were equipped with the Westinghouse Automatic Brake. In the air brake's simplest form, called the straight air system, compressed air pushes on a piston in a cylinder. The piston is connected through mechanical linkage to brake shoes that can rub on the train wheels, using the resulting friction to slow the train. The mechanical linkage can become quite elaborate, as it evenly distributes force from one pressurised air cylinder to 8 or 12 wheels.

The pressurised air comes from an air compressor in the locomotive and is sent from car to car by at rain line made up of pipes beneath each car and hoses between cars. The principal problem with the straight air braking system is that any separation between hoses and pipes causes loss of air pressure and hence the loss of the force applying the brakes. This deficiency could easily cause a runaway train. Straight air brakes are still used on locomotives, although as a dual circuit system, usually with each bogie (truck) having its own circuit.

In order to design a system without the shortcomings of the straight air system, Westinghouse invented a system wherein each piece of railroad rolling stock was equipped with an air reservoir and atriple valve, also known as a control valve.

The triple valve is described as being so named as it performs three functions: Charging air into an air tank ready to be used, applying the brakes, and releasing them. In so doing, it supports certain other actions (i.e. it 'holds' or maintains the application and it permits the exhaust of brake cylinder pressure and the recharging of the reservoir during the release). In his patent application,

Westinghouse refers to his 'triple-valve device' because of the three component valvular parts comprising it: the diaphragm-operated poppet valve feeding reservoir air to the brake cylinder, the reservoir charging valve, and the brake cylinder release valve. When he soon improved the device by removing the poppet valve action, these three components became the piston valve, the slide valve, and the graduating valve.

- If the pressure in the train line is lower than that of the reservoir, the brake cylinder exhaust portal is closed and air from the car's reservoir is fed into the brake cylinder to apply the brakes. This action continues until equilibrium between the brake pipe pressure and reservoir pressure is achieved. At that point, the airflow from the reservoir to the brake cylinder is lapped off and the cylinder is maintained at a constant pressure.
- If the pressure in the train line is higher than that of the reservoir, the triple valve connects the train line to the reservoir feed, causing the air pressure in the reservoir to increase. The triple valve also causes the brake cylinder to be exhausted to atmosphere, releasing the brakes.
- As the pressure in the train line and that of the reservoir equalise, the triple valve closes, causing the air pressure in the reservoir and brake cylinder to be maintained at the current level.

Unlike the straight air system, the Westinghouse system uses a reduction in air pressure in the train line to apply the brakes. When the engineer (driver) applies the brake by operating the locomotive brake valve, the train line vents to atmosphere at a controlled rate, reducing the train line pressure and in turn triggering the triple valve on each car to feed air into its brake cylinder. When the engineer releases the brake, the locomotive brake valve portal to atmosphere is closed, allowing the train line to be recharged by the compressor of the locomotive. The subsequent increase of train line pressure causes the triple valves on each car to discharge the contents of the brake cylinder to atmosphere, releasing the brakes and recharging the reservoirs.

Under the Westinghouse system, therefore, brakes are applied by reducing train line pressure and released by increasing train line pressure. The Westinghouse system is thus fail safe—any failure in the train line, including a separation ("break-in-two") of the train, will cause a loss of train line pressure, causing the brakes to be applied and bringing the train to a stop, thus preventing a runaway train.

Modern air brake systems are in effect two braking systems combined:

- The service brake system, which applies and releases the brakes during normal operations, and
- The emergency brake system, which applies the brakes rapidly in the event of a brake pipe failure or an emergency application by the engineer.

When the train brakes are applied during normal operations, the engineer makes a "service application" or a "service rate reduction", which means that the train line pressure reduces at a controlled rate. It takes several seconds for the train line pressure to reduce and consequently takes several seconds for the brakes to apply throughout the train. In the event the train needs to make an emergency stop, the engineer can make an "emergency application," which immediately and rapidly vents all of the train line pressure to atmosphere, resulting in a rapid application of the train's brakes. An emergency application also results when the train line comes apart or otherwise fails, as all air will also be immediately vented to atmosphere.

In addition, an emergency application brings in an additional component of each car's air brake system: the emergency portion. The triple valve is divided into two portions: the service portion, which contains the mechanism used during brake applications made during service reductions, and the emergency portion, which senses the immediate, rapid release of train line pressure. In addition, each car's air brake reservoir is divided into two portions—the service portion and the emergency portion—and is known as the "dual-compartment reservoir". Normal service applications transfer air pressure from the service portion to the brake cylinder, while emergency applications cause the triple valve to direct all air in both the service portion and the emergency portion of the dual-compartment reservoir to the brake cylinder, resulting in a 20–30% stronger application.

The emergency portion of each triple valve is activated by the extremely rapid rate of reduction of train line pressure. Due to the length of trains and the small diametre of the train line, the rate of reduction is high near the front of the train (in the case of an engineer-initiated emergency application) or near the break in the train line (in the case of the train line coming apart). Farther away from the source of the emergency application, the rate of reduction can be reduced to the point where triple valves will not detect the application as an emergency reduction. To prevent this, each triple valve's emergency portion contains an auxiliary vent port, which, when activated by an emergency application, also locally vents the train line's pressure directly to atmosphere. This serves to propagate the emergency application rapidly along the entire length of the train.

Use of distributed power (i.e., remotely controlled locomotive units midtrain and/or at the rear end) mitigates somewhat the time-lag problem with long trains, because a telemetred radio signal from the engineer in the front locomotive commands the distant units to initiate brake pressure reductions that propagate quickly through nearby cars.

Enhancements

Electro-pneumatic or EP brakes are a type of air brake that allows for immediate application of brakes throughout the train instead of the sequential application. EP brakes have been in use in German high-speed trains (most notably the ICE) since the late 1980s, and in British practice since 1949, fully described in Electro-pneumatic brake system on British railway trains. Electro-pneumatic brakes are currently in testing in North America and South Africa in captive service ore and coal trains.

Passenger trains have had for a long time a 3-wire version of the Electro-pneumatic brake, which gives seven levels of braking force. In most cases the system is not fail-safe, with the wires being energised in sequence to apply the brakes, but the conventional automatic air brake is also provided to act as a fail safe, and in most cases can be used independently in the event of a failure of the EP brakes.

In North America, WABCO supplied HSC (High Speed Control) brake equipment for several post-World War II streamlined passenger trains. This was an electrically controlled overlay on conventional D-22 passenger and 24-RL locomotive brake equipment. On the conventional side, the control valve set a reference pressure in a volume, which set brake cylinder pressure via a relay valve. On the electric side, pressure from a second straight-air trainline controlled the relay valve via a two-way check valve. This "straight air" trainline was charged (from reservoirs on each car) and released by magnet valves on each car, controlled electrically by a 3 wire trainline, in turn controlled by an "electro-pneumatic master controller" in the controlling locomotive. This controller compared the pressure in the straight air trainline with that supplied by a self lapping portion of the engineers valve, signaling all of the "apply" or "release" magnets valves in the train to open simultaneously, changing the pressure in the "straight air" trainline much more rapidly and evenly than possible by simply supplying air directly from the locomotive. The relay valve was equipped with four diaphragms, magnet valves, electric control equipment, and an axle-mounted speed sensor, so that at speeds over 60 mph full braking force was applied, and reduced in steps at 60, 40 and 20 mph, bringing the train to a gentle stop. Each axle was also equipped with anti-lock brake equipment. The combination minimised braking distances, allowing more full-speed running between stops. The "straight air" (electro-pneumatic trainline), anti-lock, and speed graduating portions of the system were not dependent on each other in any way, and any or all of these options could be supplied separately.

Later systems replace the automatic air brake with an electrical wire (in the UK, at least, known as a "round the train wire") that has to be kept energised to keep the brakes off.

More recent innovations are electronically controlled pneumatic brakes where the brakes of all the wagons (cars) and locomotives are connected by a kind of local area network, which allows individual control of the brakes on each wagon, and the reporting back of performance of each wagon's brakes.

Limitations

The Westinghouse air brake system is very trustworthy, but not infallible. Recall that the car reservoirs recharge only when the brake pipe pressure is higher than the reservoir pressure, and that the car reservoir pressure will rise only to the point of equilibrium. Fully recharging the reservoirs on a long train can require considerable time (8 to 10 minutes in some cases), during which the brake pipe pressure will be lower than locomotive reservoir pressure.

If the brakes must be applied before recharging has been completed, a larger brake pipe reduction will be required in order to achieve the desired amount of braking effort, as the system is starting out at a lower point of equilibrium (lower overall pressure). If many brake pipe reductions are made in short succession ("fanning the brake" in railroad slang), a point may be reached where car reservoir pressure will be severely depleted, resulting in substantially reduced brake cylinder piston force, causing the brakes to fail. On a descending grade, the unfortunate result will be a runaway.

In the event of a loss of braking due to reservoir depletion, the engineer (driver) may be able to regain control with an emergency brake application, as the emergency portion of each car's dual-compartment reservoir should be fully charged—it is not affected by normal service reductions. The triple valves detect an emergency reduction based on the rate of brake pipe pressure reduction. Therefore, as long as a sufficient volume of air can be rapidly vented from the brake pipe, each car's triple valve will cause an emergency brake application. However, if the brake pipe pressure is too low due to an excessive number of brake applications, an emergency application will not produce a large enough volume of air flow to trip the triple valves, leaving the engineer with no means to stop the train.

To prevent a runaway due to loss of brake pressure, dynamic (rheostatic) braking can be utilised so the locomotive(s) will assist in retarding the train. Often, blended braking, the simultaneous application of dynamic and train brakes, will be used to maintain a safe speed and keep the slack stretched as the train crests a grade.

Another solution to loss of brake pressure is the two-pipe system, fitted on most modern passenger stock and many freight wagons. In addition to the traditional brake pipe, this enhancement adds the main reservoir pipe, which is continuously charged with air directly from the locomotive's main reservoir. The main reservoir is where the locomotive's air compressor output is stored, and is ultimately the source of compressed air for all systems that use it.

Since the main reservoir pipe is kept constantly pressurised by the locomotive, the car reservoirs can be charged independently of the brake pipe, this being accomplished via a check valve to prevent backfeeding into the pipe. This arrangement helps to reduce the above described pressure loss problems, and also reduces the time required for the brakes to release, since the brake pipe only has to recharge itself.

Main reservoir pipe pressure can also be used to supply air for auxiliary systems such as pneumatic door operators or air suspension. Nearly all passenger trains (all in the UK and USA), and many freights, now have the two-pipe system.

Accidents

The air brake can fail if one of the cocks where the pipes of each carriage are joined together is accidentally closed. In this case, the brakes on the wagons behind the closed cock will fail to respond to the driver's command. This happened in 1953 to the Federal Express, a Pennsylvania Railroad train pulling in to Washington DC's Union Station, causing the train to crash into the passenger concourse and fall through the floor. Similarly, in the Gare de Lyon train accident a valve was accidentally closed by the crew, reducing braking power.

There are a number of safeguards that are usually taken to prevent this sort of accident happening. Railroads have strict government-approved procedures for testing the air brake systems when making up trains in a yard or picking up cars en route. These generally involve connecting the air brake hoses, charging up the brake system, setting the brakes and manually inspecting the cars to ensure the brakes are applied, and then releasing the brakes and manually inspecting the cars to ensure the brakes are released. Particular attention is usually paid to the rearmost car of the train, either by manual inspection or via an automated end-of-train device, to ensure that brake pipe continuity exists throughout the entire train. When brake pipe continuity exists throughout the train, failure of the brakes to apply or release on one or more cars is an indication that the cars' triple valves are malfunctioning. Depending on the location of the air test, the repair facilities available, and regulations governing the number of inoperative brakes permitted in a train, the car may be set out for repair or taken to the next terminal where it can be repaired.

Standardisation

The modern air brake is not identical with the original airbrake as there have been slight changes in the design of the triple valve, which are not completely compatible between versions, and which must therefore be introduced in phases. That said, the basic air brakes used on railways worldwide are remarkably compatible.

Vacuum Brakes

The main competitor to the air brake is the vacuum brake, which operates on negative pressure. The vacuum brake is a little simpler than the air brake, with an ejector with no moving parts on steam engines or a mechanical or electrical "exhauster" on a diesel or electric locomotive replacing the air compressor. Disconnection taps at the ends of cars are not required as the loose hoses are sucked onto a mounting block.

However, the maximum pressure is limited to atmospheric pressure, so that all the equipment has to be much larger and heavier to compensate. This disadvantage is made worse at high altitude. The vacuum brake is also considerably slower acting in both applying and releasing the brake; this requires a greater level of skill and anticipation from the driver. Conversely, the vacuum brake had the advantage of gradual release long before the Westinghouse automatic air brake, which was originally only available in the direct-release form still common in freight service.

A primary fault of vacuum brakes is the inability to easily find leaks. In a positive air system, a leak is quickly found due to the escaping pressurised air. This problem left the British railways in a terrible condition, where trains would have to be stopped at the top of grades to set the manual brakes on each car. Purchase and maintenance of a mechanical air pump on hundreds of engines is nothing compared to keeping the vacuum line in good order across a fleet of tens of thousands of freight cars.

Electro-vacuum brakes have also been used with considerable success on South African electric multiple unit trains. Despite requiring larger and heavier equipment as stated above, the performance of the electro-vacuum brake approached that of contemporary electro-pneumatic brakes. However, their use has not been repeated.

Air Brake (Road Vehicle)

Figure: Truck air actuated disc brake.

Air brakes are used in trucks, buses, trailers, and semi-trailers. George Westinghouse first developed air brakes for use in railway service. He patented a safer air brake on March 5, 1872. Originally designed and built for use on railroad train application, air brakes remain the exclusive systems in widespread use. Westinghouse made numerous alterations to improve his air pressured brake invention, which led to various forms of the automatic brake and the subsequent use on heavier road vehicles.

Compressed Air Brake System

Compressed air brake systems are typically used on heavy trucks and buses. The system consists of service brakes, parking brakes, a control pedal, an engine-driven air compressor and a compressed air storage tank. For the parking brake, there is a disc or drum brake arrangement which is designed to be held in the 'applied' position by spring pressure. Air pressure must be produced to release these "spring brake" parking brakes. For the service brakes (the ones used while driving for slowing or stopping) to be applied, the brake pedal is pushed, routing the air under pressure (approx 100-125psi) to the brake chamber, causing the brake to reduce wheel rotation speed. Most types of truck air brakes are drum units, though there is an increasing trend towards the use of disc brakes in this application. The air compressor air draws filtered air from the atmosphere and forces it into high-pressure reservoirs at around 120 PSI. Most heavy vehicles have a gauge within the driver's view, indicating the availability of air pressure for safe vehicle operation, often including warning tones or lights. Setting of the parking/emergency brake releases the pressurised air pressure in the lines between the compressed air storage tank and the brakes, thus actuating the (spring brake) parking braking hardware. An air pressure failure at any point would apply full spring brake pressure immediately.

Design and Function

A compressed air brake system is divided into a supply system and a control system. The supply system compresses, stores and supplies high-pressure air to the control system as well as to additional air operated auxiliary truck systems (gearbox shift control, clutch pedal air assistance servo, etc.).

Supply System

The air compressor is driven off of the engine either by crankshaft pulley via a belt or directly off of the engine timing gears. It is lubricated and cooled by the engine lubrication and cooling systems. Compressed air is first routed through a cooling coil and into an air dryer which removes moisture and oil impurities and also may include a pressure regulator, safety valve and a smaller purge reservoir. As an alternative to the air dryer, the supply system can be equipped with an anti freeze device and oil separator. The compressed air is then stored in a reservoir (also called a wet tank) from which it is then distributed via a four way protection valve into the front and rear brake circuit air reservoir, a parking brake reservoir and an auxiliary air supply distribution point. The system also includes various check, pressure limiting, drain and safety valves.

Control System

The control system is further divided into two service brake circuits: the parking brake circuit and the trailer brake circuit. This dual brake circuit is further split into front and rear wheel circuits which receive compressed air from their individual reservoirs for added safety in case of an air leak. The service brakes are applied by means of a brake pedal air valve which regulates both circuits. The parking brake is the air operated spring brake type where its applied by spring force in the spring brake cylinder and released by compressed air via hand control valve. The trailer brake consists of a direct two line system: the supply line (marked red) and the separate control or service line (marked blue). The supply line receives air from the prime mover park brake air tank via a park brake relay valve and the control line is regulated via the trailer brake relay valve. The operating signals for the relay are provided by the prime mover brake pedal air valve, trailer service brake hand control (subject to a country's relevant heavy vehicle legislation) and the prime mover park brake hand control.

Exposed Physical Structure

This example of the air brake consists of a physical structure on the exterior of a vehicle that will increase the vehicle's drag coefficient, and therefore slow it down. Air brakes of this sort are ineffective at normal road vehicle speeds, and therefore are reserved for vehicles which need to quickly decelerate from high speeds, such as race and high performance sports cars.

Figure: The Bugatti Veyron's hydraulic rear spoiler/air brake in the wing position.

Many high performance sports and racing cars utilise air brakes in order to slow the cars down from high speeds. The Bugatti Veyron, one of the fastest production cars in the world, features a rear spoiler which, at speeds above 200 km/h (120 mph), also acts as an air brake, snapping to a 55° angle in 0.4 seconds once the brake pedal is pressed, providing an additional 0.68 g (6.66 m/s^2) of deceleration (equivalent to the stopping power of an ordinary hatchback). Top Fuel Dragsters and other drag racing cars that routinely reach speeds greater than 150 miles per hour use a physical air brake via a parachute(s) after the completion of a race.

In 1994, NASCAR introduced roof flaps to the cars, which are designed to keep cars from becoming airborne and possibly flipping. Following Rusty Wallace's crash at Talladega, Penske Racing designed the original roof flaps. NASCAR team owner Jack Roush helped improve on the design of the roof flaps, in conjunction with Embry-Riddle Aeronautical University, Daytona, Florida, USA. During a spin, the car rotates it eventually reaches an angle where the oncoming air reacts with the profile of the vehicle in the same manner as a wing. If the speed is high enough, air flowing over this aerofoil shape will create sufficient lift to force the car to become airborne. To prevent this, NASCAR developed a set of flaps that are recessed into pockets on the roof of the car. As a car is turned around and reaches an angle where significant lift occurs, the low pressure above the flaps causes them to deploy. The first flap, oriented 140 degrees from the centreline of the car, typically deploys first. After flap deployment, higher pressure air is forced through an air tube which connects to a second flap, deploying it. This second flap ensures that, should the car continue to spin, no further lift will be created as the vehicle's angle changes. The deployment of these flaps eliminates most of the lift on the vehicle. The roof flaps generally keep the cars on the ground as they spin, although it is not guaranteed.

Bibliography

Ashby, Michael; Hugh Shercliff and David Cebon: *Materials: Engineering, Science, Processing and Design*, Butterworth-Heinemann, UK, 2007.

Ashurst, John and Nicola: *Practical Building Conservation: Mortars, Plasters, and Renders,* Gower Technical Press, UK, 2015.

Askeland, Donald R.; Pradeep P. Phule: *The Science & Engineering of Materials*, Thomson-Engineering, NY, 2015.

Bauccio, M.: *ASM Engineering Materials Reference Book,* ASM Intl., Materials Park, OH, 2014.

Beer, F.P., Johnston, Jr., E.R.: *Mechanics of Materials*, McGraw-Hill, Inc., New York, 2013.

Berry, Mary Clay: *Alaska Pipeline: The Politics of Oil and Native Land Claims*. Bloomington, IN: Indiana University Press, 2005.

Bradford, Samuel A.: *Practical Handbook of Corrosion Control in Soils*. Edmonton, Canada: CASTI Pub., 2010.

Brittain, A., and P. Morton: *Engraving on Precious Metals*, N.A.G. Press, London, 2008.

Brownell, W. E.: *Structural Clay Products,* Springer-Verlag Publisher, N.Y., 2006.

Bucciarelli, Louis L.: *Designing Engineers*, MIT Press, Cambridge, 1994.

Burns, G.; Glazer, A.M.: *Space Groups for Scientists and Engineers,* Academic Press, Boston, 2010.

Bye, G. C.: *Portland Cement Composition Production and Properties*, Pergamon Press, Oxford, 2013.

Callister, Jr., William D.: *Materials Science and Engineering – An Introduction*, John Wiley and Sons, UK, 2000.

Chen, W.F.: *Civil Engineering Handbook*. Boca Raton, FL: CRC Press, 2005.

Chesters, J. M., *Refractories: Production and Properties*, The Iron and Steel Institute, London, 2013.

China, T.: *Building Industry, Pride of the Pacific,* Honolulu, Hawaii, 2011.

Craig, Bruce D. and David S. Anderson: *Handbook of Corrosion Data. Materials Park*, OH: ASM International, 2015.

Davenport, Thomas: *Process Innovation: Reengineering Work through Information Technology*, Harvard Business School Press, Boston, 2013.

Davey, N.: *History of Building Materials*, Phoenix House, London, 1961.

David Fisher: *Rules of Thumb for Engineers and Scientists*, Gulf Pub, Houston, 2011.

David Willoughby: *Plastic Piping Handbook*. McGraw-Hill Professional, Delhi, 2002.

Deer, W. A.: *Rock-forming Minerals*, Longmans, London, 2003.

Dickenson, T. Christopher: *Valves, Piping & Pipelines Handbook*, Tulsa, OK: PennWell, 1999.

Dimes, Francis B.: *Conservation of Building and Decorative Stone*, Heineman, Butterworth, 2010.

Domminghaus, H.: *Plastics for Engineers: Materials, Properties, Applications*, Hanser Publishers, Munich, 2013.

Dorf, Richard C.: *Electrical Engineering Handbook*. Boca Raton, FL: CRC Press, 2014.

Eckel, E. C., *Cements, Limes, and Plasters*. John Wiley & Sons, New York, 1999, p. 576.

Edward H. Smith: *Mechanical Engineer's Reference Book*, Butterworth-Heinemann, Oxford, 2014.

Eichholz, D. E.: *Pliny: Natural History*, Harvard University Press, Cambridge, Mass., 2012.

Fitchen, J.: *Building Construction Before Mechanization*, MIT Press, London, 2008.

Flynn, M.J. : *Computer Architecture: Pipelined and Parallel Processor Design*, Narosa, Delhi, 2011.

Francois Cardarelli: *Scientific Unit Conversion: A Practical Guide to Metrication*, Springer, M.J. Shields. London, 2007.

Gallun, Rebecca A.: *Fundamentals of Oil & Gas Accounting*. Tulsa, OK: PennWell, 2011.

Gibson, Dan: *The Nabataeans, Builders of Petra*, CanBooks, Amman, Jordan, 2002.

Gregory K. McMillan: *Process/industrial Instruments and Controls Handbook*, McGraw-Hill, New York, 2009.

Grimwade, Mark: *Introduction to Precious Metals*, Newnes Technical Books-Butterworth, London, 2015.

Guy, N.G.: *Pipe Protection*. New York, NY: Elsevier Applied Science, 2015.

Hamlin, T.: *Architecture Through the Ages*, G. P. Putnam's Sons, New York, 2003.

Hill, Donald Routledge: *A History of Engineering in Classical and Medieval Times*, Open Court Pub., La Salle, 2014.

Hosmanek, Max: *Pipeline Construction*. Austin, TX: Petroleum Extension Service, 2014.

James Brown: *Advanced Machining Technology Handbook,* McGraw-Hill, New York, 2008.

James O. Pennock: *Piping Engineering Leadership for Process Plant Projects,* Gulf Professional Publishing, 2001.

John J. McKetta: *Piping Design Handbook.* Marcel Dekker, 2012.

Jones, Denny: *Principles and Prevention of Corrosion,* Upper Saddle River, Prentice Hall, New Jersey, 2016.

Kennedy, John. L.: *Oil and Gas Pipeline Fundamentals*, Tulsa, PennWell Books, 2013.

Kutz, M.: *Mechanical Engineers' Handbook*, John Wiley and Sons, New York, 2016.

Lamit, Louis Gary: *Pipe Fitting and Piping Handbook.* Englewood Cliffs, NJ: Prentice-Hall, 2014.

Latrobe, Benjamin Henry: *The Engineering Drawings of Benjamin Henry Latrobe,* Yale University Press, NY, 2010.

Lester, C.B.: *Hydraulics for Pipeliners,* Houston, TX: Gulf Publishing Co., 2014.

Malinowski, R.: *Concrete and Mortars in Ancient Aqueducts,* Concrete International: Design and Construction. Detroit, Michigan, 2009.

Marks, Alex: *Handbook of Oceanic Pipeline Computations.* Tulsa, OK: PennWell Books, 2010.

McHugh, Jeanne: *Alexander Holley and the Makers of Steel,* Johns Hopkins University Press, Baltimore, 2010.

McKetta, John J.: *Piping Design Handbook.* New York, NY: M. Dekker, 2012.

Mielenz, R. C.: *Concrete as a Modern Material, Modern Materials,* Academic Press, New York, 2015.

Orlandos, A. K.: *Les Materiaux de Construction,* Editions E. de Boccard, Paris, 2016.

Parsons, William Barclay: *Engineers and Engineering in the Renaissance,* Williams & Wilkins, Baltimore, 2009.

Patton, W. M.: *A Treatise on Civil Engineering,* John Wiley and Son, N.Y., 2005.

Patton, W.J.: *Materials in Industry*, Prentice-Hall, Englewood Cliffs, NJ, 2016.

Pearson, Norman: *Pipelines and Farming.* London, ON, Canada: Strathmore House, 2015.

Peckworth, H. F.: *Concrete Pipe Handbook,* American Concrete Pipe Assoc., Chicago, 2009.

Popovics, S.: *Concrete-Making Materials,* McGraw-Hill Book Co., New York, 2009.

Richard C. Dorf: *Engineering Handbook,* IEEE Press, Piscataway, NJ, 2006.

Robert C. Rosaler: *Standard Handbook of Plant Engineering,* McGraw-Hill, New York, 2006.

Roscow, James P.: *The Building of the Alaska Pipeline*. New York, NY: Prentice-Hall, 2017.

Schweitzer, Philip A.: *Handbook of Corrosion Resistant Piping*. Malabar, FL: R.E. Krieger, 2015.

Shigley, J.E.: *Mechanical Engineering Design,* McGraw-Hill, Inc., New York, 2012.

Singer, C.: *A History of Technology,* The Clarendon Press, Oxford, 2007.

Singh K & Soler A: *Mechanical Design of Heat Exchangers*, Arcturus, London, 2014.

Sosnin, H.A.: *Procedures for Pipewelding*. Washington, DC: National Association of Plumbing, 2013.

Stanley, Frank A.: *Punches and Dies: Layout, Construction and Use*, McGraw-Hill, New York, 2006.

Taylor, W. H.: *Concrete Technology and Practice,* Angus and Robertson Publisher, 2015.

Thomas Sixsmith & R. Hanselka: *Handbook of Thermoplastic Piping System Design*. Marcel Dekker Ltd, London, 2007.

Timoshenko, S., MacCullough, G.H.: *Elements of Strength of Materials,* Van Nostrand Co., New York, 2007.

Tullis, J. Paul: *Hydraulics of Pipelines: Pumps, Valves, Cavitation, Transients*. New York, NY: John Wiley & Sons, 2009.

Turner, W.: *Introduction to Industrial and Systems Engineering*, Prentice Hall, NY, 2012.

Van Deman, E. B.: *The Building of the Roman Aqueducts*, Carnegie Institution of Washington, Washington, DC, 2004.

Van Vlack, L. H.: *Elements of Materials Science,* Addison-Wesley Publishing Co., Reading, Mass., 2007.

Wachtman, John B.: *Mechanical Properties of Ceramics*, Wiley-Interscience, John Wiley & Son's, New York, 2002.

Waddell, J. J.: *Construction Materials Ready Reference Manual,* McGrawHill Book Co., New York, 2015.

Ward-Perkins, J. B.: *Roman Imperial Architecture,* Penquin Books, New York, 2013.

Warring, R.H.: *Handbook of Valves, Piping, and Pipelines*. Houston, TX: Gulf Publishing Co., 2012.

White, G.F.: *Natural Hazards: Local, National, Global*, Oxford University Press, New York, 2004.

Williams, Peter J.: *Pipelines and Permafrost: Physical Geography and Development in the Circumpolar North*. New York, NY: Longman, 2009.